AF806128

Smart Innovation, Systems and Technologies

Volume 53

Series editors

Robert James Howlett, KES International, Shoreham-by-sea, UK
e-mail: rjhowlett@kesinternational.org

Lakhmi C. Jain, University of Canberra, Canberra, Australia;
Bournemouth University, UK;
KES International, UK
e-mails: jainlc2002@yahoo.co.uk; Lakhmi.Jain@canberra.edu.au

About this Series

The Smart Innovation, Systems and Technologies book series encompasses the topics of knowledge, intelligence, innovation and sustainability. The aim of the series is to make available a platform for the publication of books on all aspects of single and multi-disciplinary research on these themes in order to make the latest results available in a readily-accessible form. Volumes on interdisciplinary research combining two or more of these areas is particularly sought.

The series covers systems and paradigms that employ knowledge and intelligence in a broad sense. Its scope is systems having embedded knowledge and intelligence, which may be applied to the solution of world problems in industry, the environment and the community. It also focusses on the knowledge-transfer methodologies and innovation strategies employed to make this happen effectively. The combination of intelligent systems tools and a broad range of applications introduces a need for a synergy of disciplines from science, technology, business and the humanities. The series will include conference proceedings, edited collections, monographs, handbooks, reference books, and other relevant types of book in areas of science and technology where smart systems and technologies can offer innovative solutions.

High quality content is an essential feature for all book proposals accepted for the series. It is expected that editors of all accepted volumes will ensure that contributions are subjected to an appropriate level of reviewing process and adhere to KES quality principles.

More information about this series at http://www.springer.com/series/8767

Huapu Lu
Editor

Proceedings of the Second International Conference on Intelligent Transportation

Springer

Editor
Huapu Lu
Department of Civil Engineering
Tsinghua University
Beijing
China

ISSN 2190-3018 ISSN 2190-3026 (electronic)
Smart Innovation, Systems and Technologies
ISBN 978-981-10-2397-2 ISBN 978-981-10-2398-9 (eBook)
DOI 10.1007/978-981-10-2398-9

Library of Congress Control Number: 2016948262

© Springer Science+Business Media Singapore 2017
This work is subject to copyright. All rights are reserved by the Publisher, whether the whole or part of the material is concerned, specifically the rights of translation, reprinting, reuse of illustrations, recitation, broadcasting, reproduction on microfilms or in any other physical way, and transmission or information storage and retrieval, electronic adaptation, computer software, or by similar or dissimilar methodology now known or hereafter developed.
The use of general descriptive names, registered names, trademarks, service marks, etc. in this publication does not imply, even in the absence of a specific statement, that such names are exempt from the relevant protective laws and regulations and therefore free for general use.
The publisher, the authors and the editors are safe to assume that the advice and information in this book are believed to be true and accurate at the date of publication. Neither the publisher nor the authors or the editors give a warranty, express or implied, with respect to the material contained herein or for any errors or omissions that may have been made.

Printed on acid-free paper

This Springer imprint is published by Springer Nature
The registered company is Springer Nature Singapore Pte Ltd.
The registered company address is: 152 Beach Road, #22-06/08 Gateway East, Singapore 189721, Singapore

The original version of the book was revised: Acknowledgment has been included in Chaps. 14 and 15. The Erratum to the book is available at 10.1007/978-981-10-2398-9_29

Contents

A New Feature Evaluation Algorithm and Its Application to Fault of High-Speed Railway

Jing Du, Weidong Jin, Zhenzhen Cai, Fei Zhu and Zhidan Wu

Abstract Multi-criterion feature ranking algorithms can ease the difficulty on selecting appropriate ranking criterion caused by single-ranking algorithms, and improve the reliability of feature ranking results. However, the issue of conflict between different single-ranking algorithms is often overlooked. By treating this task as a search and optimization process, it is possible to use the D-S theory and evidence conflict to reduce conflicts between different single-criterions and improve the stability of feature evaluation. This work presents a new multi-criterion feature ranking algorithm based on D-S theory and evidence conflict theory combining different criteria improving classification performance of feature selection results. Comparison between the new algorithm and Borda Count, Fuzzy Entropy, Fisher's Ratio and Representation Entropy methods are done on train fault dataset. The obtained results from the experiment demonstrate that the new algorithm has highest classification accuracy than the other four criterions on all cases considered.

Keywords Feature · Evaluating · Multi-criterion fusion · D-S evidence theory · Evidence conflict theory · Fault classification

1 Introduction

Feature selection aims at finding a feature subset that has the most discriminative information from the original feature set [10, 12, 16]. There are two major issues related to current feature selection. The first one is to define new single-criteria and

---

J. Du (✉) · W. Jin · Z. Cai · Z. Wu
Department of Electrical Engineering, Southwest Jiaotong University, Chengdu 610031, China
e-mail: dawn_on_way@163.com

F. Zhu
Department of Electrical Engineering, Southwest Jiaotong University, Chengdu 610031, China

© Springer Science+Business Media Singapore 2017

H. Lu (ed.), *Proceedings of the Second International Conference on Intelligent Transportation*, Smart Innovation, Systems and Technologies 53,
DOI 10.1007/978-981-10-2398-9_1

another is how to fuse a series of different feature evaluating criteria to solve the redundancy and irrelevance among the features.

To solve the first problem: Many scholars have made a lot of studies with different kinds of feature selection techniques to evaluate features. Mahalanobis Distance, Fuzzy Entropy, Fisher's ratio, Representation Entropy etc. are the examples of this category. Different feature selection methods can be generally classified into wrapper model and filter model [6]. It is also known that the single feature ranking criterion cannot reflect characteristics of features completely.

To solve how to fuse a series of different ranking criteria problem, Weizhong Yan proposed the multi-criterion feature ranking scheme—MCFR to integrate all the single ranking criteria used in the study [14]. With the aim of utilizing individual ranking criteria, both two fusion models are used in this work to compare with each other. In the multiple rankings fusion model, the Borda count method is used to implement ranking vectors fusion [13]. Feng developed a multi-criterion fusion-based recursive feature elimination (MCF-RFE) algorithm to improve classification accuracy of ranking results [15].

In the fusion process, how we should combine different criteria, which we refer to as the fusion rule is the most important step. As mentioned above, using different evaluating criteria will yield different ranking orders, that is, the same feature may have different positions in different ranking orders. This implies that there are conflicts among different single-ranking algorithms. If the conflicts are not taken into account adequately, it might produce counter-intuitive results when fusing these different ranking orders. What's more, the counter-intuitive results will lead to low classification accuracy. Fortunately, the D-S based evidence conflict theory can effectively solve the uncertain and conflict existing in fusion [3, 11]. Recently, many improved D-S theory have proposed to solve the uncertain and conflict existing in fusion, i.e., Zhou proposes an optimal model to learn discounting reliability based evidence distance criterion which considers improving focusing degree and reducing conflict simultaneously [17]. Therefore, to solve the reliability of feature ranking fusion problem, we propose an innovative multi-criterion feature selecting method which is based on the improved D-S theory by Zhou.

In the experimental section, the classification accuracy of the new method and the accuracy of four single evaluation criteria (*Fisher's ratio, Fuzzy Entropy, Representation Entrop* (*RE*), *MD*), the accuracy of multi criteria feature evaluation method (Borda Count) are compared. The results obtained from experiment indicate that the classification accuracy of the new method is superior to the above method, and it also shows that taking D-S theory as the fusion rule can effectively reduce the conflict in the process of fusion, more reliable results are obtained.

2 Basic Feature Ranking Criterion

A great deal of feature evaluation criteria have been presented in the document. Obviously, it is not necessary to fuse all feature selection criterion, which is also impractical [18]. As mentioned above, different feature selection criteria can be generally classified into wrapper model and filter model. The wrapper methods often outperform than filter because of the better results, while filter methods work generally much faster than wrapper methods. This specialty requires a principle for basic criterion choosing: basic criterion should be selected from both models. For computational simplicity and criteria from both models debated above, *Mahalanobis distance* (*MD*), *Fisher's ratio*, *Fuzzy Entropy* and *Representation Entropy* (*RE*) are the basic criteria used in this paper. *Mahalanobis distance* (*MD*) and *Fisher's ratio* are wrapper strategies which use an induction algorithm to estimate the merit of feature subsets. *Fusion Entropy* and *Representation Entropy* (*RE*) are information theory-based filter methods. Details of the four individual basic criteria are presented below.

2.1 Representation Entropy, RE

Assume that $\lambda_i (i = 1, \ldots, n)$ are the eigenvalues of the $n \times n$ covariance matrix of a feature space with n-dimensional features. Then, the representation entropy can be defined as [5]

$$\lambda_i' = \lambda_i \Big/ \sum\nolimits_{i=1}^{n} \lambda_i \tag{2.1}$$

$$RE = -\sum_{i=1}^{n} \lambda_i' \log \lambda_i' \tag{2.2}$$

where $0 \le \lambda_i' \le 1$ and $\sum_{i=1}^{n} \lambda_i = 1$. RE is the representation entropy of the feature set.

Removing a feature from the original feature space, we can obtain a new RE value. With the RE value of the original feature space minus the new one, the representation entropy of the removed feature can be acquired. The eliminated feature will be positioned higher, if the new RE value has a larger decrease compared with the original one. Eliminate only one feature at a time, we can obtain the RE values of all features from the feature space. Then, according to the decreases, the feature ranking can be acquired.

2.2 Fisher's Ratio

Fisher's ratio is an individual feature selection criterion which can assess the influence of a particular feature on classification. The Fisher's ratio of a feature i can be defined as the ratio of interclass variance (TE) and infraclass variance (TA) [1]:

$$F_i = TE_i / TA_i \tag{2.3}$$

$$TE_i = \frac{1}{C}\sum_{i=1}^{C}\left(v_{ij} - \sum_{j=1}^{C} v_{ij}/C\right) \tag{2.4}$$

$$TA_i = \frac{1}{C}\sum_{i=1}^{C}\left(\sum_{n=1}^{N_j}(f_{ijn} - v_{ij})^2/N_j\right) \tag{2.5}$$

According to the above formulate C indicates the number of class. v_{ij} represents the mean of feature i within class j. f_{ijn} is the ith feature of the nth sample from the jth class. And N_j is the number of samples from the jth class. The larger the Fisher's ratio value, the more representational the feature is.

2.3 Fuzzy Entropy

The particularity of fuzzy sets is to seize the concept of partial membership. Considering the idea of fuzzy set, the measure of fuzzy entropy corresponding to Shannon probabilistic entropy proposed by De Luca and Termini should be [4]:

$$H(A) = -\sum_{j=1}^{n}\left(\mu_A(x_j)\log\mu_A(x_j) + (1-\mu_A(x_j))\log(1-\mu_A(x_j))\right) \tag{2.6}$$

According to formulate 2.6 $\mu_A(x_j)$ indicates the fuzzy values, and x_j represents all samples of the feature set. The Fuzzy Entropy describes the fuzzy degree of a fuzzy set. So $H(A)$ has the minimal value, if the feature A is good, otherwise, the feature A is bad.

2.4 Mahalanobis Distance

Mahalanobis distance is the covariance distance between different data and it is an effective method to calculate the similarity between two sample sets. Let v_i and v_j

represent respectively the mean value of features vector for class i and j, the feature set co-variance is $\sum$. The Mahalanobis distance between class i and j can be described as formulate 2.7 [14]

$$M\_dis_{i,j} = \sqrt{(v_i - v_j)\sum^{-1}(v_i - v_j)^T}. \tag{2.7}$$

For a feature set with more than two classes, the Mahalanobis distance for the ith class can be expressed as:

$$M\_dis_i = \sum_{i=1}^{C}\sum_{j=i+1}^{C}\sqrt{(v_i - v_j)\sum^{-1}(v_i - v_j)^T} \tag{2.8}$$

where C is the number of classes. Eliminating a feature from the original feature space, we can acquire a new Mahalanobis distance value. With the Mahalanobis distance value of the original feature space deduct the new Mahalanobis distance value, we acquired the Mahalanobis distance of the removed feature. The eliminated feature will be positioned higher, if the new value has a larger decrease compared with the original one.

3 Multi-criterion Feature Evaluation Based on D-S Theory

The reason for utilizing multi-criteria is obvious. Using different feature selection criteria will achieve different feature subsets and ranking order. Among these different selection results we do not have the priori knowledge to decide that which is the best. It is a reasonable solution to integrate different 'opinions' from single feature selection criteria to yield a consensus ranking order. Multi-criterion feature selection has been a topic of recent interest.

3.1 The Motivation of Using Theory

Two reasons motivate us to utilize the D-S theory as the fusion imperative. Firstly, using different feature ranking algorithms will produce different ranking orders. Table 1 indicates the individual rankings of the standard dataset Wine. The individual rankings are obtained through four different single feature ranking criteria, containing MD, Fisher's ratio, Fuzzy Entropy and RE. Obviously, the results of the individual rankings are quite different. That is to say it existing high conflict among different criteria, and the ranking of the same feature can be totally different. If the conflicts are not taken into account appropriately, it might produce counter-intuitive

Table 1 Ranking order based on four different feature selection criteria on wine data set

Criterion	Feature												
	1	2	3	4	5	6	7	8	9	10	11	12	13
Fisher's ration	7	2	13	1	3	4	12	10	9	6	5	11	8
Fuzzy entropy	1	3	13	7	4	11	2	12	6	9	8	10	5
MD	13	7	10	1	3	4	12	2	11	8	6	9	5
RE	13	8	11	3	6	9	1	12	7	2	10	4	5

results when fusing these different ranking orders. Fortunately, the evidence theory based on D-S theory can effectively deal with the uncertainty and conflict between individual rankings, preventing the production of fusion paradox in the process of integrating single rankings.

Secondly, since the D-S theory has obvious advantages in the fusion of different information, it has been successfully applied in multi-sensor fusion. Motivated by these two reasons, this paper presents a multi criteria feature evaluation method based on D-S theory.

3.2 *Dempster-Shafer (D-S) Evidence Theory*

The Dempster-Shafer (D-S) evidence theory, firstly proposed by Dempster in 1967, and then developed by Shafer in 1976 [2, 8], attracts more and more attention as an uncertainty reasoning method. In the DS evidence theory, let $\Theta = \{A_1, A_2, A_3 \ldots A_n\}$ consists of N hypotheses that is independent each other. Θ is called the frame of discernment including all possible propositions and assumptions. Θ and its subsets constitute the power set $2^\Theta = \{A_1, A_2, A_3 \ldots A_2^N\}$. Defining the mapping $m : 2^X \to [0, 1]$ is the basic probability assignment functions, respectively $m_1, m_2, m_3 \ldots m_n$. The corresponding Dempster combination rule is:

$$m(A) = \begin{cases} 0, & A = \Phi \\ \frac{\sum_{\cap A_j = A} \prod_{1 \le i \le N} m_i(A_i)}{1-K}, & A \ne \Phi \end{cases} \tag{3.1}$$

$$k = \sum_{\cap A_j = A} \prod_{1 \le i \le N} m_i(A_i) \tag{3.2}$$

According to the formulate 3.2, K represents the conflict coefficient among the evidences.

Though D-S evidence theory is attractive, counterintuitive and invalid conclusions are produced for highly conflicting evidence. The reliability of evidence is generally difficult to estimate, unreliable evidence tends to affect the fusion process, even lead to the wrong result. Therefore, it is necessary to add reliability information of evidence in the integration process in order to improve the fusion result. Shafer introduced discount operator to handle the situation where the information sources is not reliable [8]. Supposing the reliability of information sources is α ($0 < \alpha < 1$), then the Unreliability is $1 - \alpha$, also called as the discount factor. For inhibiting unreliability and the conflict of evidence, Shafer purposes discounting evidence based on reliability α, the unreliable part will be assigned to entire collection. The corresponding formula is:

$$m_j^d(A) = \begin{cases} \alpha_j \times m_j(A) & A \neq \Phi \\ 1 - \alpha_j + \alpha_j \times m_j(A), & A = \Phi \end{cases} \tag{3.3}$$

Such methods are often acquiring the reliability of each evidence firstly, relying on a certain conflict or distance measurement criteria between evidences, and then fusing the discounted evidences based on obtained reliability. Selecting the appropriate measurement criteria is so crucial that will have a great effect on the discounting factor α. Although there are many different ways of calculating α, those methods are only applicable to some special cases. So in the fusion process we employ a method about calculating the reliability α proposed by Zhou.

3.3 Multi-criterion Fusion Based on D-S Theory

In order to get more reliable and more stable feature subset, this paper proposes a multi criteria feature evaluation method based on D-S theory and evidence conflict theory. The multi criteria evaluation method combines different individual evaluation criteria, and can acquire feature subsets containing more information. In the following, we introduce the overall fusion procedure firstly. Then, the details of the fusion criterion are illustrated.

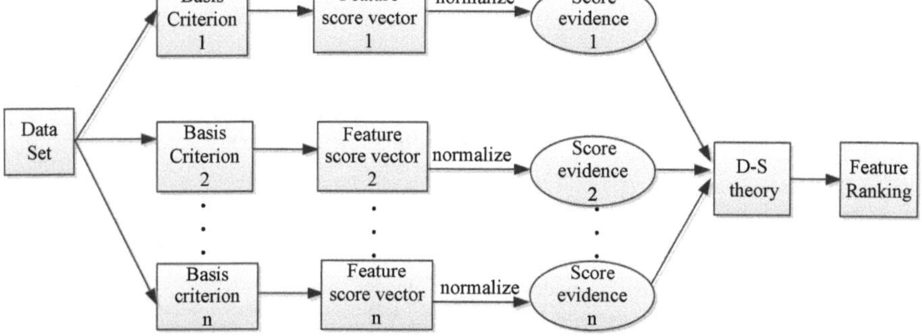

Fig. 1 Multi criteria fusion based on D-S theory

In this paper, we employ a score-based multi-criterion fusion method. As it is described in Fig. 1, given a feature dataset, each individual basic criterion first yields the corresponding feature score vector. The value of the feature score decides whether it is important. Then, feature score vector is normalized to be a score evidence vector which contains normalized scores of all features. Thirdly, the multi-criterion algorithm is utilized to fuse the score evidence vector into a comprehensive score vector. Finally, ranking the comprehensive score vector can obtain a comprehensive feature rank.

Next, the detail of the fusion procedure is illustrated. In order to make sure that the score vector which obtained from four individual basic criteria are comparable, normalizing four score vectors is the first thing that should be done. In *MD*, *RE*, *Fisher's ratio*, a feature will be ranked higher if the score is greater, however, in Fuzzy Entropy method, the smaller the score of a feature is, the greater the contribution to classification is. To be consistent with the first three methods, we take the reciprocal of the score based on fuzzy entropy of each feature. Then, the score vectors which acquired from four basic criterion are normalized to be a vector, and the value of each vector element range from 0 to 1. Suppose existing N single criteria, and s_i represents the score vector obtained by the ith $i(1 \leq i \leq N)$ individual criterion, then the vector s_i can be normalized as follows:

$$e_i = (s_i - s_{i\min}) \Big/ \sum (s_i - s_{i\min}) \tag{3.4}$$

where $s_{i\min}$ is the minimum values in vector s_i, and $\sum e_i = 1$. Here we get the basic probability distribution functions $e_1, e_2, e_3 \ldots e_N$.

The identification framework $\Theta = \{F_1, F_2, \ldots, F_M\}$ of fusion system contains M complete incompatible assumption propositions. M is equal to the number of the features and F_i is the ith feature. In the fusion process we employ an optimal method for combining conflicting evidence by Zhou. Zhou proposes an optimal model to learn discounting reliability α based evidence distance criterion which considers improving focusing degree and reducing conflict simultaneously. The fusion procedure can be briefly summarized as follows.

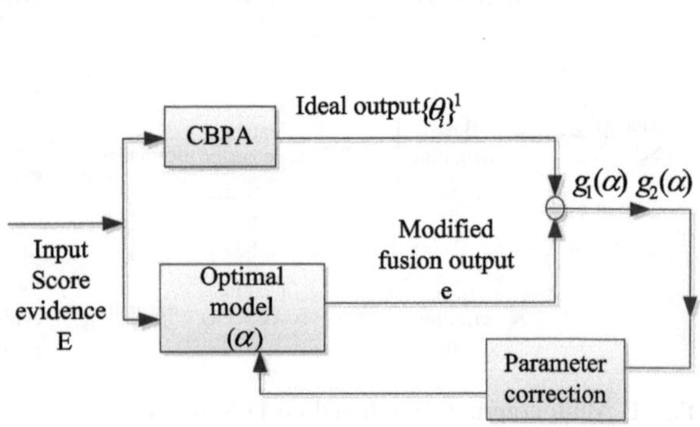

Fig. 2 The process of parameters learning

3.3.1 Fusion Procedure

1. Calculate the reliability of the evidence α:

There are N pieces evidence e_j, $j = 1, \cdots N$. The corresponding reliability vector $\alpha = [\alpha_1, \cdots \alpha_j \cdots, \alpha_N]$ is the unknown parameter which is learned by optimization model as follows in Fig. 2:

Where the input score evidence E is the evidence to be fused. α is the reliability parameter vector to be optimized; e is the integration of the results with the optimized α;$\{\theta_i\}^1$ is the ideal output; Here an ideal output $\{\theta_i\}^1$ is a CBPA [9], namely in the identification framework the value of elements θ_i is 1, the other elements assigns to zero. $g_1(\alpha)$ is the difference between the ideal output $\{\theta_i\}^1$ and modified fusion output e, $g_2(\alpha)$ is the value conflict K among the modified evidence. And the optimal model expressed as formula (3.4).

$$\min\{g_1(\alpha), g_2(\alpha)\} = \begin{cases} g_1(\alpha) = \underset{i,\alpha}{\min} d_J(\overset{N}{\underset{j=1}{\oplus}}\{e_j^d\}, \{\theta_i\}^1), \quad i = 1, \cdots, n \\ g_2(\alpha) = \overset{N}{\underset{j=1}{\oplus}}\{e_j^d\}(\phi) \end{cases} \tag{3.5}$$

where $\{\theta_i\}^1$ represents $e(\theta_i) = 1$. e_j^d is the e modified evidence e_j by adopting the formula (3.7). $\oplus_{j=1}^N\{e_j^d\}$ is the result of fusing e_j^d, $j = 1, \cdots N$ according to Dempster combination rule. $\oplus_{j=1}^N\{e_j^d\}(\Phi)$ is the value of conflict K when $m_j^d, j = 1, \cdots N$ fused

(1) And then the formula 3.4 should be subject to

$$0 < \alpha_j < 1, \quad j = 1, \cdots N \tag{3.6}$$

(2) The reliability of the normal evidence should be greater than abnormal evidence, therefore the value of the reliability of evidence should satisfy certain order relations.

$$\alpha_{\sigma(1)} \geq \alpha_{\sigma(2)} \geq \cdots \geq \alpha_{\sigma(j)} \geq \cdots \alpha_{\sigma(N)} \tag{3.7}$$

where $\{\sigma(1), \cdots \sigma(N)\}$ is an order which relates with $\{1, \cdots N\}$. The reliability of evidence relies on the falsity proposed by Schubert of evidence [7]. The above optimization problem is a multi-objective optimization problem with linear constraints, solved by the FGOALATTAIN function.

2. Fuse rule

$$e_j^d(\mathrm{A}) = \begin{cases} \alpha_j \times e_j(\mathrm{A}) & \mathrm{A} \neq \Phi \\ 1 - \alpha_j + \alpha_j \times e_j(\mathrm{A}), & \mathrm{A} = \Phi \end{cases} \tag{3.8}$$

$$e(\mathrm{A}) = \bigoplus_{j=1}^{N} e_j^d(\mathrm{A}), \quad \forall \mathrm{A} \subset \theta \tag{3.9}$$

Then we can modify the evidence according to formula (3.7) by using the obtained reliability α. Fusing the rectified evidence $e_j^d(\mathrm{A})$ complies with the Dempster combination rule. The obtained score vector is $[e_1, e_2, \ldots, e_N]$. The elements of score vectors arrange in descending order, obtaining comprehensive feature ranking.

4 Experiment and Results

For proving the validity and superiority of the proposed multi criteria algorithm, the measurements fault data of high speed trains are evaluated. In this paper, we utilize the new multi criteria method, *Borda Count* and four single evaluation criteria—*Fuzzy Entropy, Fisher's ratio, RE, MD* to evaluate the characteristics. Removing a redundant feature each time, the classification accuracy is obtained by using the rest of the feature subset to classify. The classification accuracy of the above five methods is compared.

4.1 High Speed Train Data Experiment Design

In order to validate the validity of the new method in the diagnosis of high speed rail faults, the new method is used to simulate the measure data of a certain type of high speed train. On the high-speed train four conditions (normal, lateral damper failure, failure of anti hunting damper, air spring loss of gas), eight dimensional feature including wavelet coefficient of mean value, and fast Fourier transform of the mean, variance etc. are extracted from the measure data. Each condition has 20 sets of samples, with a total sample sets of 80. A set of samples were selected respectively from four operating conditions, and the remaining 76 groups were used as test sample.

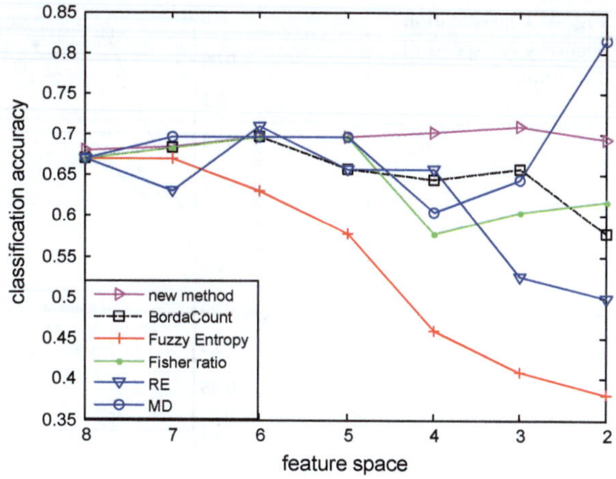

Fig. 3 Classification accuracy (at speed of 120 km/h)

Table 2 Average classification accuracy under different speed of feature space

Evaluation criteria	120 km/h	140 km/h	160 km/h	200 km/h	220 km/h
New method	0.6956	0.8176	0. 8780	0.9624	0.9492
B-count	0.6560	0.8064	0.7744	0.8985	0.9248
Entropy	0.5434	0.6823	0.6804	0.7293	0.7800
F-ratio	0.6503	0.7029	0.7575	0.9586	0.9361
RE	0.6221	0.6278	0.6203	0.8308	0.7986
MD	0.6898	0.8045	0.8778	0.9454	0.9323
Original feature space	0.6710	0.6315	0.7763	0.9605	0.9210

Note using B-count, F-entropy, F-ratio represent Borda count, fuzzy entropy, Fisher's ratio method

4.2 Experimental Result Analysis

Figures 3 etc., shows the accuracy of the 6 methods under different speeds in each feature space. Table 2 shows the average value of 6 methods under different speeds in each feature and the classification accuracy of the original feature space. Available from figures and Table 2, compared with other methods, the new method has higher classification accuracy for all five kinds of speed in each feature space. At the process of removing redundant features, the classification accuracy rate shows up first and then down. Moreover, the average classification accuracy obtained by the new method is the highest of every feature space. And other methods can only be used to evaluate the characteristics of a certain speed, but it is not suitable for other speed. Such as the Borda Count method only has a better evaluation results at the speed of 140, 220 km/h, but doesn't apply to other speed. Even though fisher's ratio method has a good evaluation only for the speed of 200 and 220 km/h, the accuracy is lower than the value of the new method, and the evaluation of the other speed is very poor. The accuracy of the new method improves 22.76 %, compared with the original feature space at

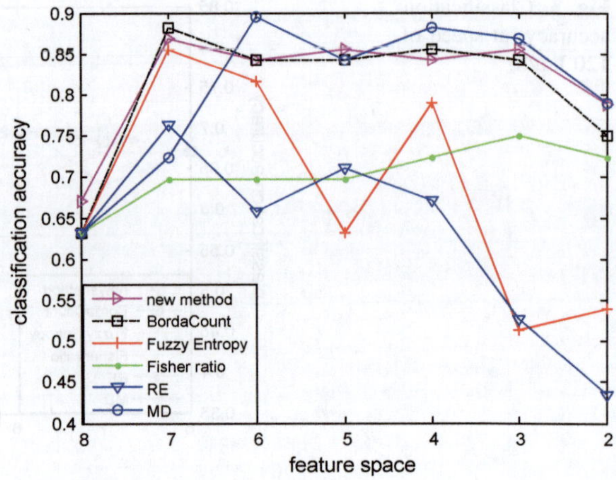

Fig. 4 Classification accuracy (at speed of 140 km/h)

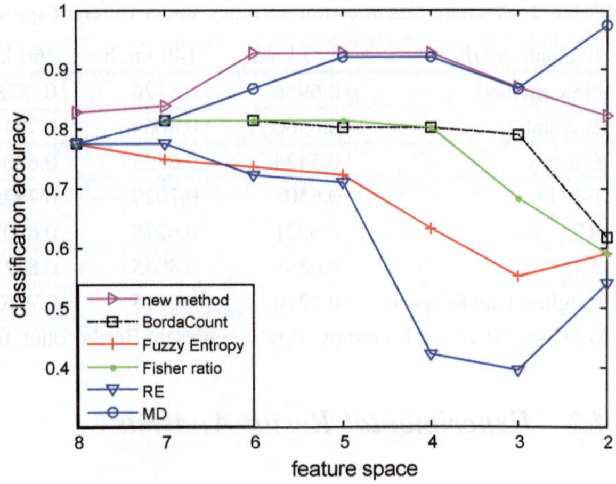

Fig. 5 Classification accuracy (at speed of 160 km/h)

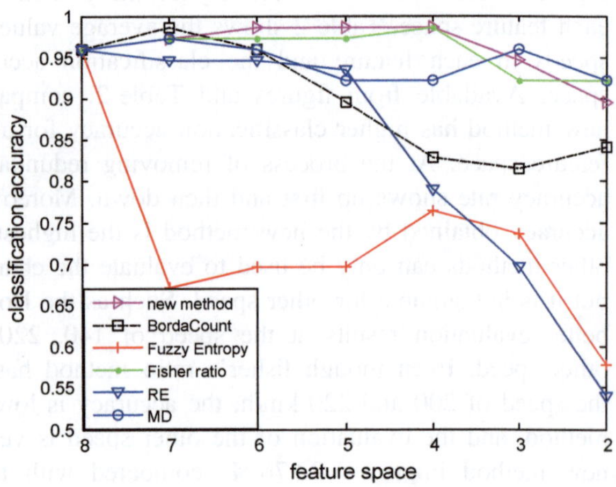

Fig. 6 Classification accuracy (at speed of 200 km/h)

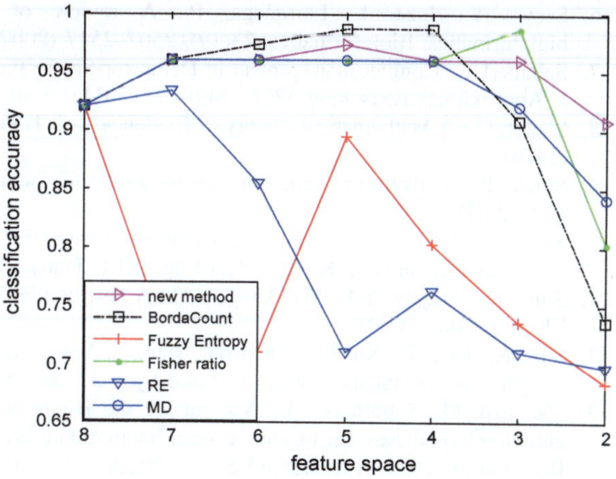

Fig. 7 Classification accuracy (at speed of 220 km/h)

140 km/h speed, improving 11.58 % at the speed of 160 km/h. All these shows that the new method can better evaluate the features, and has a universal property (Figs. 4, 5, 6 and 7).

5 Conclusion

In this paper, we have analyzed and discussed multi-criterion based on the improved D-S theory for feature selection on the fault high speed train data. Experiment study of the new multi-criterion method with the Borda Count, Fisher's ratio, Fuzzy Entropy, RE based on the performance of classification accuracy has been calculated. The obtained result indicates that the proposed method can effectively evaluate the characteristics of the feature set, remove the redundant features and cut down the complexity of the classifier, effectively improve the classification accuracy of the high speed train fault.

References

1. Abdulla, W.H., Kasabov, N.: Reduced feature-set based parallel CHMM speech recognition systems. Inf. Sci. **156**(1), 21–38 (2003)
2. Dempster, A.P.: Upper and lower probabilities induced by a multivalued mapping. Ann. Math. Stat. pp. 325–339 (1967)
3. Li, B., Wang. B., Wei. J., Huang, Y., Guo, Z.: An efficient combination rule of evidence theory. J. Data Acquisition Proc. (2002)
4. Luukka, P.: Feature selection using fuzzy entropy measures with similarity classifier. Expert Syst. Appl. **38**(4), 4600–4607 (2011)
5. Mitra, P., Murthy, C.A., Pal, S.K.: Unsupervised feature selection using feature similarity. IEEE Trans. Pattern Anal. Mach. Intell. **24**(3), 301–312 (2002)

6. Saeys, Y., Inza, I., Larrañaga, P.: A review of feature selection techniques in bioinformatics. Bioinformatics, **23**(19), 2507–2517 (2007)
7. Schubert, J.: Conflict management in Dempster-Shafer theory using the degree of falsity. Int. J. Approximate Reasoning **52**(3), 449–460 (2011)
8. Shafer, G.: A Mathematical Theory of Evidence, vol. 1. Princeton University Press, Princeton (1976)
9. Smets, P.: Analyzing the combination of conflicting belief functions. Inf. Fusion **8**(4), 387–412 (2007)
10. Song, Q., Ni, J., Wang, G.: A fast clustering-based feature subset selection algorithm for high-dimensional data. Knowl. Data Eng. IEEE Trans. **25**(1), 1–14 (2013)
11. Sun, Q., Xiu qing, Y.E., Wei kang, G.U.: A new combination rules of evidence theory. Acta Electron. Sin. (2000)
12. Sun, X., Liu, Y., Xu, M., Chen, H., Han, J., Wang, K.: Feature selection using dynamic weights for classification. Knowl. Based Syst. **37**, 541–549 (2013)
13. Van Erp, M., Schomaker, L.: Variants of the Borda count method for combining ranked classifier hypotheses. In: In The Seventh International Workshop on Frontiers in Handwriting Recognition. 2000. Amsterdam Learning Methodology Inspired By Human's Intelligence Bo Zhang, Dayong Ding, and Ling Zhang (2000)
14. Yan, W.: Fusion in multi-criterion feature ranking. In: IEEE International Conference on Information Fusion, 10th July 2007, pp. 1–6, (2007)
15. Yang, F., Mao, K.Z.: Robust feature selection for microarray data based on multicriterion fusion. IEEE/ACM Trans. Comput. Biol. Bioinfor. (TCBB) **8**(4), 1080–1092 (2011)
16. Yu, L., Liu, H.: Efficient feature selection via analysis of relevance and redundancy. J. Mach. Learn. Res. **5**, 1205–1224 (2004)
17. Zhou, Z., Xu, X.B., Wen, C.L., Lv, F.: An optimal method for combining conflicting evidences. Acta Autom. Sin. **38**(6), 976–985 (2012)
18. Zhu, J., Fei, Z.: Feature selection for high-dimensional and small-sized data based on multi-criterion fusion. J. Convergence Inf. Technol. **7**(19) (2012)

Steering Angle Balance Control Method for Rider-Less Bicycle Based on ADAMS

Yuanyuan Feng, Runjia Du and Yuping Xu

Abstract Bicycle is a typical unstable system. Although it is partly similar to the inverted pendulum systems, the complex mechanical characteristics cause that it is difficult to be analyzed and controlled. In this paper, Constrained Lagrangian method is used to building a rider-less bicycle dynamic model. Through the analysis of the dynamic model, the control system is designed to be divided into two parts: the forward control and the balance control. The rear wheel gets a stable speed by controlling the torque of the motor. Steering Angle Balance Control method is designed to keep the system balance. Use ADAMS to simulate a rider-less bicycle control system. The simulation shows that using Steering Angle Balance Control method can balance the bicycle to run forward and swerve stably.

Keywords Rider-less bicycle · Constrained lagrangian method · Steering angle balance control · ADAMS

1 Introduction

The rider-less bicycle has advantages like small size, flexibility, and environment friendly etc. It draws a plenty of attention because it has a bright future of applying to forest patrol, field rescue, entertainment and many other respects. However, since the system of bicycle is non-linear and non-holonomic system, there are lots of precarious and dubious factors. Thus, using mechanism to control bicycle and make it run steadily is not easy.

Through the neoteric research that aimed at riding bicycle, Getz [1] set the torque of handelbar and the driving torque of rear wheel as the input of a system,

Y. Feng (✉) · R. Du · Y. Xu
School of Railway Tracks and Transportation, East China Jiaotong University, Nanchang, Jinagxi Province 330013, China
e-mail: 452955812@qq.com

© Springer Science+Business Media Singapore 2017

H. Lu (ed.), *Proceedings of the Second International Conference on Intelligent Transportation*, Smart Innovation, Systems and Technologies 53,
DOI 10.1007/978-981-10-2398-9_2

and built a simple forepart bicycle-robot-dynamics model with an inner balance control machine. Iuchi [2] remodeled and redesigned bicycles through input the control handelbar, driving rear and motor driving which can control roll single pendulum angle. And they successfully realized that bicycle running in a static speed at 2.5 km/h. Tanaka [3, 4] built the non-linear dynamics model of bicycle system, besides, they used partial linear theory to simplify the model. They brought up a design concept that use position controller to stabilize the bicycle body and implement track control using track controller. Domestic professor Wei Shi Min and his research team [5] are working at the analysis of a front-wheel drive bicycle robot.

This article builds up a rider-less bicycle model, and using Constrained Lagrangian method to deduce the dynamic equation. Through the analysis of the dynamic equation, design a bicycle model which the driving control and balance control are separated. According to the different running-speed commands, the bicycle gets a stable forward speed by adjusting the torque of the rear driving motor. In the condition of having a stable speed, the body of the bicycle can keep balance through actualize the feedback of bicycle body roll angle to adjust the steering angle by using the Steering Angle Balance Control method.

2 Rider-Less Bicycle Dynamic Model

2.1 Model Building

The rider-less bicycle is considered as a two-part platform: a rear frame and a steering mechanism. Before analyzing the bicycle mode, a few consume conditions are settled as following: (1) the contact of the tread and the ground is point-contact, thus the thickness of the tire are supposed to be ignored; (2) the frame of the bicycle is regard as a point mass; (3) the ground is horizontal; (4) there is no side sliding when the bicycle is running [6–8].

Like what is shown in Fig. 1, using three coordinate systems: fixed coordinate system $(\mathrm{X}, \mathrm{Y}, \mathrm{Z})$, this coordinate system is fixed on the ground; bicycle body moved coordinate system (x, y, z), the axis x of this coordinate system is in the direction of the connect between point C_2 which is the contact point of rear wheel and ground and point C_1 which is the contact point of front wheel and ground. The coordinate origin O is in the position of C_1. The axis x and the axis y are orthogonal, and parallel with the ground; bicycle body coordinate system $B(x_B, y_B, z_B)$. The x_B axis and z_B axis of this coordinate system are fixed on the bicycle frame plane, the origin O_B is the centroid. As the bicycle body coordinate system relative to the fixed coordinate system, ψ is the roll angle, φ is the list angle. If define the unit vectors of these three coordinate systems are: $(\mathbf{I}, \mathbf{J}, \mathbf{K})$, $(\mathbf{i}, \mathbf{j}, \mathbf{k})$, $(\mathbf{i_B}, \mathbf{j_B}, \mathbf{k_B})$, the conversion of these three coordinate system is:

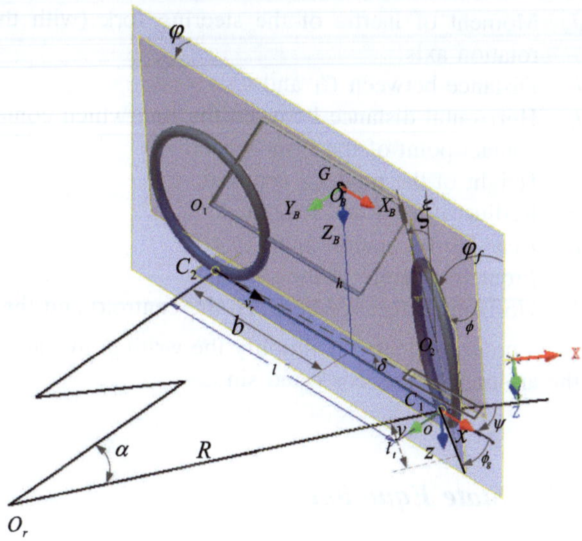

Fig. 1 The rider-less bicycle system model

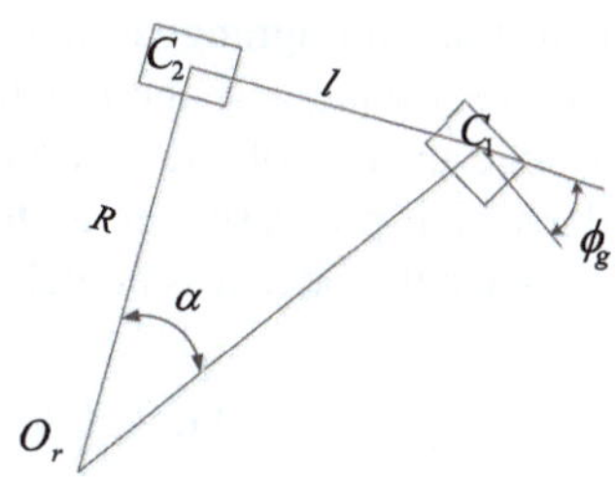

Fig. 2 Rider-less bicycle model overlook-diagram

$$\begin{bmatrix} \mathbf{i_B} \\ \mathbf{j_B} \\ \mathbf{k_B} \end{bmatrix} = \begin{bmatrix} \cos\psi & \sin\psi & 0 \\ -\cos\varphi\sin\psi & \cos\varphi\cos\psi & \sin\varphi \\ \sin\varphi\sin\psi & -\sin\varphi\cos\psi & \cos\varphi \end{bmatrix} \begin{bmatrix} \mathbf{I} \\ \mathbf{J} \\ \mathbf{K} \end{bmatrix} \tag{1}$$

Let O_r denote the instantaneous rotation center R is represent the rotation radius. ϕ_g is the projection of the steering angle on the ground. According to the geometry connection between these, like the diagram which is shown in Fig. 2:

$$\alpha = \phi_g. \tag{2}$$

Other parameters in the model:

v_r Speed of the rear wheel and the ground contact point. Since tire has no side sliding, the direction of the v_r is along the axis x_B.

ϕ Angle of the steering.

m Mass of the bicycle system.

J_s Moment of inertia of the steering fork (with the front wheel set) about its rotation axis.
l Distance between C_1 and C_2.
l_t Horizontal distance between the line which contains the rotation axis and the contact-point of the front wheel.
h Height of the model's centroid.
r Radius of front and rear tires.
δ Rear frame rotation angle.
ξ Front fork angle of the steering.
b Horizontal distance between the centroid and the center of the rear wheel.

In order to clear and simplify the writing, this article use x, C_x and S_x to present the arbitrary angle, $\cos x$ and $\sin x$.

2.2 State Equation

Aiming at the non-holonomic system which stated as above, this article imposes the Constrained Lagrangian method to get the dynamic equation conveniently. The first step gets separately the kinetic energy of the rider-less bicycle system, the rotation kinetic energy of the steering fork, and the potential energy of the system. The derivation processes are as following:

when the rider-less bicycle system is running forward, the speed of the centroid is:

$$\mathbf{v_G} = \left(v_r + h\dot{\psi}S_\varphi\right)\mathbf{i} + \left(b\dot{\psi} + h\dot{\varphi}C_\varphi\right)\mathbf{j} + h\dot{\varphi}S_\varphi \quad (3)$$

When turning the steering, the steering fork produces the rotation kinetic energy. The ground projection of the steering angle ϕ is ϕ_g. And give the definition of the dynamic steering variable σ [9]:

$$\sigma := \tan\phi_g = \frac{l}{R} \quad (4)$$

Through the geometry shape of the steering mechanism get the geometry connection as follow:

$$\tan\phi_g C_\varphi = \tan\phi C_\xi \quad (5)$$

When the roll angle and steering angle are small, Eq. (5) can be replaced by Eq. (6) approximately.

$$\dot{\sigma}C_\varphi = \dot{\phi}C_\xi \quad (6)$$

Besides, the rotation of the steering also affects the potential energy of the system. When the steering angle is zero, the projection of the bicycle body likes the dotted line which is shown in Fig. 1 because of the height of the body roll centroid is $h\cos\varphi$. And when the steering rotated an angle ϕ, the projection of the bicycle body rotated an angle of δ, meanwhile, the height of the centroid get short by Δh_G. The δ can be approximate to:

$$\delta = \frac{l_t \phi_g C_\xi}{l} \tag{7}$$

Using an approximate of small angle $\sigma \approx \phi_g$, Δh_G can be calculated as:

$$\Delta h_G = \delta b S_\varphi \approx \frac{b l_t \sigma C_\xi}{l} S_\varphi \tag{8}$$

As what has been listed above, the Lagrangian L of the system is:

$$L = \frac{1}{2} J_s \dot{\phi}^2 + \frac{1}{2} m \mathbf{v}_\mathbf{G}^2 - mg(h\cos\varphi - \Delta h_G) \tag{9}$$

According to the Euler-Lagrange equation:

$$\frac{d}{dt}\left(\frac{\partial L}{\partial \dot{\mathbf{q}}}\right) - \frac{\partial L}{\partial \mathbf{q}} = \mathbf{u} \tag{10}$$

In the equation:

$\boldsymbol{q} = [\varphi \quad s \quad \sigma \quad \psi]^T$; $\dot{\boldsymbol{q}} = [\dot{\varphi} \quad v_r \quad \dot{\sigma} \quad \dot{\psi}]^T$; $\mathbf{u} = \begin{bmatrix} 0 \\ \tau \\ \tau_\sigma \\ 0 \end{bmatrix}$ and s is the remove of the bicycle. τ is the torque which throw on the rear wheel's axle and make the bicycle to move forward. τ_σ is a turning torque of handlebar.

When bicycle is running forward, the rear wheel's speed and the roll angle have a relation as the following equation, which immediate the non-holonomic constraint of the system:

$$v_r = R\dot{\psi} = \frac{l}{\sigma}\dot{\psi} \tag{11}$$

The non-holonomic constraint can be simplified as follows:

$$\mathbf{A_c}\dot{\mathbf{q}} = 0 \tag{12}$$

where $\mathbf{A_c} = \begin{bmatrix} 0 & 1 & 0 & -\frac{l}{\sigma} \end{bmatrix}$

Using non-holonomic constraint combine with Lagrange multipliers to resettle the dynamic Eq. (10), the equation will be in the format as follows:

$$\mathbf{H}(\mathbf{q})\,\ddot{\mathbf{q}} + \mathbf{V}(\dot{\mathbf{q}},\mathbf{q}) = \mathbf{u} + \mathbf{A}_\mathbf{c}^{\mathrm{T}}\lambda \tag{13}$$

The $\mathbf{H}(\mathbf{q})$ in the equation is a 4 × 4 matrix, and $\mathbf{V}(\dot{\mathbf{q}},\mathbf{q})$ is a 4 × 1 matrix. The λ in the equation is the Lagrange multipliers. Then the equation can be farther trimmed into a new format like following:

$$\dot{\mathbf{q}} = \mathbf{C}(\mathbf{q_r})\dot{\mathbf{q}}_\mathbf{r} \tag{14}$$

where $\dot{\mathbf{q}}_\mathbf{r} = [\,\dot{\varphi} \quad v_r \quad \dot{\sigma}\,]^{\mathrm{T}}$. The $\mathbf{C}(\mathbf{q_r})$ is a 4 × 3 matrix that each line of which is an new space vector of $\mathbf{A_c}$.

Calculi Eq. (14), and there is a new equation as follows:

$$\ddot{\mathbf{q}} = \mathbf{C}(\mathbf{q_r})\,\ddot{\mathbf{q}}_\mathbf{r} + \dot{\mathbf{C}}(\mathbf{q_r})\dot{\mathbf{q}}_\mathbf{r} \tag{15}$$

The dynamic equation with three variables can be obtained by putting Eq. (15) into Eq. (13) and multiplies with $\mathbf{C}^{\mathrm{T}}(\mathbf{q_r})$:

$$\begin{bmatrix} mh^2 & \frac{mbh\sigma}{l}C_\varphi & 0 \\ \frac{mbh\sigma}{l}C_\varphi & (1-\frac{h\sigma}{l}S_\varphi)^2 + \frac{b^2\sigma^2}{l^2} & 0 \\ 0 & 0 & \frac{J_sC_\varphi^2}{C_\xi^2} \end{bmatrix} \ddot{\mathbf{q}}_\mathbf{r} - \begin{bmatrix} -m(1-\frac{h\sigma}{l}S_\varphi)\frac{h\sigma C_\varphi}{l}v_r^2 + mg(hS_\varphi + \frac{l_tbC_\xi}{l}\sigma C_\varphi) - \frac{bh}{l}C_\varphi v_r\dot{\sigma} \\ 2m(1-\frac{h\sigma}{l}S_\varphi)\frac{h\sigma}{l}C_\varphi\dot{\varphi}v_r + \frac{mbh\sigma}{l}S_\varphi\dot{\varphi}^2 + B_\omega\dot{\sigma} \\ \frac{2J_s}{C_\xi^2}S_\varphi C_\varphi\dot{\sigma}\dot{\varphi} + \frac{mbgl_tC_\xi}{l}S_\varphi \end{bmatrix} = \begin{bmatrix} 0 \\ \tau \\ \tau_\sigma \end{bmatrix} \tag{16}$$

And in Eq. (16): $\mathrm{B}_\omega = 2\left[(1-\frac{h\sigma}{l}S_\varphi)\frac{h}{l}S_\varphi - \frac{b^2\sigma}{l^2}\right]v_r - \frac{b}{l}v_r - \frac{bh}{l}C_\varphi\dot{\varphi}$

When the J_s is small and smooth enough, get $\dot{\sigma}$ as one of the input capacity. Thus the equation can be simplified as follows:

$$\mathbf{M}\ddot{q}_1 = \mathbf{K}_m + \mathbf{B}_m\begin{bmatrix}\dot{\sigma} \\ \tau\end{bmatrix} \tag{17}$$

And in Eq. (17):

$$\dot{q}_1 = [\,\dot{\varphi} \quad v_r\,]^T;$$

$$\mathbf{M} = \begin{bmatrix} \mathrm{M}_{11} & \mathrm{M}_{12} \\ \mathrm{M}_{21} & \mathrm{M}_{22} \end{bmatrix} = \begin{bmatrix} h^2 & \frac{bh\sigma}{l} C_\varphi \\ \frac{bh\sigma}{l} C_\varphi & \left(1 - \frac{h\sigma}{l} S_\varphi\right)^2 + \frac{b^2\sigma^2}{l^2} \end{bmatrix};$$

$$\mathbf{K_m} = \begin{bmatrix} -\left(1 - \frac{h\sigma}{l} S_\varphi\right) \frac{h\sigma C_\varphi}{l} v_r^2 + g(hS_\varphi + \frac{l_t b C_\xi}{l} \sigma C_\varphi) \\ 2\left(1 - \frac{h\sigma}{l} S_\varphi\right) \frac{h\sigma}{l} C_\varphi \dot{\varphi} v_r + \frac{bh\sigma}{l} S_\varphi \dot{\varphi}^2 \end{bmatrix};$$

$$\mathbf{B_m} = \begin{bmatrix} -\frac{bh}{l} C_\varphi v_r & 0 \\ B_\omega & \frac{1}{m} \end{bmatrix}.$$

3 Control System Design

Equation (17) indicates that the torque τ of the rear wheel has no effect on the bicycle body's rolling dynamic. Thus design to separately control the bicycle running forward and the bicycle balance. In the forward control, the running speed can get stable through adjusting the torque of the motor and stabilize the rotational speed of the rear wheel according to the different command of forward running speed. When balance controlling the system, using the Steering Angle Balance Control method. When the bicycle body is incline, the angle sensor can sense the roll angle as well as the angular velocity. Using the signal processing to get the accurate roll angle, then use the PI controller to adjust the steering angle which can keep the balance of the bicycle. After the system gets balanced, if the reference angle φ_d of the bicycle body is changed, the bicycle curve can be achieved using the Steering Angle Balance Control method which stated above (Fig. 3).

When the forward running speed is an invariable, which means $\dot{v}_r = 0$, the dynamic equation of the system can be simplified as:

$$\ddot{\varphi} - \frac{g}{h} S_\varphi = \frac{\sigma^2 S_\varphi C_\varphi}{l^2} v_r^2 + \left(\frac{g l_t b C_\xi}{l h^2} C_\varphi + \frac{v_r^2 C_\varphi}{l h}\right)\sigma - \frac{b C_\varphi v_r}{l h} \dot{\sigma} \tag{18}$$

When φ is a small angle, the linearization of the system make the transfer function of the system is:

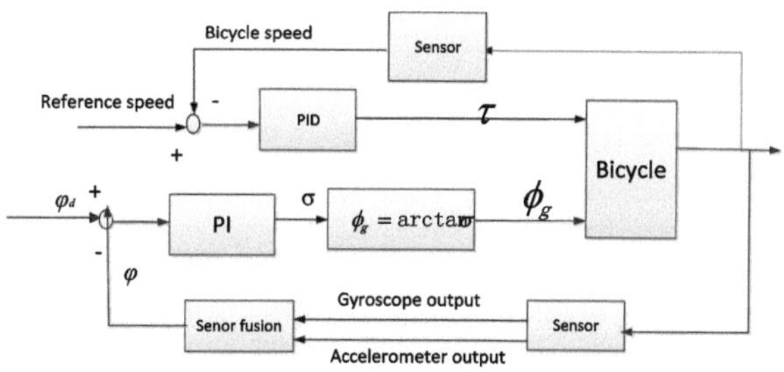

Fig. 3 The block diagram of control system

Table 1 The parameters of rider-less bicycle

Mass of tires (m_W)	0.5 kg
Mass of bicycle M	3.1 kg
Moment of inertia of the steering fork (J_s)	3.675E −002 kg/m^2
Steering front fork angle ξ	15°
Horizontal distance between the line which contains the rotation axis and the contact-point of the front wheel (l_t)	0.018 m
The horizontal distance between the centroid and the center of the rear wheel (b)	0.15 m
Radius of tires(r)	0.07 m
Distance between the center of two tires (l)	0.29 m
Height of the bicycle centroid (h)	0.25 m
The horizontal distance between the centroid and the center of the rear wheel (b)	0.15 m

$$G_{\varphi\sigma}(s) = -\frac{\frac{bv_r}{lh}s + \frac{hv_r^2 + gbl_tC_\xi}{lh^2}}{s^2 - g/h} \tag{19}$$

Design and make the small rider-less bicycle for laboratory use, the detail parameters are as Table 1:

The Open loop transfer function is:

$$G_k s = \frac{-2.069v_r s - (1.41 + 13.793v_r^2)}{s^2 - 39.2}$$

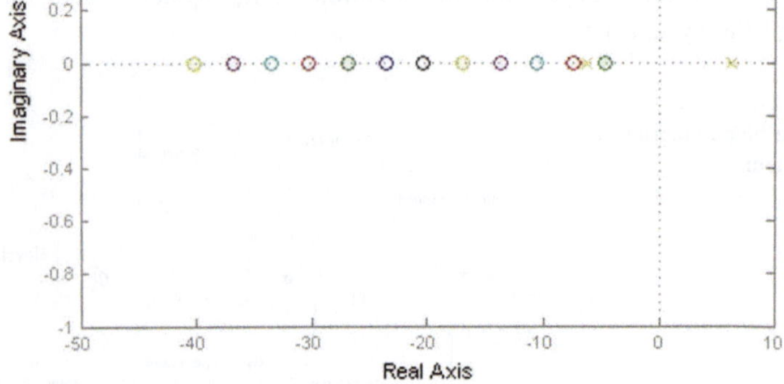

Fig. 4 The zeros and poles while v_r is increasing

Figure 4 shows that with the increasing of v_r, a pole of the system is always stays in right half planc. This is an unstable system, therefore, through the feedback of the real-time roll angle φ, the excogitation that using PI controller to control the motor rotation angle of the steering can maintain the steady of the system. The angular amount of the steering angle can be expressed as follows:

$$\sigma = -K_P(\varphi - \varphi_d) - K_I \int (\varphi - \varphi_d)dt \tag{20}$$

Using Laplace transform can get the expression of the controller's transfer function D:

$$D(s) = K_P + \frac{K_I}{s} = \frac{sK_P + K_I}{s} \tag{21}$$

And the Closed loop transfer function of the system can be ensured by Eq. (20):

$$Gs = \frac{G_k sD(s)}{1 + G_k sD(s)} \tag{22}$$

As an example, the preform of the reference roll angle φ_d is a square wave. By final-value theorem, the state of the roll angle can be obtained after the system becomes stable.

$$\lim_{t \to \infty} \varphi(t) = \lim_{s \to 0} sG(s)\varphi_d(s) = 0 \tag{23}$$

where $\varphi_d(s) = \delta(\frac{1-e^{-\varepsilon s}}{s})$, δ is the peak value of the square wave, ε is the duration time of the square wave.

The result indicated that Steering Angle Balance Control method can make the roll angle to be convergent, and that a square wave preform can be used as the reference roll angle φ_d to realize the bicycle's steering.

4 Simulation by ADAMS

This article uses the ADAMS (Automatic Dynamic Analysis of Mechanical Systems) software to imitate the control method which is indicated above. Through building a motor model of rider-less bicycle in the ADAMS software environment (Fig. 5) to observe the actual effect of the Steering Angle Balance Control method that indicated in this article.

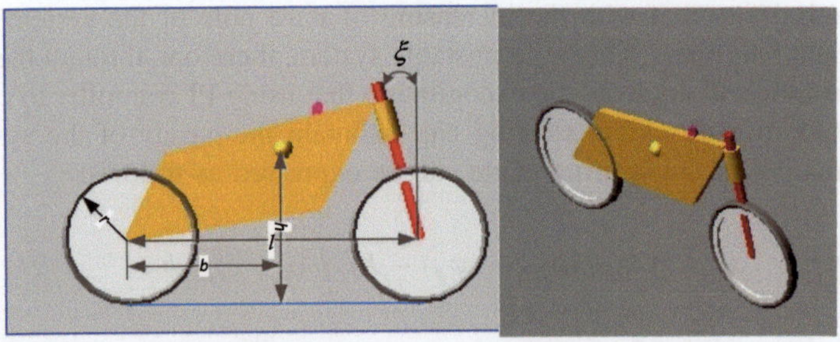

Fig. 5 The front elevation and the three-view drawing of ADAMS model

4.1 The Simulation of Balance Control Under Different Target Speed

Put different torques to the rear wheel as the impetus for the bicycle to move forward. The rotational speed of the rear wheel is increased from zero to 410, 820, 1240 degree/s, and let the eventually theoretical forward speed be 0.5, 1, 1.5 m/s. After the speed being stable, adding a disturbing torque at centroid. Design the value of K_I, K_P of the balance controller separately to make sure that the bicycle steadily moving forward (Figs. 6, 7, 8, 9, 10, 11 and 12; Table 2).

From the simulation data which listed above, some conclusions can be attained as follows:

1. Because the bicycle tires will have small distortion when the bicycle is running forward, the radius of gyration will be smaller than the actual radius. Therefore, the actual speed of running steadily forward will be less than the value that rotational speed multiplies with r.

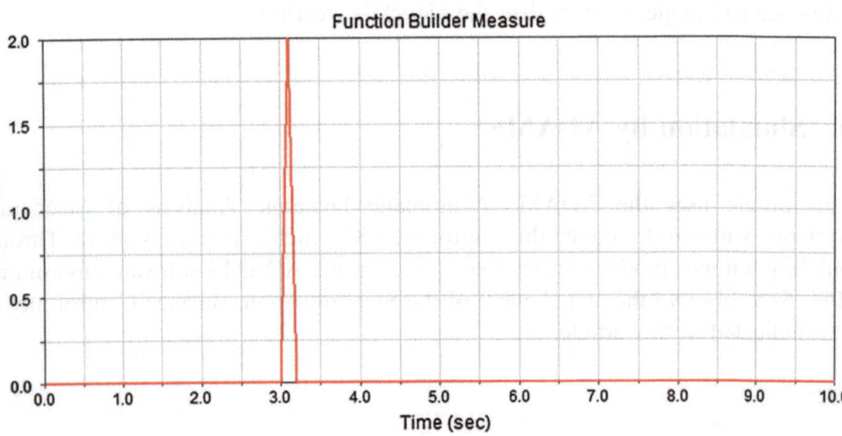

Fig. 6 The disturbing torque which added to the centroid after the system become stable

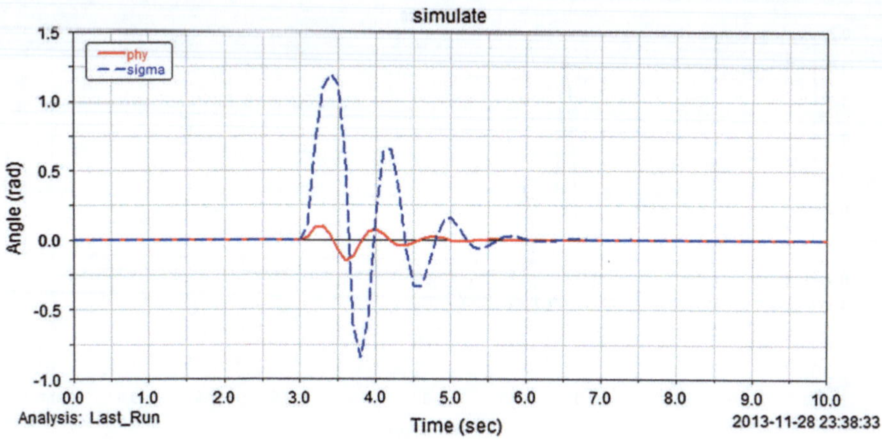

Fig. 7 The result of ϕ and σ when the target rotational speed of ADAMS model is 410 degree/s

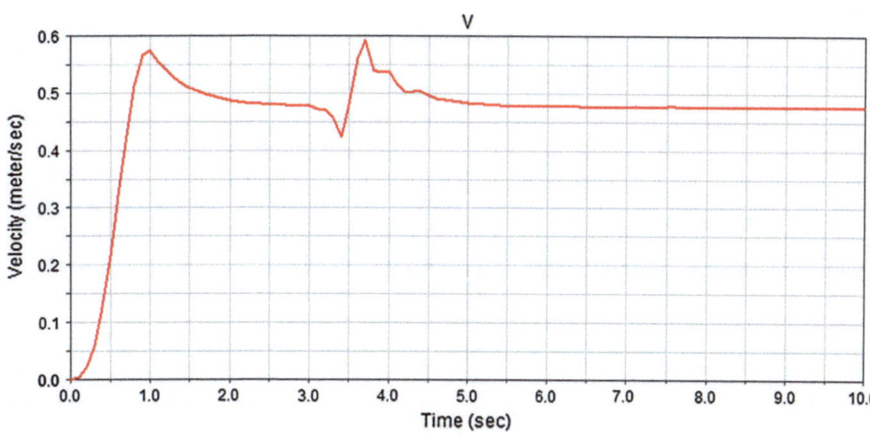

Fig. 8 The speed of the bicycle when the target rotational speed of ADAMS model is 410 degree/s

2. When the bicycle body incline to the right, the steering will turn right. It will make the weight of the body shift to the left and engender a gravity torque. And the gravity torque will pull the system back to the balance point again, Vice versa.
3. The speed v_r of the system has negative correlation with K_P and K_I of the balance controller: when the speed v_r increases gradually, the control coefficient $K_P(3-2.5)$ and $K_I(100-30)$ that make the system stable and the stable time is always 3 s (from the third sec to the sixth sec) are both decrease.
4. The smaller target speed that given to the rear wheel, the easier the running forward of the bicycle will be affected by the torque of the steering; according to

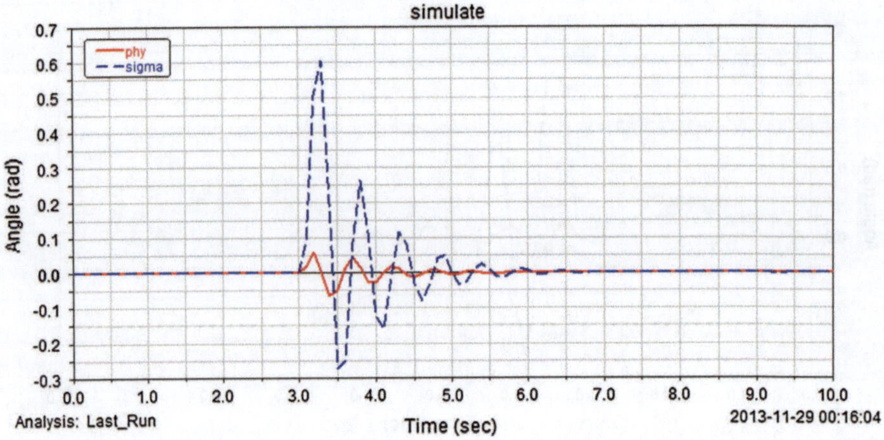

Fig. 9 The result of ϕ and σ when the target rotational speed of ADAMS model is 820 degree/s

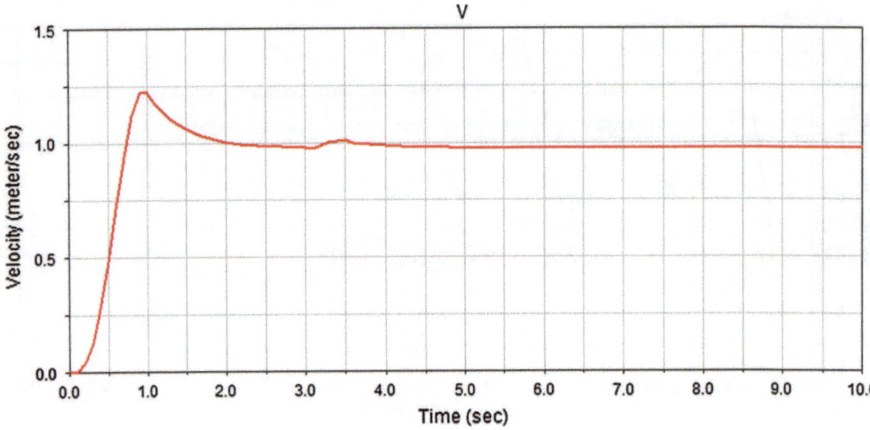

Fig. 10 The speed of the bicycle when the target rotational speed of ADAMS model is 820 degree/s

this assumption, this is one of the reason why the balance control is difficult when the speed approaches to zero.

5. When the command of the rotational speed which given to the rear wheel accumulates from zero to the stable speed, the speedup-static-friction that tires get from the ground is heavier than the uniform velocity-static-friction. When the command suddenly changed from accumulate speed state to the stable speed, however, the bicycle still get a strong effect due to the speedup-static-friction which makes the overshoot phenomenon exists in the bicycle's speed.

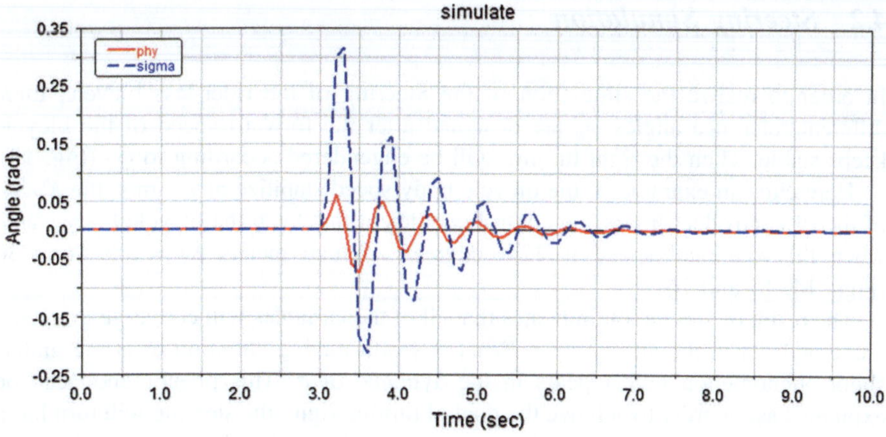

Fig. 11 The result of ϕ and σ when the target rotational speed of ADAMS model is 1240 degree/s

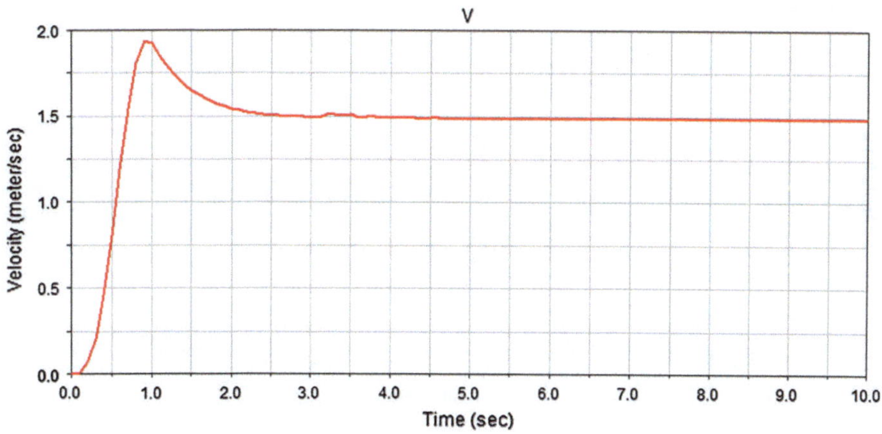

Fig. 12 The speed of the bicycle when the target rotational speed of ADAMS model is 1240 degree/s

Table 2 The relationship between PI parameter and target speed in the balance controller

v_r	K_P	K_I
0.5	3	100
1	2.8	80
1.5	2.5	30

4.2 Steering Simulation

In order to realize the simulation of the Steering of the rider-less bicycle, three different reference angles φ_d are designed after the forward speed of the bicycle keeps stable. Then the right turning will be engendered according to φ_d (Fig. 13).

Here cites an example of the bicycle-body speed stabilize at 1.5 m/s, the K_P, K_I of the PI controller choose the control coefficient that keep the bicycle body stable when the rotational speed is 1240 degree/s. Which means $K_P = 2.5,\ K_I = 30$ (Figs. 14, 15 and 16).

The result of simulation indicates that all of the variables will converge to a fixed value and get in the steady-state. What is worth paying attention to is the undershoot phenomenon will appears in the dynamic of σ. This phenomenon can be explained as, in order to achieve the goal of turning right, the steering will turn left a little at first, which makes the weight center engenders a right incline gravity torque.

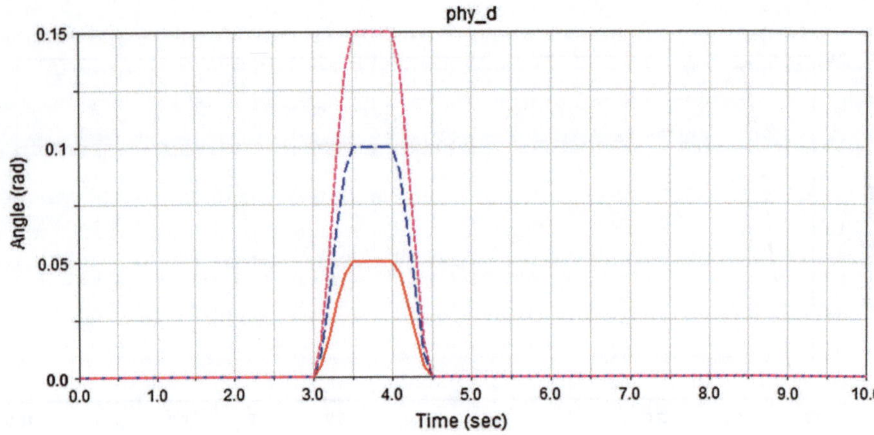

Fig. 13 Three different reference angles φ_d of ADAMS

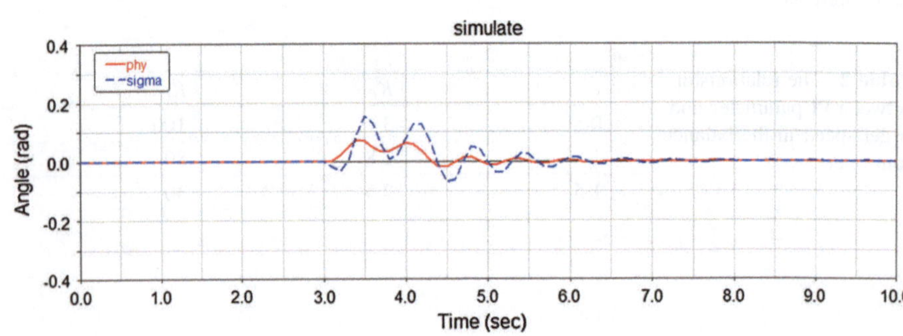

Fig. 14 The result of ϕ and σ when the reference angle φ_d of ADAMS is 0.05 rad

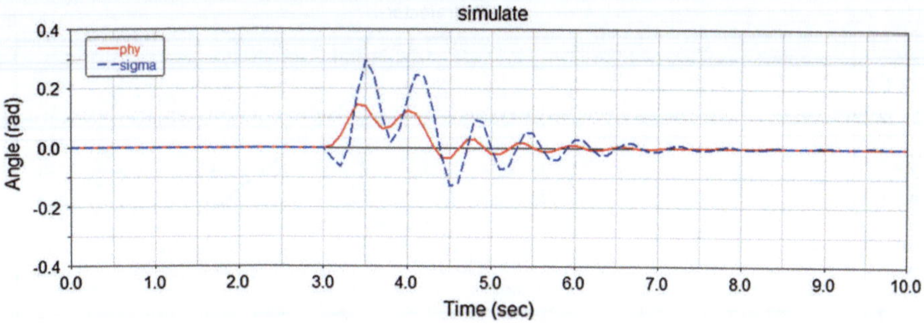

Fig. 15 The result of ϕ and σ when the reference angle φ_d of ADAMS is 0.1 rad

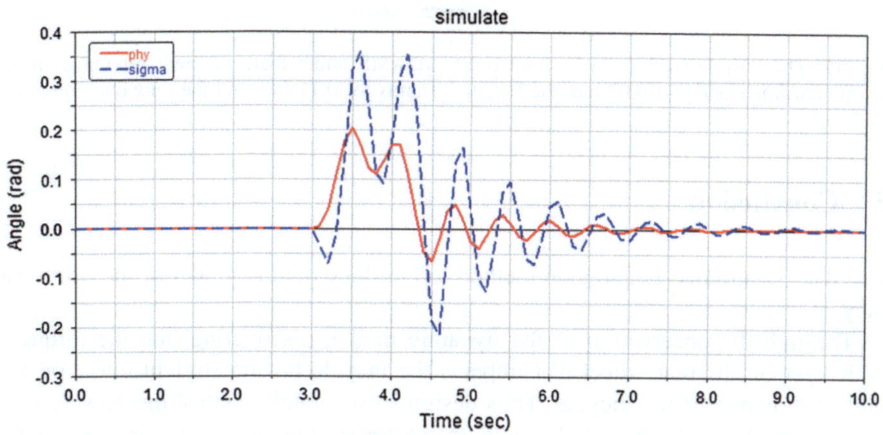

Fig. 16 The result of ϕ and σ when the reference angle φ_d of ADAMS is 0.15 rad

Then the steering will turn right immediately for keeping balance. Until the reference angle return-to-zero, the balance controller finally pushed the bicycle back into vertical and running along a straight line.

Figure 17 contains the bicycle tracks under three different reference angle, the original point is the beginning of the bicycle, the red line is the turning centroid's track while the reference roll angle is 0.05 rad, the blue line is the turning centroid's track while the reference roll angle is 0.1 rad, and the pink one is the turning centroid's track while the reference roll angle is 0.15 rad. These lines show that the bigger the bicycle body's reference angle, in the same time, the smaller the turning radius of the bicycle; the bigger the angle that the bicycle turns, also the bigger the angle which the steering is supposed to turn.

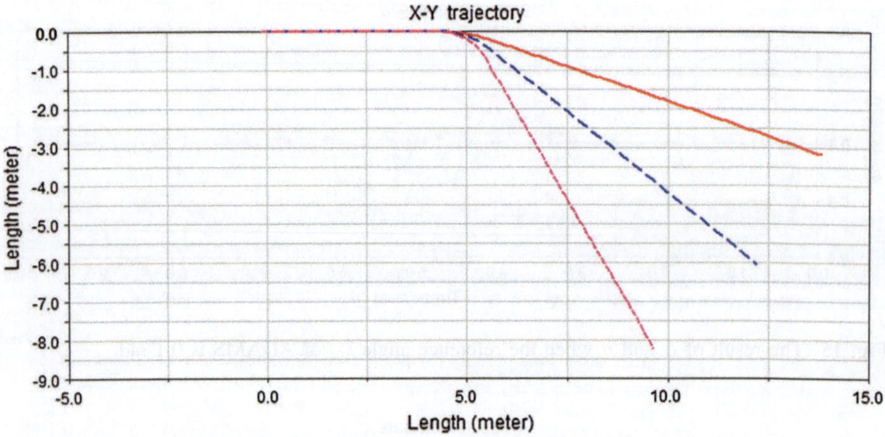

Fig. 17 The bicycle tracks which simulate by ADAMS under there different reference angles φ_d(the *red line* stands for 0.05 rad, the *blue line* stands for 0.11 rad, and the *pink* one represents 0.15 rad)

5 Conclusion

Use the Constrained Lagaragian method to build a dynamic model of the rider-less bicycle.

Through the observation of the dynamic model, discovering that the torque τ which set on the rear wheel that impetus the bicycle has no straight effect on the incline dynamic of the bicycle. Thus design to separately control the bicycle running forward and the bicycle balance. In the forward control, the rear wheel get a stable speed through adjust the torque of the motor under different forward speed commands. In the condition of the bicycle's speed is stable, design a steering-angle-balance-control method to keep the bicycle system in balance. Through feedback the real-time roll angle of the bicycle and control the curve of the steering by controller.

Use the simulation of the ADAMS, simulating the balance control and steering of the rider-less bicycle separately. The simulation indicates that the Steering Angle Balance Control method can realize the bicycle balance control and the curve control effectively. Meanwhile, the bigger the speed of the bicycle, the smaller the control coefficient (K_P, K_I) of the balance controller PI. Thus the bicycle body is easier to control; besides, after the bicycle get to the target speed.

References

1. Getz, N.H., Marsden, J.E.: Control for an autonomous bicycle. In: Proceedings of 1995 IEEE International Conference on Robotics and Automation, vol. 2, pp. 1397–1402 (1995)
2. Iuchi, K., Niki, H., Murakami, T.: Attitude control of bicycle motion by steering angle and variable COG control. In: 31st Annual Conference of IEEE (IECON 2005), 6 pp. Industrial Electronics Society (2005)
3. Tanaka, Y., Murakami, T.: A study on straight-line tracking and posture control in electric bicycle. Industr. Electron. IEEE Trans. **56**(1), 159–168 (2009)
4. Tanaka, Y., Murakami, T.: Self sustaining bicycle robot with steering controller. In: The 8th IEEE International Workshop on Advanced Motion Control, 2004 (AMC'04), pp. 193–197 (2004)
5. Yonghua, H., Qizheng, L., Shimin, W., Lei, G.: Control system design for a bicycle robot based on double DSP. Comput. Meas. Control, **19**(12), 2965–2968, 2988 (2011) (in Chinese)
6. Guo, L., Huang, Y., Liao, Q., Wei, S.M.: Kinemics analysis for self-balancing bicycle robot. J. Beijing Univ. Posts Telecommun. **34**(6), 99–102 (2011) (in Chinese)
7. Zhong W.: Design of embedded control system for the riderless bicycle. Beijing University of Posts and Telecommunications (2010) (in Chinese)
8. Yu X.: A study on the control and realization for bicycle robot nonlinear system. Beijing University of Posts and Telecommunications (2010) (in Chinese)
9. Jingang, Y., Yizhai, Z, Dezhen, S.: Autonomous motorcycles for agile maneuvers, part I: dynamic modeling. In: Proceedings of the 48th IEEE Conference on Decision and Control, 2009 held jointly with the 2009 28th Chinese Control Conference. CDC/CCC 2009, pp. 4613–4618 (2009)

The Management of Slight Traffic Accident Based on the Internet

Desheng Yu, Laimei Fu and Gong Qin

Abstract With the rapid development of social economy, the number of motor vehicles is increasing rapidly, especially the number of private cars, which leads to many urban traffic congestion. The rate of traffic accident is increasing, therein slight traffic accidents account for the most. Using mobile Internet technology, remote guidance without casualties and accidents by the two sides to negotiate and determine the responsibility, shorten the time of the accident waiting and processing, can greatly ease the traffic congestion problem, optimization and promotion based on the Internet traffic minor accident processing mode, is an effective measure to achieve intelligent traffic.

Keywords The internet · Intelligent transportation · Slight accidents

1 Introduction

With the development of the social economy, People's living standards and income are rising. More and more people choose to buy private cars in order to improve the convenience of travel. As a result, the number of motor vehicles increase rapidly, which lead to the shortage of road resources, increasingly contradictions of traffic supply and demand, urban traffic problems increasingly serious etc. In another way, the unreasonable road network layout and structure, the development lagging of bus

D. Yu · L. Fu (✉) · G. Qin
East China JiaoTong University, #330013 Nanchang, China
e-mail: 503263137@qq.com

D. Yu
e-mail: 545617186@qq.com

D. Yu
School of Economics and Management, East China JiaoTong University, Nanchang, China

© Springer Science+Business Media Singapore 2017

H. Lu (ed.), *Proceedings of the Second International Conference on Intelligent Transportation*, Smart Innovation, Systems and Technologies 53,
DOI 10.1007/978-981-10-2398-9_3

and the inadequate traffic management level all can be the main direct factors, which lead to the bad condition of urban traffic. Data show that the average speed of the majority of large urban motor vehicles in the peak period of travel is low to 10 km/h, the residents can get out of travel efficiency.

Traffic congestion not only affects the travel efficiency, but also consumes a lot of energy, which hinders the rapid development of economy and society, meanwhile, increases the probability of road traffic accidents in a large level. According to the current traffic accident processing system, the accident vehicle is instructed to remain where it were, as to waiting for the traffic police or insurance personnel to deal with. Obviously, the procedure would take the road for a long time, which probably lead to more serious congestion, even part of the roads traffic paralysis. Therefore, this paper proposes a kind of traffic accident processing mode, which is based on the Internet, to achieve the purpose of rapid processing of traffic accidents.

2 The Current Slight Traffic Accident Processing Mode

At present, there are four small modules in the process of dealing with the traffic accident. It is the module of the accident, the traffic police brigade, the insurance company and the vehicle maintenance (Fig. 1).

2.1 The Traffic Police Brigade Module

When the traffic accident happened, and got an accident alarm call, the traffic police should be carried out to deal with it as soon as possible. In accordance with the relevant provisions of the slight traffic accident treatment should assigned at least 1–2 police officers to do the prospecting and responsibility in the scene [1]. With such a high incidence of accident and shortage of police, the traffic police can't arrived in time, or the accident owners don't cooperate, at same time the parties push to take

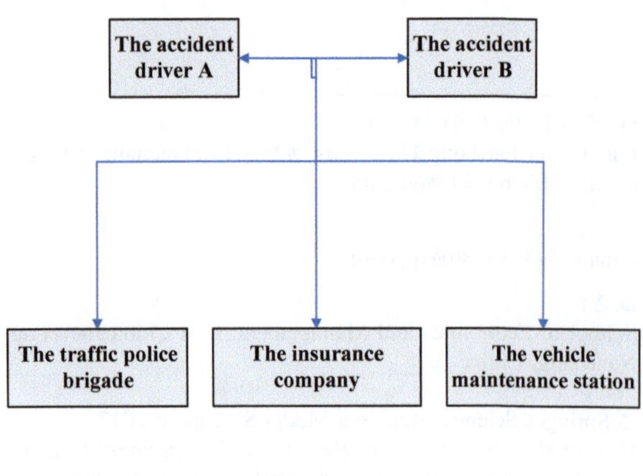

Fig. 1 The flow chart of current slight traffic accident processing

off the accident responsibility, which leads to the accident vehicle could not in time to evacuate the scene of the accident that situation occurs frequently, and which causes the traffic jams.

2.2 The Accident Driver's Module

Owners in a traffic accident, should ensure that the accident scene, and then the two sides of the vehicle to determine the handling of the accident. There are generally two ways: first, the two sides negotiated settlement; the two is to contact the police and insurance companies to the scene [2].

2.3 The Insurance Company Module

Insurance companies in the accident after receiving the phone call to the scene, the division of responsibilities, and confirm the issue of compensation. After the division of responsibility and confirmation of compensation for the incident, they should prepare the relevant information for compensation.

2.4 The Vehicle Maintenance Station Module

At present, most of the vehicle maintenance point have not joined the Internet platform, and their customers are mainly from nearby residents' vehicles and the passing vehicles, who using the Internet platform to attract customers rarely. According to the visiting survey, there have more business in the section of good vehicle maintenance points. Where almost without using of network platform to drum up business, only some of the location is not good shop through the addition of such as sticky web site to attract customers, but the effect is not ideal.

3 The Processing Pattern of Slight Traffic Accident Based on the Internet

3.1 Network Topology

Network platform is the information control center that deals with the minor traffic accident, the analysis and responsibility division, insurance compensation and vehicle maintenance. When a minor traffic accident happens, the parties involved take out mobile phone or digital camera or video camera to photograph the scene of the accident according to the requirement and then send these photos to the network terminal. The traffic police through the network terminal assess the scene photos whether they are valid or not. If the scene photos are valid then the accident can be accepted and then submitted to the analysis and responsibility division. If the photos are invalid then a request is made for the parties to retake the photos. Insurance company through the Internet, make the insurance settlement according to the photos uploaded by the parties, at the same time they also communicate with the traffic police, after that they can make determination. Vehicle maintenance station through the network platform make plans for the owners to choose from. These plans are made according to the uploaded photos.

Based on Internet slight accident handling information network topology is shown in Fig. 2.

3.2 Accident Treatment System Module Partition

The detailed partition of system module is not only need to consider the setting of the accident processing jobs, but also take the accident treatment procedures into

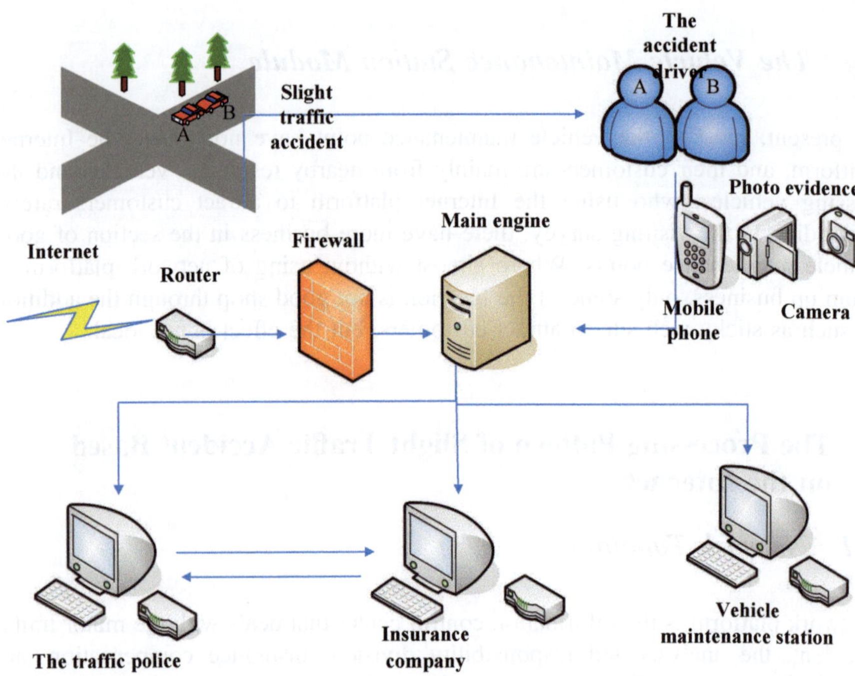

Fig. 2 Based on the Internet slight traffic accident treatment topology

account. Slight accidents processing system which is mainly related to the content of four modules, as shown in Fig. 3.

3.2.1 The Traffic Police Brigade Module

The traffic police don't need to the scene of the accident any more, for based on Internet slight traffic accident treatment, who directly through the network platform to accept online. The slight accidents which can be processing rapidly that the traffic police request the parties to take at least three photos (The first one is the location of the collision; The second one is to stand in the accident, the photos taken before and after the position in each 5 meters away. What must to take a road lines, the yellow line and dotted line in pictures; the third one is using panoramic model to catch the background photo) to upload online. If the photos are effective, then the traffic police accept this accident online, and the parties can evacuate away the scene of the accident in a timely. For conditions in retreat quickly, but the parties don't withdraw from the scene in time, at the same time cause traffic jams, traffic administration can drag the accident vehicle mandatory away from the scene, and the driver should be fined. After making photos are effective, if one side not go to the fast center to handle matters of the accident what is the party's joint agreement. The traffic administration will hold accountable the party who doesn't abide by the agreement, which is guaranteeing the effective implementation of the retreat quickly, and improving the efficiency of slight traffic accident treatment. Reducing the traffic congestion, meanwhile, saving manpower and financial resources [3].

3.2.2 Owners Module

It is more convenient and efficient for the parties to process the slight traffic accidents by network. Only taken forensic photographs do as the traffic regulations, then waiting for the traffic police to accept online. After accepting then the owners can be evacuated away the scene of the accident immediately, without quarrelling at

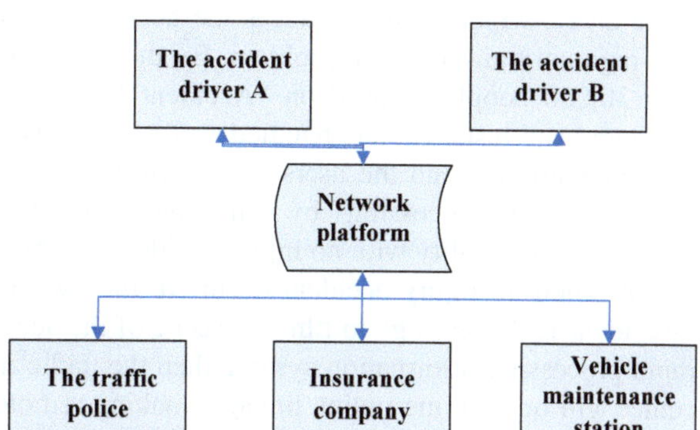

Fig. 3 Accident processing system module partition

the scene of the accident, or waiting for the traffic police and coming the insurance company to perform the responsibility, which also saved a lot of procedures and time, and improved efficiency [4].

3.2.3 Insurance Company Module

According to the parties provided photos of the scene of the accident and the defined responsibility by the traffic police department, then the Insurance Company processing the compensation. That can save personnel rushed to the scene to confirm responsibility, and also save manpower, material and financial resources and that making processing program more concise and convenient.

3.2.4 Vehicle Maintenance Station Module

Through the network platform, displaying the price of the vehicle maintenance station and customer reviews, providing transparent price and reputation, and holding some online discount activities to attract customers. At the same time, that also allows owners to find stores nearby more convenient, so that to realize win-win situation between vehicle maintenance station and the owner.

3.3 The Example of APP—"Accident E Processing" Trying in Parts of Beijing

In May 2015, "Accident E Processing" of mobile phone APP tried out in Zhongguancun boundary of Haidian District and Shahe boundary of Changping District in Beijing City. It is estimated there are over 90 % of the traffic accident which is no injury accident currently in Beijing. According to the traditional processing mode, the accident from the police to move away from the vehicle takes about half an hour. However, using of rapid processing App just needs 5 min to make away, which greatly reduces the accident processing time and ease up to 4 into regional transportation problem. By the end of July 2015, it already has more than 30,000 people focused on "Accident E Processing" and has 10,313 people registered and installed, and handled traffic accidents 185, which obtained the traffic police department and the users' consistent high praise.

"Accident E Processing" by using the mobile Internet technology to remotely confirm responsibility with no injury accident and be negotiated by both sides. That means when no injury accident occurred, the owner could under the direction of "Accident E Processing" to film the scene of the accident photos and upload to the rapid processing information system, then the traffic management bureau command center will be realtime online linkage, making remote determine the responsibility of the accident, and directing the two sides to make their vehicles evacuating lanes and fill out the electronic agreement of self negotiation protocol for traffic accident online. The insurance company will provide a more quickly and conveniently claims service, which according to photos by the scene of the accident and rapid processing of electronic agreement. The "Accident E Processing" software interface as shown in Fig. 4.

According to the Fig. 4 "Accident E Processing" software interface, there has four modules: basic information, instructions, historical records and accident treatment. Basic information has the weather, date, temperature, and Vehicle tail restriction or not, etc. The use of the software process is shown in Fig. 5. At first, entering accident processing module and getting the pop-up warm prompt: when it occurs traffic accident, please open the danger alarm flashlight, also must open outline marker lamps and rear position lamps, and place warning signs in the vehicle rear at night; Two, if there is any party without vehicle license, either party driving without driver's license, or either party driving people suspected of drinking, which must give the alarm call 122 to process in the scene of the accident; Three, we will make the following steps to guide you to deal with traffic accidents: A, taking picture for evidence; B, responsibility identification; C, insurance report. The second step, photo forensics. When taking photos, the "Accident E Processing" will guide you how to obtain evidence in detail. If it occurs the single vehicle accident then photographed forensics directly. In more than two cars and accidents, according to the different types of accident guidance you photo evidence: collision, and lines, retrograde, reverse, slide, switch doors, violation of traffic signal and other. The third step, this APP would check definition of the pictures uploaded after photographing. Photographing again if there is photograph that definition can't

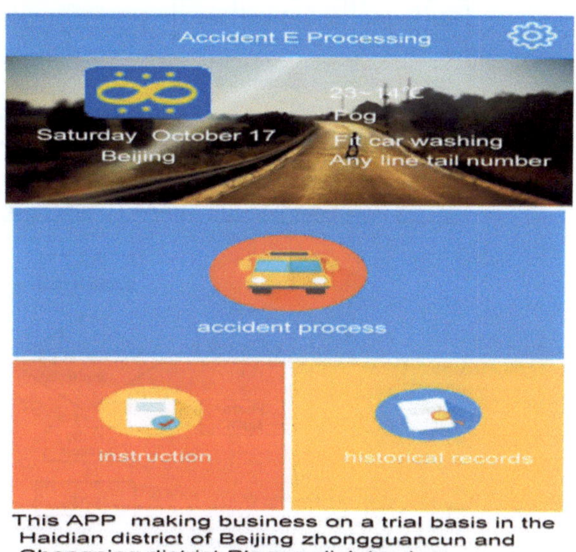

Fig. 4 "Accident E Processing" software interface

afford to requirement. If check successfully, traffic police department accepts and carries out the responsibility confirmation, then the parties put away the cars. The fourth step, insurance reports. After responsibility confirmation, the parties can report self-help insurance through APP. Photographs and responsibility confirmation information would be passed to insurance company automatically, and the APP will finish insurance report. Procedures of self-help insurance report at the scene: self-help alarm, telephone alarm, and not report. If self-help alarm is selected, confirm insurance company. After insurance report successful, insurance company will audit and connect with parties, and vehicles of both sides can voluntarily leave. Satisfaction evaluation is available after finishing the procedures of this traffic accident case [5].

4 Optimization "Accident E Processing" Mobile Phone APP

According to the traditional traffic accident processing mode, it takes about half an hour to have the vehicle moved away from the accident's scene from the calling the police. Compared with it, the fast processing App only takes about 5 min or so to

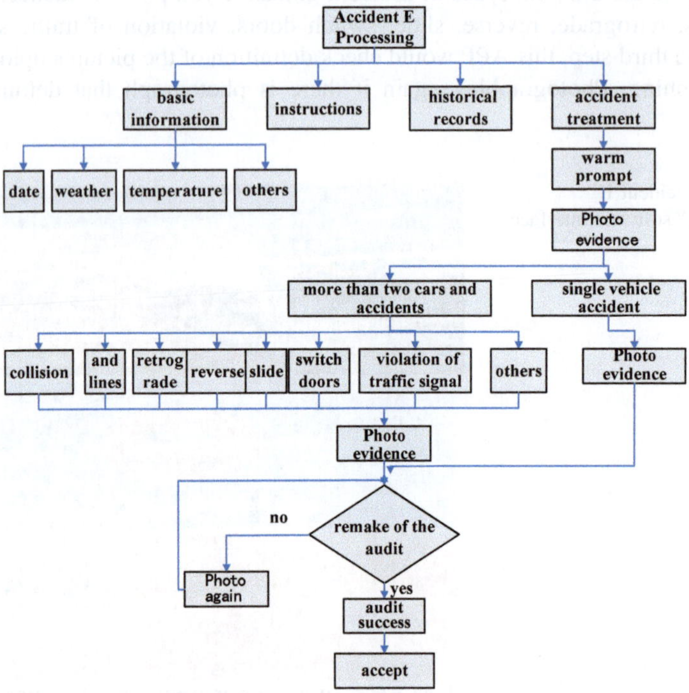

Fig. 5 "Accident E Processing" flow chart of accident

free the road. The app, by its remote guidance on negotiation the responsibility of the two parties on minor traffic accident, eases traffic congestion and greatly shortens the processing time of accident's scene.

In the earlier introduction of the app's fourth steps of the insurance fraud cases, party's insurance must be reported after the ending of the responsibility determination which can not be synchronized with the traffic police, extending the accident processing time. This paper puts forward an optimization suggestion that insurance report and the traffic police's responsibility determination can be determined at the same time. Insurance companies and the traffic police department are on the same network platform and can receive photos and respectively carry out insurance compensation and liability identification at the same time. In case of complex uncertain traffic cases, Insurance can be carried out after the police departments responsible for determining compensation. Meanwhile, we can also add a function module vehicle maintenance module. In accordance with the vehicle damage, vehicle maintenance, the transparent price provided by the repair shop and customer reviews, car owners can choose a satisfactory repair shop nearby which is facilitate the owners and help vehicle maintenance shop to attract customers and achieves a mutual benefit and a win-win situation [6].

5 Conclusion

To quickly handle minor traffic accident, to alleviate urban traffic congestion and improve the traffic environment to create infinite possibility, developed countries represented by the United States has been basically established intelligent transportation platform in urban. Intelligent transportation platform used in practice in our country is very limited. The former mentioned in the article Beijing "Accident E Processing" software, also only in the trial stage in the local area in Beijing. As the traffic laws and regulations to establish and perfect continuously, and the coverage of the Internet and software technology research and development, based on the Internet and software to realize traffic accident treatment needs further research.

Acknowledgments This work has been supported by Youth Foundation of China Ministry of Education and Social Science Project (grant 15YJAZH091) and Social science planning project of Jiangxi province (grant 14YJ24).

References

1. Huang, J.: Road traffic accidents, mechanical identification. Open J. Transp. Technol. **2**, 175–178(2013)
2. Gupta, P., Jain, N., Sikdar, P.K., Kumar, K.: Geographical information system in transportation planning. India Map India Conf. **12**(23), 76–79 (2003)

3. Thobias, Renatus, Richards, et al.: Development of a cost effective GIS crash analysis tool for highway safety improvement. 30th International Traffic Records Forum, 2004
4. Guo, R., Hassan, A.H., Hu Y., et al.: Road traffic accident data collection and analysis for road safety research. Proc. Infants **22**(16), 134–136 (2005)
5. Memit, E.: Calibration and Validation of CORSIM for Swedish Road Traffic Condition [J]. Transp. Res. Board Ann. Meet. **24**(13), 89–92 (2004)
6. Polese, G., Chang, S.K.: Towards a theory of normalization for multimedia databases. Human-Centric Comput. Lang. Environ. **21**(9), 93–95 (2002)

Application of FBG Sensors in High-Pile Wharf Structure Monitoring System

Jiashuai Xu and Guangshuang Ge

Abstract Structure monitoring technology is gradually applied in the field of hydraulic structure; However, the traditional electrical sensor technology has some limitations in the field of hydraulic structure monitoring system, especially for long-term and real-time structure monitoring. In order to achieve long-term and real-time monitoring of hydraulic structures, Fiber Bragg Grating (FBG) sensor technology is adopted in engineering practice in this paper, and first used in a high-pile wharf structure monitoring system of a supported project, the results of loading test and finite analysis verified that FBG sensors have great advantages in hydraulic structure monitoring.

Keywords FBG sensors · Structure monitoring · Real-time

1 Introduction

In structure monitoring fields, vibrating wire sensors are often used to monitor the strain and displacement of the structure components, accelerometers are used to monitor the vibration parameters, and they have the common features that the input and output signals are all electric signal. However, the sensors are usually working in bad environment, electrical signal in this environment is usually influenced by Electromagnetic, simultaneously, the weaves and corrosive of the seawater make the electrical sensors been more easily damaged, thus, the structure monitoring systems based on electrical sensors are weak on long-term and real-time structure monitoring. Compared with electrical sensors, Fiber Bragg Grating (FBG) sensors have great advantages in hydraulic structure monitoring filed; they have a dual function of sensing and transmission. Besides, FBG sensors have some other performances, such as anti-interference ability, corrosion resistance, safety and reliability, durability and high sensitivity, making them very suitable for long-term

J. Xu (✉) · G. Ge
Tianjin Research Institute of Water Transport Engineering, Tianjin 300456, China
e-mail: xxjjssinfo@163.com

© Springer Science+Business Media Singapore 2017

H. Lu (ed.), *Proceedings of the Second International Conference on Intelligent Transportation*, Smart Innovation, Systems and Technologies 53,
DOI 10.1007/978-981-10-2398-9_4

monitoring of hydraulic structures. In this paper, FBG sensors are first used in the structure monitoring system to monitor the strain, displacement and structure vibration of a hydraulic structure.

2 The Performance of FBG Sensors

2.1 FBG Strain Sensors

FBG strain sensors must consider the cross sensitivity between temperature and strain. The sum of the strains caused by the temperature and the load of the object can be given as Eq. 1 (Fig. 1).

$$\varepsilon_0 = K(\lambda_1 - \lambda_0) + B(\lambda_{t1} - \lambda_{t0}) \tag{1}$$

The strain only caused by the change of the load is calculated by Eq. 2.

$$\varepsilon = K(\lambda_1 - \lambda_0) + B(\lambda_{t1} - \lambda_{t0}) - \alpha \times \Delta T \tag{2}$$

In Eq. 2, ε_0 Represents the total strains (unit: $\mu\varepsilon$); ε Represents the strain caused by load (unit: $\mu\varepsilon$); K is the coefficient of the strain gauge strain (unit: $\mu\varepsilon/\text{nm}$, $K = 816.33514455$); B is the temperature correction coefficient of the FBG sensor (unit: $\mu\varepsilon/\text{nm}$, $B = -877.5708325$); λ_1 is the current wavelength value of the grating(unit: nm); λ_0 is the initial wavelength value of the grating (nm, $\lambda_0 = 1558.827$); λ_{t1} is the current wavelength value of the temperature compensation grating, (unit: nm); λ_{t2} is the initial wavelength value of the temperature compensation grating, (unit: nm); α is the thermal expansion coefficient of the object being measured, (unit: $\mu\varepsilon/°\text{C}$); $\Delta T = 100 \times (\lambda_{t1} - \lambda_{t0})$ (unit: °C).

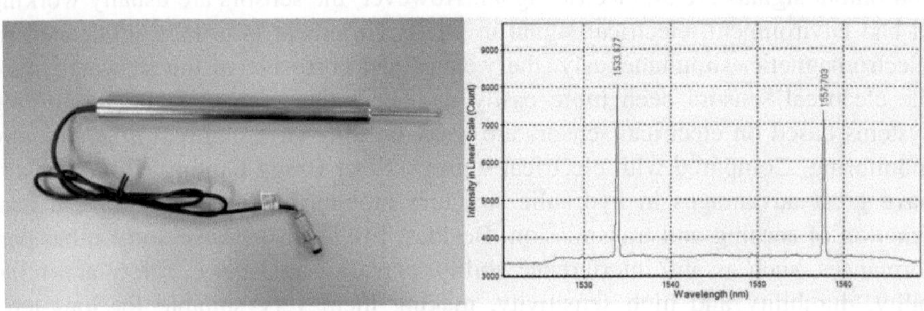

Fig. 1 FBG strain sensor and wavelength characteristic

2.2 FBG Displacement Sensors

The displacement can be given as Eq. 3 (Fig. 2).

$$\Delta L = A((\lambda_1 - \lambda_2) - (\lambda_{10} - \lambda_{20}))^2 + B((\lambda_1 - \lambda_2) - (\lambda_{10} - \lambda_{20})) + C \quad (3)$$

A, B, C is the coefficients of the polynomial ($A = 3.769703159$, $B = 20.49708772$, $C = 0.027215851$); ΔL is the change in length of the FBG displacement sensor(unit: mm); λ_1, λ_2 is the current wavelength values of the grating(unit: nm); $\lambda_{10}, \lambda_{20}$ is the initial wavelength value of the grating(unit: nm, $\lambda_{10} = 1551.657$ nm, $\lambda_{20} = 1549.438$ nm).

2.3 FBG Displacement Sensors

The accelerometer can be calculated by Eq. 4 (Fig. 3).

$$a = \frac{\Delta\lambda}{K} \quad (4)$$

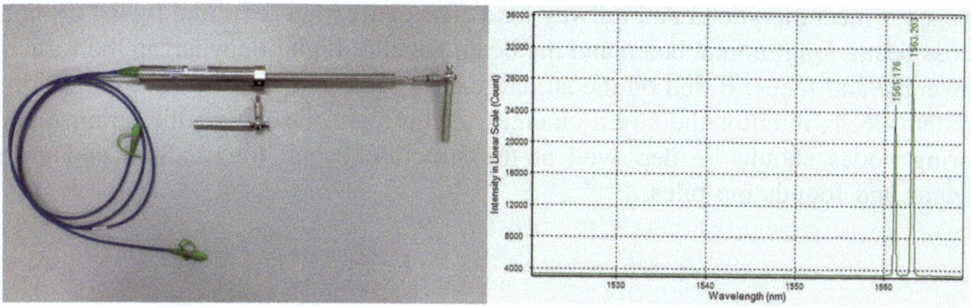

Fig. 2 FBG displacement sensor and wavelength characteristic

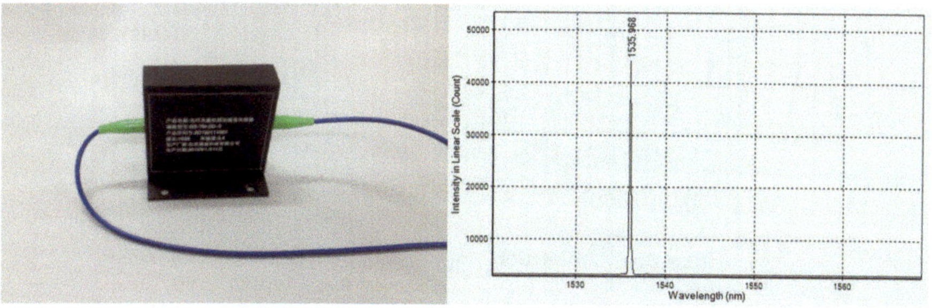

Fig. 3 FBG accelerometer and wavelength characteristic

In Eq. 4, *a* represents acceleration; $\Delta\lambda$ represents the wavelength variation of the peaks and valleys. *K* represents the sensitive of the sensor.

3 The Monitoring System

3.1 *Support Project*

The total length of the wharf is 371.66 m, including the wharf platform length 330 m, width 34 m, distance between the rows is 10.4 m, divided into 5 segments, each row has 6 piles, 2 straight piles and 2 pair of fork piles. The wharf bears important task of production, port average tonnage of the ship continue rise, in order to meet the needs of production, waters on the dock were repeatedly dredged by owners, improved to 20 million tons berths production transformation (Figs. 4 and 5).

3.2 *Monitoring Items*

3.2.1 Strain Monitoring

Vertical load of high pile wharf includes structure self gravity, bulk cargo load, transport machinery load and railway load. The entire load is carried by the ground, cross beam, longitudinal beam and the beam panel of high pile wharf in the form of coverage and focus. Based on the structure stress analysis, the structure monitoring system needs monitor the strain changes of the high pile wharf; the strain monitoring nodes should be deployed at the structure panel, longitudinal and cross beams and foundation piles.

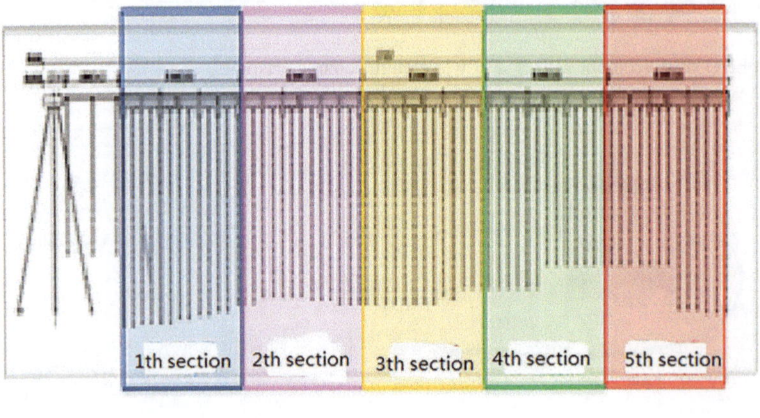

Fig. 4 Structural facade of supported project

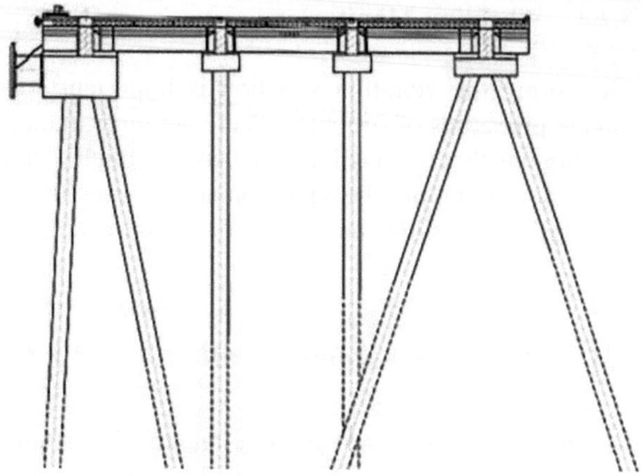

Fig. 5 Structural section of supported project

3.2.2 Relative Displacement Monitoring

Component damage is most likely to occur on the high pile wharf, the front cap of structure contains main panel, pile cap, pile, and shipping components and so on, and all the components are made a continuous structure by reinforced concrete.

Analyzing the damage factories, relative displacement of pier of is caused directly by slope deformation, horizontal displacement is caused by silt clay layer at the wharf, and horizontal displacement can caused the dislocation between the components at the back cap, the increasing load of the yard is the mainly reason of the wharf slope deformation. Corrosion of external environment creates a condition for relative dislocation between the piles and the beams.

After analyze the damage of the components and the relative displacement between the components, according to the relevant specifications and standards of the structure detection, structure health monitoring system must monitor the relative displacement of the components.

3.2.3 Terminal Displacement Monitoring

Due to the inhomogeneous geological structure, the plastic deformation of sub grade and the load of the structure and the working load, the displacement and settlement of the wharf can be occurred; large displacement and settlement may result in the dislocation of the component's cracking or the dislocation of the joint, finally leading to the reduction of the structure durability. The observation of the continuous deformation of the pier is an effective way to grasp the safety condition of the wharf and find out the problem in time. The overall displacement monitoring of the wharf mainly includes horizontal and vertical displacement.

3.2.4 Vibration Monitoring

Monitoring the vibration situation is important; vibration attribute is one of the unique properties of structure. There are three main aspects. First of all, monitoring the ship whether has a collision behavior, timely detection the emergencies. Second, the system monitors the vibration of the structure under the working load. Third content is to monitor the development and change of structure vibration trend.

3.3 Monitoring Items Installation of FBG Sensors

In supported project, the sampling rate of FBG strain sensor and FBG displacement sensor is 1 Hz, the sample rate of FBG accelerometer is 50 Hz. Data is stored in a unified format into the industrial computer and finally into the database (Figs. 6, 7, 8, 9 and 10).

4 Monitoring Data and Simulation Data of FBG Sensors

4.1 Finite Analysis of Strain Changes

In order to prove the correctness of data acquisition, a step by step load test of the wharf structure is carried on, Due to space limitations, this paper only analysis the strain data (Table 1).

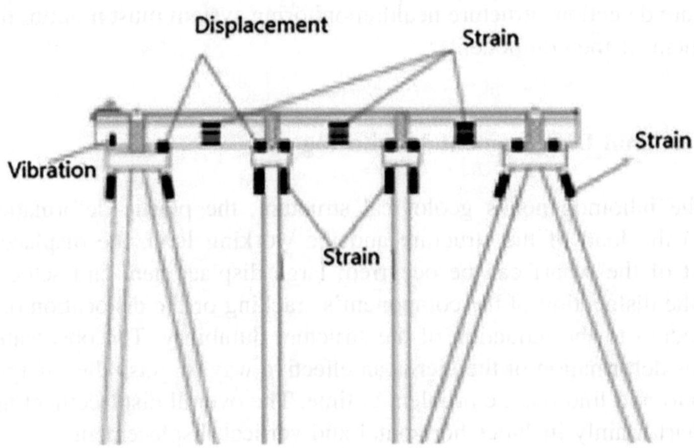

Fig. 6 Deployment of FBG sensors in high piled wharf

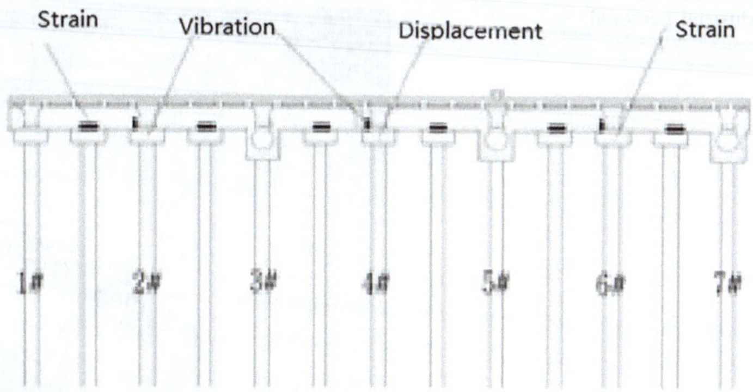

Fig. 7 Deployment of FBG sensors in high piled wharf

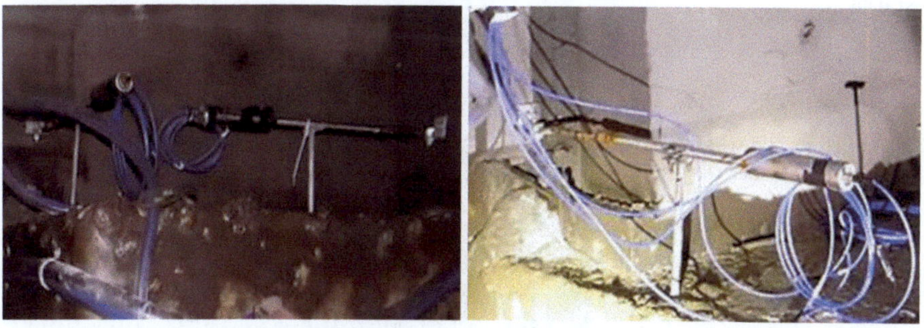

Fig. 8 The installation of the FBG sensor

Fig. 9 FBG sensor demodulation

Fig. 10 Industrial personal computer

Table 1 Step by step load test data

	1st	2nd	3rd	4th	5th	6th	7th	8th	9th
Load	200	100	100	100	100	100	100	100	100
Overall	200	300	400	500	600	700	800	900	1000

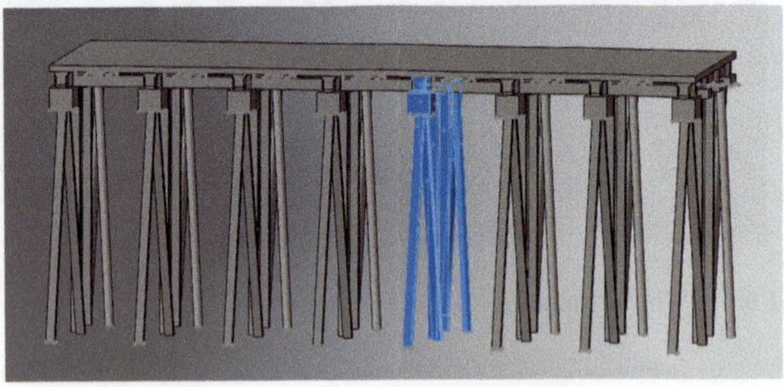

Fig. 11 High pile wharf finite element model

Model dimensions are based on actual size of the project. Finite element analysis tool ANSYS is adopted, contact unit types are targe170 and contac173. High pile wharf is a kind of linear elastic constitutive model. Ground is classic Drucker-Prager model. Figure 11 shows the high pile wharf finite element model.

In calculating progress, simulated load is carried on. The load values ranges from 0 to 200 tons, obtaining the load-strain relationship at the observation points as the leaning samples. Parts of statistical results of numerical analysis are shown in Table 2.

4.2 *Load Test on Support Project*

In supported project, the FBG strain sensors are arranged on a row of a certain structure, the positions of the FBG strain sensors are shown in Fig. 12.

Table 2 Calculated results by finite element method

Load	1st	2nd	3rd	4th	5th	6th
200	64	69	85	87	79	90
300	76	80	96	97	85	94
400	85	88	109	108	93	116
500	97	101	102	115	96	126
600	99	89	126	126	102	134
700	102	110	138	130	115	142
800	104	110	153	140	120	151
900	115	134	167	165	124	162
1000	127	152	176	170	136	174

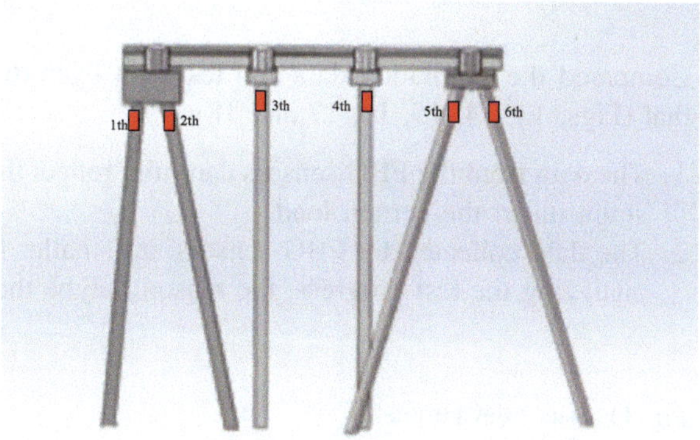

Fig. 12 Positions of FBG strain sensors

For bearing capacity prototype test of high-piled wharf piles, there is no related technical specifications, in the test, loading and unloading processes referenced to the existing relevant technical regulations. The test load is 10 MN; loading process is shown in Table 3. In order to ensure the accuracy of the data, the sensor data is collected after each load when the sensor data is stable. The initial data collected by FBG sensors are wavelength, by formula (1) and (2), the strain parameters can be obtained, the calculated results of the test are shown in Table 3.

Table 3 Test results of pile's strain

Load	1st	2nd	3rd	4th	5th	6th
200	58	60	85	87	79	90
300	68	75	96	97	85	94
400	77	80	109	108	93	116
500	90	99	102	115	96	126
600	94	100	126	126	102	134
700	101	105	138	130	115	142
800	104	109	153	140	120	151
900	113	132	167	165	124	162
1000	116	141	176	170	136	174

4.3 *Performance of FBG Sensors*

Compared the simulation data and test data from different piles, we can conclude that (Figs. 13, 14, 15, 16, 17 and 18):

1. The data from the FBG sensors can fully reflect the change trend of the structure strain under the certain load.
2. The data collected by FBG sensors are smaller than the simulation data, after analyzing the test progress, the reason maybe the installation of the sensors.

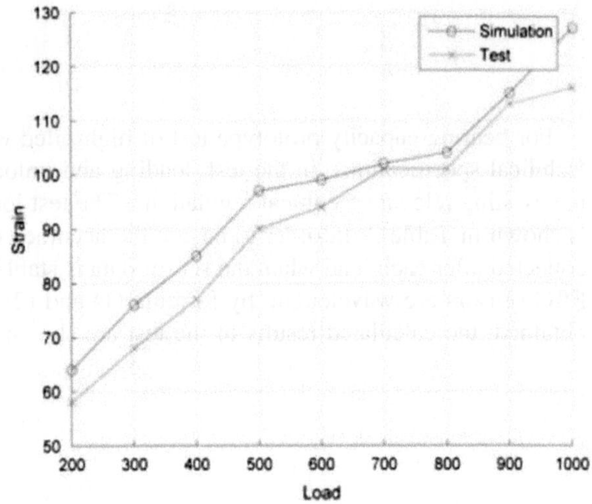

Fig. 13 Data from 1st pile

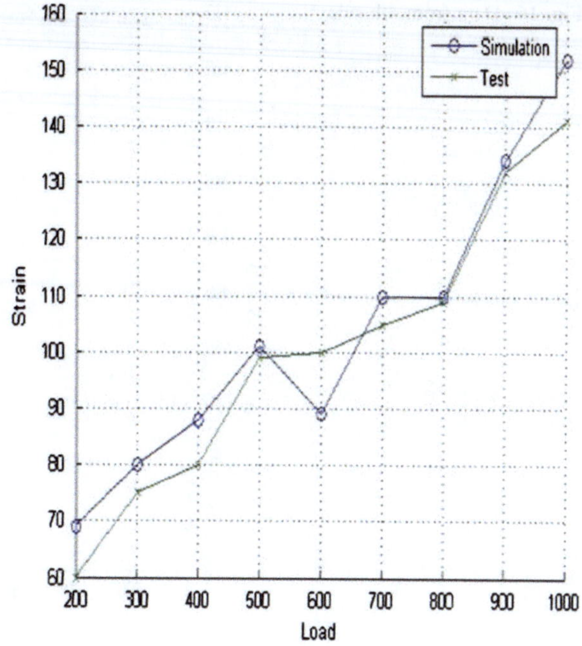

Fig. 14 Data from 2nd pile

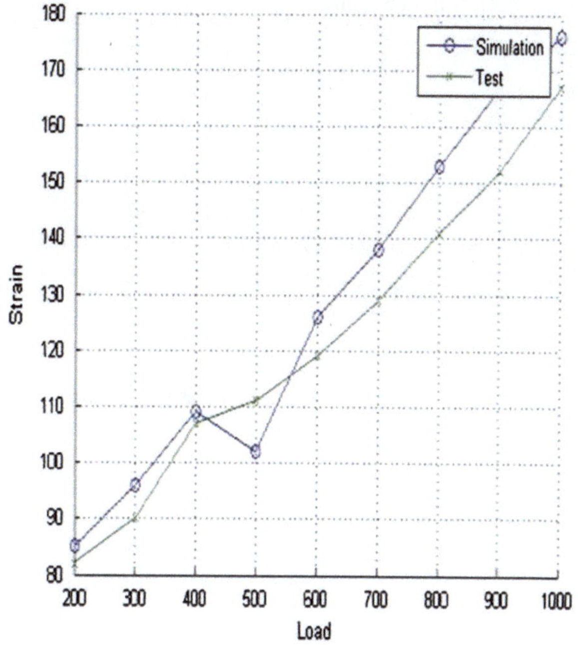

Fig. 15 Data from 3rd pile

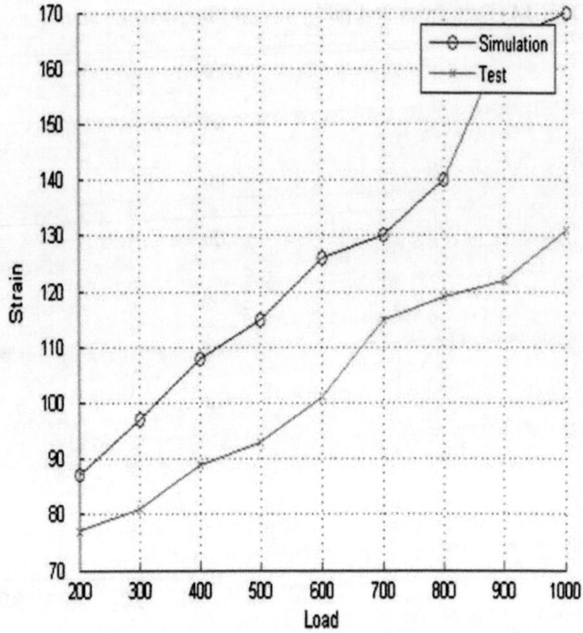

Fig. 16 Data from 4th pile

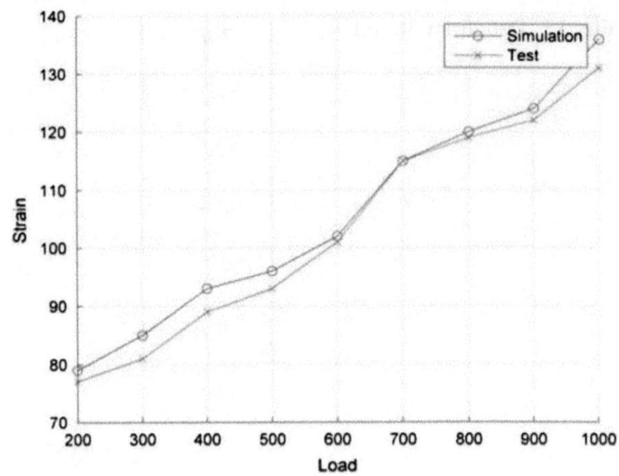

Fig. 17 Data from 5th pile

Fig. 18 Data from 6th pile

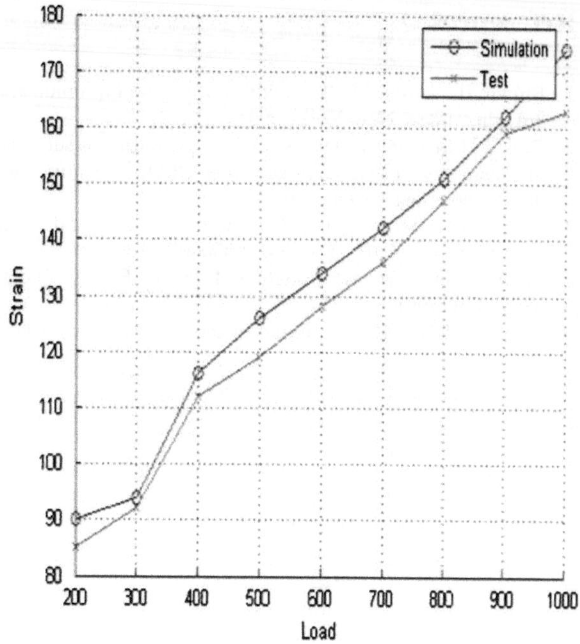

5 Conclusions

In this paper, FBG sensor is introduced into the structure monitoring system of high pile wharf structure, the characteristics of FBG sensors are analyzed. In support project, FBG sensors are adopted, in order to verify the accuracy of FBG sensors, loading test and finite analyze are carried on, simulation data was obtained by finite analysis, test data was obtained through supported project, after analyze the results, we can draw the following conclusions.

1. In order to verify the accuracy of the data from the structure monitoring system, load test and simulation method can be taken.
2. In the fields of hydraulic structures, FBG sensors have great advantages compared with other sensors considering the durability.
3. The installation of the FBG sensors must take correct methods and carry out effective protections, making the data from the sensors being more accurate.

References

1. Brown, B., Aaron, M.: The politics of nature. In: Smith, J. (ed.) The rise of modern genomics, 3rd edn. Wiley, New York (2001)
2. Dod, J.: Effective substances. In: The dictionary of substances and their effects. Royal Society of Chemistry (1999). Available via DIALOG. http://www.rsc.org/dose/title of subordinate document. Cited 15 Jan 1999
3. Slifka, M.K., Whitton, J.L.: Clinical implications of dysregulated cytokine production. J. Mol. Med. (2000). doi:10.1007/s001090000086
4. Smith, J., Jones Jr., M., Houghton, L., et al.: Future of health insurance. N. Engl. J. Med. **965**, 325–329 (1999)
5. South, J., Blass, B.: The future of modern genomics. Blackwell, London (2001)

The Rail Transportation and Bus Intelligent Connection Under the Comprehensive Transportation System

Yong Hu, Min Huo and Gong Qin

Abstract In order to alleviate traffic pressure, most cities in the world using the development of rail transportation as backbone, buses as the main body of the urban transportation network structure. This article puts forward to building the integrated intelligent scheduling system based on Internet platform in order to improve the efficiency, unblocked and consistency of the transfer. This way can realize the goal of the rail transportation and public traffic intelligent connection. This system includes passengers and scheduling command two sub-systems, two subsystems contact with each other and cooperation.

Keywords Rail transportation · Public vehicle · Intelligent connection

1 Introduction

With the development of the society progress, the phenomena of urban traffic congestion problem becomes more and more serious. In order to relieve the traffic pressure, most cities in the world utilizing the development of rail transportation as backbone [1], buses as the main body of the municipal transportation network structure. As an important part of the urban public transportation, rail transportation and public traffic connection will be more harmonious and efficient. Both of them should be more coordinated in order to enhance the efficiency of travel. Therefore, different kinds of comprehensive transportation system intelligent transportation connection is the strategic research direction of the future transport interchange. This article suggests an idea to build the integrated intelligent scheduling system based on the internet platform.

Y. Hu (✉) · M. Huo · G. Qin
School of Railway Tracks and Transportation, East China Jiaotong University, Nanchang 330013, Jiangxi, China
e-mail: 987090508@qq.com

G. Qin
e-mail: 474467408@qq.com

© Springer Science+Business Media Singapore 2017

H. Lu (ed.), *Proceedings of the Second International Conference on Intelligent Transportation*, Smart Innovation, Systems and Technologies 53,
DOI 10.1007/978-981-10-2398-9_5

2 The Analysis of Connection Status

What is called connection? It means one method of transportation change to another. In other words can say it transfer. Being dependent on the traditional significance, transfer including transportation between inner cities and urban transportation between inside and outside.

Developed and in the world, such as New York, Tokyo, London, Paris [2], as early as last century had completed track traffic infrastructure construction. Meanwhile, these countries formed a set of underground, ground and the ground three distinct spatial location effects of three-dimensional public transportation network. The phenomena of urban traffic congestion problems become more and more serious. At the same time provides the further advance opportunity for connection among the unusual vehicle. Bates, who takes part in, has to provide explanations of different types of interchange facilities, on the basis of its service scope and classification, connection of different sites according to the different scale and put forward relevant measures to effectively improve docking.

Katherine F. Turnbull according to the time needed for park-and-ride summed up the rail transit site set up relevant park-and-ride facilities, to ensure the use efficiency not less than 80 %. As for Hall, R, based on the most European countries urban rail transport interchange change to the status qua, analysis of factors which influence the transfer and transfer facilities, then gives appropriate advice on how to set up the tubing.

3 The Analysis of Rail Transit and Bus Interchange Requirements

Bus as the urban traffic system which be used frequency highest, the most widely and the users' largest transportation. It has advantages of a large capacity, network coverage area broadly and the per capita road area is lesser. Especially in today, traffic congestion increasing more serious, many major cities in the world are planning and construction given the priority to building with rail transit skeleton [3], conventional public transport as the main body, under the background of the mode of transportation, urban transportation integration and makes in connection with rail transit vehicles is preferred.

When urban residents travel from location to rail transit site, there has a certain distance. The survey shows that when the travel distance between the location and traffic site more than 600 m, it is too tired to arrive. Meanwhile, the utilization time rate of walking is low, in this case, residents have a tendency travel by bus to transfer go to the site. Therefore, in order to meet the demand of a large number of passengers' transfer, improve the efficiency of public transport overall, to realize the integration of passenger transportation organization, we should optimize the transfer.

4 Discuss About Rail Transportation and Public Traffic Intelligent Connection Mode

With the development of the social economy, urbanization process accelerated and the population agglomeration. Urban residents travel demand growth sharply, traditional single rail transit and bus interchange are too difficult to meet the demand for the rapid progress of society. In the background of intelligent transportation advance, we should use information technology to optimize traffic connection link, exploring the rail transit and bus intelligent connection mode.

4.1 The Analysis of Intelligent Connection Patterns

At present, the bus is fixed departure intervals, more delays due to traffic jams, cars moving, the minimum rate of punctuality and time of arrival in the rail transit site is uncertain. So on the basis of the existing connection should be emphatically considered two cooperate departure and arrival of the means of transportation, reduce transfer passengers waiting time, and improve the efficiency of urban public transportation system of the whole transportation, rail transportation and intelligent shuttle buses coherence.

Rail transportation has its own characteristics, such as fast, large volume, high reliability and automation, which are mainly laid on the principal roads of the city. Especially when the connection sites of passengers from the departure distance greater than 600 m [4], passengers have a tendency transfer by bus. During the transfer process, the bus departure of uncertainty and unexpected road congestion caused by vehicle delay phenomenon will impact transfer efficiency, reduce the overall efficiency of urban transportation. The Internet platform construction of comprehensive transportation system, bus system and rail transportation system integration, so that passengers get the transfer of bus operation information and the connection of the site before the trip, planning ahead travel mode and route, and buy electronic tickets, reduce queue up for the bus coin credit card and waiting time, to improve the transfer efficiency, realize the Internet platform of intelligent feeder. At the same time, the system according to each connection site real-time feedback feeder traffic scheduling shuttle bus vehicles, the deployment of effective organization of vehicles, when passenger flow peak stage, deployed shuttle bus, rapid transit passengers, transfer passenger flow, to ensure the smooth connection of the site; when the low peak traffic period, in a timely manner to reduce shuttle vehicle deployment, reduce empty running.

4.2 Intelligent Feeder Systems

The existing condition of city rail transit and bus connecting the schematic shown in Fig. 1:

From the diagram one, that the existing mode of rail transit and bus feeder is split into six nodes [5], train in the rail transit-rail transit station-the track traffic transfer channel-rail transit access gateway station- bus stations-bus vehicles. Each node is linked to each other, and the traffic flow is bidirectional. In order to make it more coordinated and integrated, the transport process is more coordinated:

Figure 2 integrated scheduling intelligent feeder system is separated into two subsystems, respectively is the passenger system and urban comprehensive transportation dispatch and command system. The two subsystems are for use in collective masses of passengers and urban transportation staff. Between the two systems through the Internet for information exchange, rail transit and bus intelligent feeder, the specific operation is as follows:

(1) Passenger system
Passenger system is a system for the public. This system is comprised of a query system and a ticketing system. When the passengers for daily travel, the system can be carried out through the system to check the bus in the vicinity of the site and the transfer station of the railway station, meanwhile the system

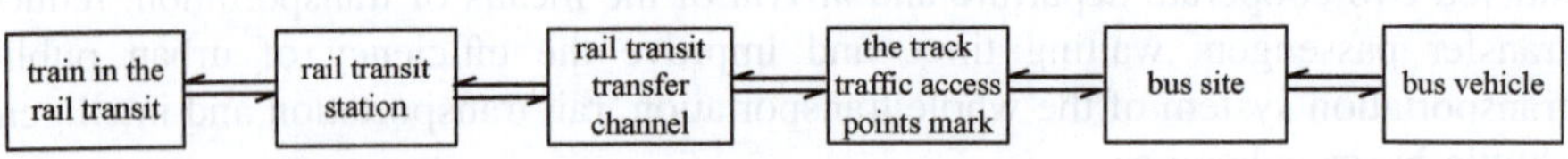

Fig. 1 The rail transit and bus connection process existing mode

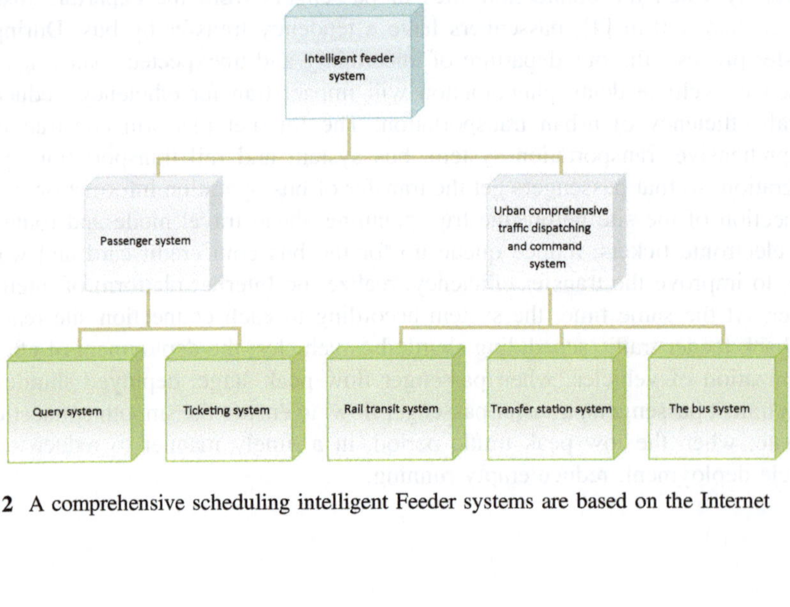

Fig. 2 A comprehensive scheduling intelligent Feeder systems are based on the Internet

through the corresponding program design to provide passengers a detailed travel time plan, including waiting for the bus to the time T1, the bus generally running time T2, the bus stops at the arrival time of T3 and the transit passage of the T4, which constitutes the time required to reach the transfer station T (T = t1 + t2 + t3 + t4). In addition, the query system can also give a rail transit train to arrive at the exact time of the transfer station, the passengers according to the time of the inquiry to arrange their travel plans, planning travel time. Also in order to reduce the time of passengers wait in line to buy a ticket on the train and resulting traffic site traffic congestion phenomenon, ticketing system will passengers ride the bus and rail transportation train two stroke unified ticket system, providing online enough to buy services, system use of mobile phone Bluetooth technology and wireless sensor technology, passengers don not need to buy paper tickets, through the network to buy electronic tickets at each site without queuing, directly on the car, just take out the phone, the use of the platform for the use of electronic tickets to buy tickets, so as to reduce the queuing time to achieve high efficiency and intelligent connection.

(2) Urban comprehensive traffic dispatching and command system
The system receives and processes the traffic information of shared feedback, and arranges the transportation organization reasonably. Its concrete operation process is: When passengers use the query system, according to the time plan given by the system to decide whether to buy tickets, when the passenger ticket system have done, the system will be included in the passenger's travel routes and transfer records and specific travel time, and to bring together all the passenger information submitted to the site of the integrated dispatch command center. Comprehensive dispatch command center in real-time to update the passenger information, and analysis of the different sites of the traffic flow and passenger flow, as well as the transfer of the peak time and the truncated peak period. When transfer flow or access to stop flow is higher than the average threshold, or transfer time in a day in the busy periods of time point, dispatch command center system passes command scheduling to the public traffic system and the transfer station system, increase investment in transport vehicles to bus system, many additional buses commute to the high flow transfer station, transfer of stranded passengers quickly, ease the site's transfer congestion, reduce the waiting crowd. At the same time scheduling command center passes the instructions to the transfer site, the site on duty staff according to this large passenger flow situation enabled station intelligent guiding equipment, effective evacuation, guide passenger flow orderly evacuation, the transfer channel access station crowd density decreased. When the transfer station is depressed, the site can be transferred to the public transport command center of the system. The bus is dispatching center can arrange the vehicle transportation, reduce the number of the bus, improve transportation efficiency and save energy, protect the environment and achieve sustainable development. Passenger system and urban comprehensive transportation

dispatching system interaction, mutual influence, real-time information exchange and update traffic real news, making rail and bus connection in real time intelligent.

5 Conclusion

With the development of information society, Internet is applied more and more widely, based on Internet, the rail transit and bus intelligent feeder is the inevitable trend of future research In this paper, we propose an intelligent feeder system to optimize the allocation of traffic resources, improve the efficiency of public transportation, and improve the public transportation efficiency, and alleviate the growing problem of urban traffic congestion.

Funding The work described in this paper was supported by Youth Foundation of China Ministry of Education and Social Science Project (15YJAZH091) & Social science planning project of Jiangxi Province (14YJ24).

References

1. Wu, J.: The discussion of the definition and operation mode of community buses. Pub. Util. **6**, 6–7 (2011)
2. Barden, S.A., Seneviratne, J.P.V.: Public light bus operation surveys 1972. Hong Kong: Transport and road research laboratory, 1975 and road traffic—a case study of Nanchang, Jiangxi Teacher College, Nanchang (2007)
3. Hu, X.P., Wu, L.R., Wang, W. et al.: A value—function—based model for evaluating the passenger satisfaction in the flexible community bus problem. In: International Conference on management science and engineering, 2010. Lv, Urban spatial structure Hu based on land use research of Nanchang perspective, Jiangxi Teacher College, Nanchang (2013)
4. Qin, G., Pingfang, H.: On the spatial structure of Nanchang's Tourism. J. Jiangxi Agric. Univ. Nanchang **9**(2) (2010). Tang jie. The brief analysis of operation and development mode of the community buses. Annual China Urban Planning Conference Paper (2012)
5. Wang, S.: Functions of transfer nodes and their role in the rail transit network. Metro Light Rail **4** (2003)

Development of a Surrogate Conflict Indicator for Freeway Exits Using Trajectory Data

Tangyi Guo, Xuejiao Jiang and Wei Fan

Abstract Crash occurrences are random rare events which are difficult to detect and reconstruct. Evaluating traffic safety using crash surrogate measures instead of historical crash data is attracting more and more attention. The trajectory, a continuous function of a vehicle's temporal-spatial kinestate, is capable of depicting the process of crash potentials by means of theoretical analysis or simulation. The trajectory provides easy access to accurate calculation of the commonly used safety surrogate indicators like TTC and PET, which are difficult to obtain in practice. TTC and PET are the time differences between two vehicles during a quasi accident process. Compared with the situation in the final conflicting point, two vehicles may encounter a smaller distance or a bigger speed variation while approaching the conflict point. TTC and PET are deficient in describing abreast driving, or distinguishing the severity levels for approximately equal TTC (PET) cases. To remedy these shortcomings, this paper proposes a surrogate indicator Kj, the ratio of conflicting distance divided by relative speed. And an exponential model is developed to predict the conflict probability. Kj and the conflict probability are both time frame based, illustrating the changing process at each time frame during a conflict phase. The indicator Kj and conflict probability make it easier to describe the conflict mechanism and distinguish the levels of conflict severity.

Keywords Conflict · Surrogate safety indicator · Trajectory · Freeway exit

1 Introduction

The freeway exit is the most important connection between a freeway mainline and a crossroad. Most geometric designs change right before the deceleration lane or the ramp taper begins. Because the curved exit ramp are unexpected, drivers may need

---

T. Guo (✉) · X. Jiang · W. Fan
School of Automation, Nanjing University of Science and Technology,
#200 Xiao Ling Wei, Nanjing 210094, China
e-mail: transtor@njust.edu.cn

© Springer Science+Business Media Singapore 2017

H. Lu (ed.), *Proceedings of the Second International Conference on Intelligent Transportation*, Smart Innovation, Systems and Technologies 53,
DOI 10.1007/978-981-10-2398-9_6

to make a lane change, weave, or diverge to the desired lane, or even brake sharply to avoid a collision. Drivers have an elevated crash risk when they drive on freeway entrances and exits as compared with other sections of freeways. According to crash data from the U.S. Fatality Analysis Reporting System and General Estimates System in 2001, 18 % of all Interstate crashes occurred at interchanges, although such locations constituted less than an estimated 5 % of total freeway mileage. Of these interchange-related crashes, 83 % occurred on entrance or exit ramps [1].

Accidents are rare random events which are difficult to capture or reconstruct. Furthermore, more than 90 % accidents are caused by sophisticated human errors. Therefore, the accident statistics alone are of little use to support the research purpose to reduce accident frequency and severity. Many researchers have turned to surrogate safety measures to identify "quasi accidents" or "near accidents" as indicators of inappropriate behaviors potential risks. Besides traffic conflict technology, the trajectory-based methods have been treated as an increasingly effective tool to analyze the relationship between different traffic participants, vehicles, roadways, and environment.

As for driving trajectory, there are mainly two categories of research focuses, namely the trajectory observatories and trajectory analysis and application. Trajectory observatories intend to provide, process, and store objective and extensive measures and vehicle trajectory data [2]. In the last couple of decades, efforts to collect vehicle trajectory parameters through vehicles-based and/or site-based observations have been deployed. Vehicle-based methods make use of probe vehicles equipped with Global Positioning System (GPS) [3, 4], Aerial Photography and Remote Sensing [5], vehicle-to-vehicle (V2V) and vehicle-to-infrastructure (V2I) communication technology [6] that travel in the traffic stream and collect trajectory information of a test vehicle and/or adjacent vehicles. These methods are capable of collecting long distance trajectory data, however, one of the main shortcomings is that only a small samples could be recorded currently [5]. Site-based methods make use of the local installed devices including high-resolution video cameras, CCTV, radars [7] and infra-red [8] for detailed road traffic trajectories collection. In contrast to vehicle-based method, roadside-based methods are able to collect satisfactory amount of trajectory data, however, their coverage length are limited to 200 m, and they have no insight into the in-vehicle naturalistic driving behavior observation. Therefore, in practice, in vehicle observation and site recording are combined.

Another research field is trajectory analysis and application, which are undertaken after the field trajectory observatory process. The most important task relies on image processing algorithms: identification of moving objects, and filtering and classification of the road users of interests [9]. These tasks are mainly solved by professionals majoring in computer science, applied mathematics, or automatic control. For transportation researchers, more efforts are put into how to use these trajectory data to serve the purposes of developments on traffic flow theory [10], driving behavior analysis [11–13], transportation systems impacts modeling [14, 15] and surrogate safety measurement [6, 16, 17].

A variety of safety indicators including Time to Collision (TTC), Post-Encroachment Time (PET), Deceleration Rate (DR), Gap Time (GT) have been adopted in safety assessment and micro safety simulation studies [18, 19]. Taking the full course of vehicles over space and time into account, Minderhoud et al. (2001) developed two extended time-to-collision indicators, named TET (Time Exposed Time-to-collision) and TIT (Time Integrated Time-to-collision). The developed TET indicator expresses the exposition time to safety-critical approach situations, whereas the developed TIT indicator additionally takes into account the encountered TTC-values during these safety-critical approaches [20]. Both TET and TIT give a more complete and comprehensive picture of the safety level on a particular stretch of road during a particular period of time. Another research emphasis on surrogate safety measures is determining the threshold value of each measure. The default maximum TTC threshold value in SSAM (Surrogate Safety Assessment Model) is 1.50 s and the default maximum PET threshold value is 5.00 s; however, the user may override these with preferred alternate values ranging up to 9.95 s [18]. Taking these safety surrogates as explanatory variables, the trajectory-based safety models like crash rate, frequency, severity, and crash probability are derived. Oh and Kim (2010) proposed a methodology to estimate the rear-end crash potentials, which consists of two components. The first used a binary logistic regression (BLR) to predict a vehicle's trajectory belonging to either 'changing lane' or 'going straight'. The second derived crash probability by an exponential decay function using time-to-collision (TTC) between the subject vehicle and the front vehicle [18]. Additionally, an aggregated measure, crash risk index (CRI) was developed to accumulate rear-end crash potential for each subject vehicle. Besides safety assessment and estimation, the real-world driving trajectory data are also helpful in geometric design [21–23] and barrier design [24, 25].

This paper is organized as follows: Sect. 2 discusses the conflict detection and identification. And the surrogate safety indicator Kj is proposed for each time frame. Section 3 develops an exponential form of conflict probability model, and a sample case is illustrated.

2 Conflict Detection and Identification

2.1 Model Assumption

While analyze traffic conflict, many previous research treat a vehicle as a mass point. This assumption simplifies the task of modeling; however, it is incapable of describing the situation of two infinite approaching vehicles. In contrast, the rectangle vehicle model is more close to the real word vehicles, but it increases the difficulty of modeling the spatial relationship between vehicles. We hereby assume a 'point plus radius' vehicle model as a compromise. In Fig. 1, the mass points of two moving vehicles are Pi and Pi + 1. The closer the two mass points are, the

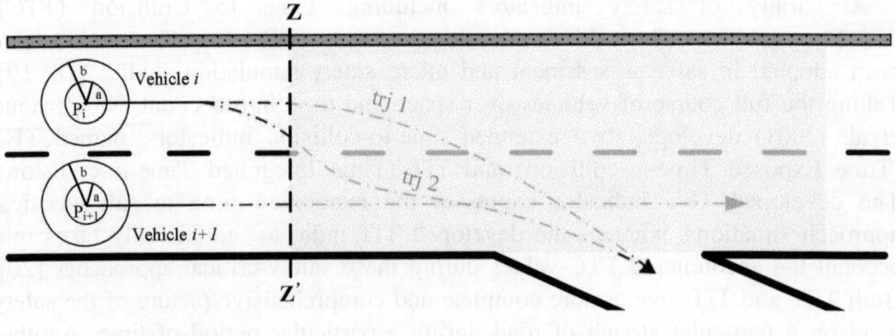

Fig. 1 Conflict scenario at a freeway exit

greater the risk. The area around each mass point is divided into three parts. While the circle with radius a is defined as a severe conflict zone, the ring with radius b–a and the area beyond radius b are defined as a general conflict zone and a safe zone, respectively. Assigning different values of a and b can distinguish different vehicle types. Figure 1 illustrates the conflict scenario between an exiting vehicle and a going straight vehicle at a two lane freeway exit.

The kinestate of vehicle i at time frame j is denoted by Si, j:

$$S_{i,j} = \left(x_{i,j}, y_{i,j}, v_{x,ij}, v_{y,ij}\right)^T \tag{1}$$

where, $(x_{i,j}, y_{i,j})$ represents the coordinates of vehicle i at time j, and (vx, vy) are the velocity components in the direction x and y.

Before section Z-Z′ the driving trajectories of both vehicles can be extracted from recorded video. After m time frames, vehicle i's trajectory $Q_i^{(m)}$ is composed of m points. $Q_i^{(m)} = (q_{i,1}, q_{i,2}, \cdots, q_{i,m}) \in H$, where H is the known space before section Z-Z′. $q_{i,m} = (x_{i,m}, y_{i,m})$ is the coordinates of the mth point on trajectory $Q_i^{(m)}$. After passing section Z-Z′, some vehicles' states are unknown due to limited camera coverage. In this case, there is need to predict vehicles' trajectories based on the historical moving data. The time interval between two adjacent frames is the reciprocal of frame rate F (fps). So for the j + 1 frame, in other words, after 1/F second, vehicle i's coordinates become $(x_{i,j+1}, y_{i,j+1})$.

$$\begin{cases} x_{i,j+1} = x_{i,j} + v_{x,ij}/F + 0.5a_{x,ij}/F^2 \\ y_{i,j+1} = y_{i,j} + v_{y,ij}/F + 0.5a_{y,ij}/F^2 \end{cases} \tag{2}$$

where, a_x, a_y are the acceleration components.

2.2 Space-Based Conflict Detection

We define traffic conflict as: A conflict occurs if and only if the time gap that two vehicles arriving at a confluence is smaller than a certain threshold. In other words, a conflict should satisfy the temporal-spatial criteria simultaneously. The threshold is different from the concept of TTC and PET. There is difficulty for the threshold in identifying conflict severity. The circle model provides a convenient method in computing the distance between two interaction vehicles.

In each time frame, the distance between i and i + 1 is calculated based on the kinestate Si, j and Si + 1, j. In time frame j, the distance dij is given as:

$$d_{ij} = \sqrt{\left(x_{i,j} - x_{i+1,j}\right)^2 + \left(y_{i,j} - y_{i+1,j}\right)^2} \tag{3}$$

According to the model assumption, the space-based conflict identification criteria is:

$$\begin{cases} d_{ij} \le a_i + a_{i+1} & \text{severe conflict} \\ a_i + a_{i+1} < d_{ij} \le b_i + b_{i+1} & \text{general conflict} \\ d_{ij} > b_i + b_{i+1} & \text{none conflict} \end{cases} \tag{4}$$

2.3 Time-Based Conflict Detection

TTC and PET are both safety surrogate indicators with smaller minimum values during a conflict event indicating a higher probability of or nearness to a collision. TTC is defined as "The projected time until two road users would collide if they continue on their collision course with unchanged speeds and direction". Whereas, PET is defined as "The elapsed time between the departure of an encroaching vehicle and the actual arrival of a trailing vehicle at the same location" [18]. Figure 2 illustrates the conflict scenario at a freeway exit by means of TTC.

The general equation to calculate TTC of vehicle i is given as [20]

$$TTC_i = \frac{X_i(t) - X_{i+1}(t) - l_{i+1}}{v_{i+1}(t) - v_i(t)} \quad \forall v_{i+1}(t) > v_i(t) \tag{5}$$

where v denotes the speed, X the position, and l the vehicle length.

At freeway exits, the potential of side-swipe and rear-end are the main types of conflicts, as depicted in Fig. 3.

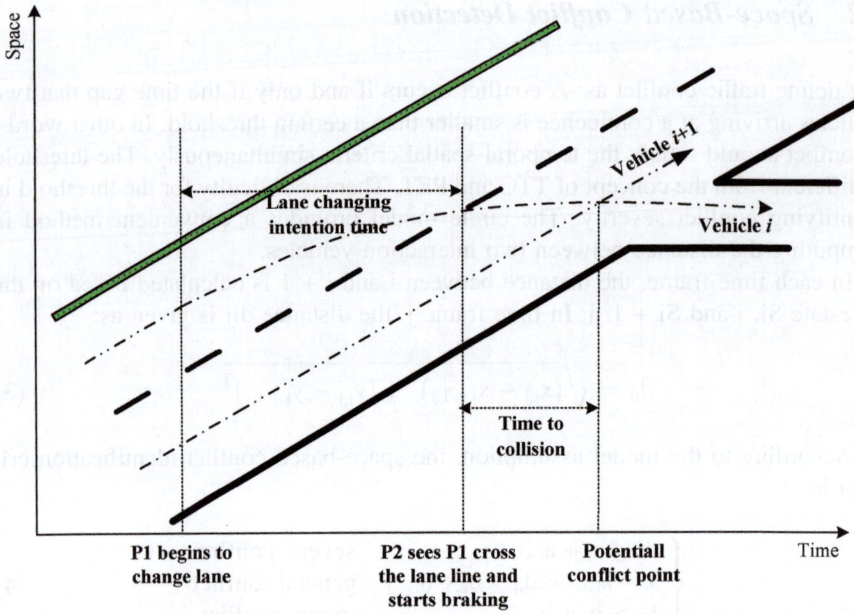

Fig. 2 Conflict scenario explained by time to collision at a freeway exit

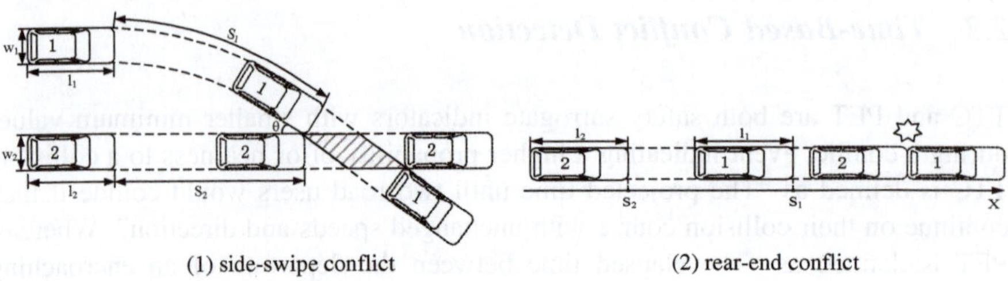

Fig. 3 Two common conflict scenarios at freeway exits

(1) for the side-swipe case,

$$\begin{cases} \mathrm{TTC} = \frac{S_1}{V_1}, \text{ if } \frac{S_2}{V_2} \le \frac{S_1}{V_1} < \frac{S_2}{V_2} + \frac{\frac{W_1}{\sin\theta} + \frac{W_2}{\tan\theta} + l_2}{V_2} \\ \mathrm{TTC} = \frac{S_2}{V_2}, \text{ if } \frac{S_1}{V_1} < \frac{S_2}{V_2} \le \frac{S_1}{V_1} + \frac{\frac{W_2}{\sin\theta} + \frac{W_1}{\tan\theta} + l_1}{V_1} \end{cases} \tag{6}$$

(2) for the rear-end case,

$$\mathrm{TTC} = \frac{S_1 - S_2 - l_1}{V_2 - V_1}, \quad \forall V_2 > V_1 \tag{7}$$

where, S denotes the distance from vehicle front to the conflict zone, W vehicle width, l vehicle length, and θ the conflict angle.

PET is much easier to calculate than TTC. PET is the difference of time that two vehicles encroaching the conflict zone.

$$PET = t_1 - t_2 \tag{8}$$

The default maximum TTC threshold value in SSAM is set as 1.50 s and the default maximum PET threshold value is 5.00 s; however, the user may override these with preferred alternate values ranging up to 9.95 s [18, 20]. Furthermore, the severity levels of conflict are not clearly defined by specific TTC or PET threshold.

2.4 *Distance-ΔV Combined Indicator*

In case 1, two vehicles passing abreast on adjacent lanes (Fig. 4), may cause a slight or even serious conflict if assessed using the space based indicator. However, in reality, the risk is low. The value of TTC or PET in case 2 is approximately equal to that in case 3. The time based indicators produce the same severity levels for both case 2 and 3. Again, in reality, the levels of severity should be different.

To explain and distinguish the levels of conflict severity, we introduce a Kj indicator, which is ratio of dij divided by relative speed ΔV at frame j.

$$K_j = \frac{d_{ij}}{\Delta V} \tag{9}$$

The smaller Kj is, the severe the conflict occurs. For case 1, $\Delta Vy \rightarrow 0$, $Kj \rightarrow \infty$, the risk is low; for case 2 and 3, the levels of severity depend. It is meaningless to calculate Kj for a single frame. In practice, we calculate the K values of all the frames that a conflict period covers. And then set the minimum of Kj as the conflict indicator during this period.

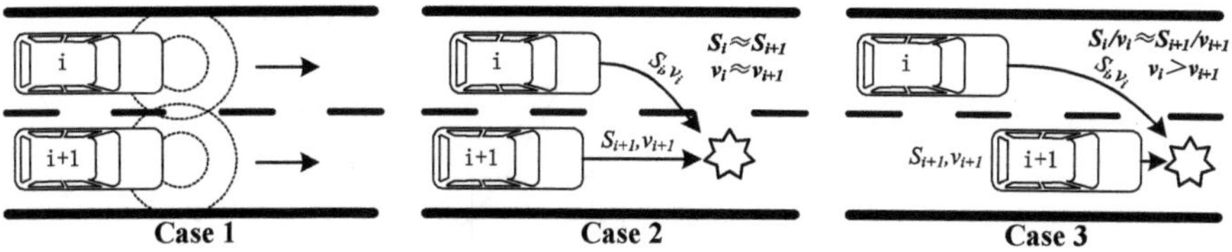

Fig. 4 The cases that TTC and PET could not explain clearly

3 Conflict Probability

A generalized exponential decay function (function form: $y = a+b \times \exp(x/c)$) can be used to predict the conflict potential. Since the potential or probability ranges from 0 to 1, set a = 0 and b = 1[16]. So the conflict probability at time frame j is given by Eq. (10). The model parameter c decides the exponential curve shape, as shown in Fig. 5. Given Kj, a bigger c value produces a higher conflict probability.

$$P(\text{conflict} \mid j) = \exp(-K_j/c) \tag{10}$$

3.1 Procedure

In this paper, vehicles' trajectories are extracted from high-resolution videos recorded at freeway exit areas. The traffic operational condition on freeway is much simpler than that on urban roadways, which makes it possible and easier for data screening and training using the neural network toolbox in Matlab. The trajectory extraction and prediction procedure are as follows.

(1) Take traffic videos at the target freeway exit for a certain time period.
(2) In Matlab, through image preprocessing, object extraction, target tracking, extract all the vehicles trajectories by means of background differencing method. And the trajectories are smoothed using Kalman filtering.
(3) Predict trajectories using neural network method.
(4) Extract the kinestate variable Si, j for time frame j.
(5) Calculate the value of surrogate conflict indicator K j using Eq. (9).
(6) Calculate the conflict probability using Eq. (10), and then choose the peak value as the conflict probability (Fig. 6).

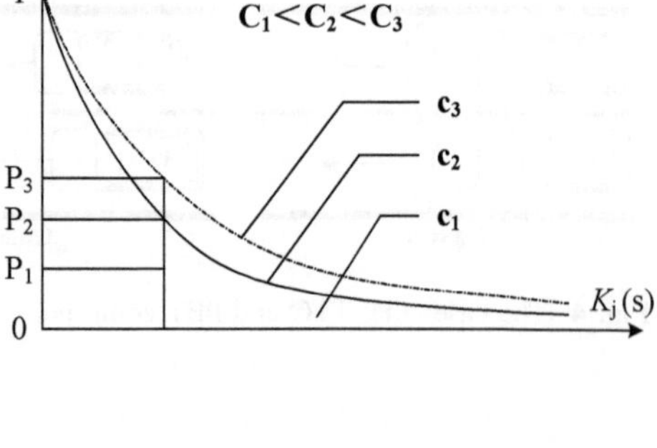

Fig. 5 The exponential probability curve with different parameter c

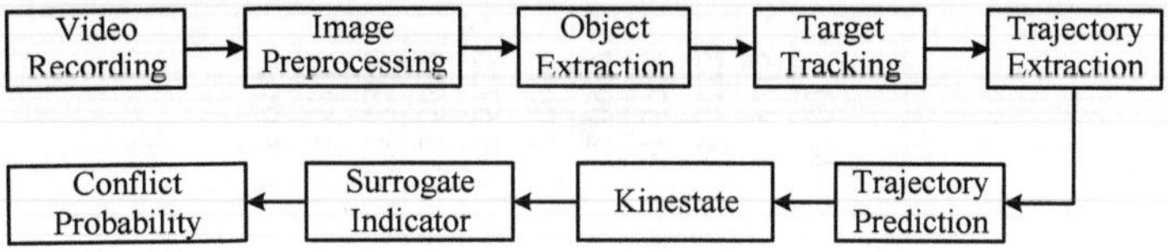

Fig. 6 The procedure for trajectory analysis and conflict prediction

3.2 *Sample Case*

Figure 7 shows both the real and predicted trajectories that two vehicles cover. The predicted trajectories are close to the real ones, indicating an acceptable accuracy. Table 1 lists 10 frames (from frame 357 to 366) of the kinestate data including vehicle coordinates, speed components, and the K values.

From Fig. 8 and Table 1, it is clear that the two vehicles approach closer to each other from frame 357 to 363, and leave away from frame 363 to 366. The K value reaches the peak of and then drops down. At the peak, the conflict probability is 59.93 %, which is also the highest compared with the other 9 frames. While approaching the conflict point, there are three options for both vehicles: ① vehicle i + 1 perceives an unacceptable gap and gives up lane changing; ② vehicle i + 1 notices vehicle i's lane changing intention, and then slows down; ③ both vehicles do not take any action and a collision occurs. According to the real trajectory in Fig. 7, the collision is avoid by option ②.

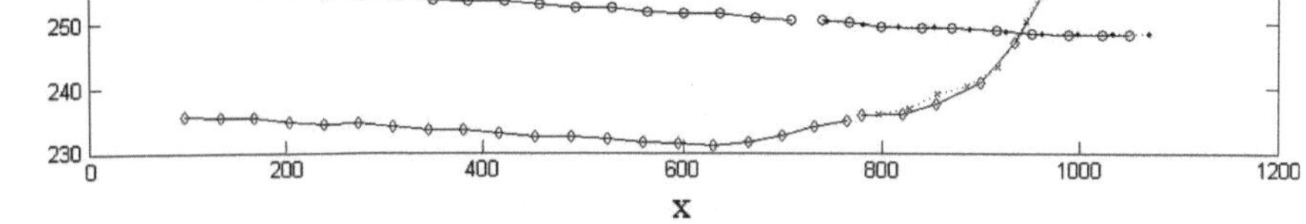

Fig. 7 Real and predicted trajectories

Table 1 The Kinestate variables and K values (10 frames)

Frame	Vehicle i				Vehicle i + 1				K		
	x_i	y_i	v_{xi} (m/s)	V_{yi} (m/s)	x_{i+1}	y_{i+1}	v_{xi+1} (m/s)	V_{yi+1} (m/s)	K_{jx} (s)	K_{jx} (s)	K_j (s)
357	744.5	250.4	18.00	−0.05	796.5	235.9	15.75	0.45	23.1111	29.0000	23.4215
358	780.6	249.9	18.05	−0.25	826.9	236.9	15.20	0.50	16.2456	17.3333	16.3183
359	816.6	249.5	18.00	−0.20	856.3	239.1	14.70	1.10	12.0303	8.0000	11.5708
360	**852.5**	**249.5**	**17.95**	**0.00**	**886.5**	**240.3**	**15.10**	**0.60**	**11.9298**	**15.3333**	**12.0937**
361	888.6	249.1	18.05	−0.20	916.6	242.2	15.05	0.95	9.3333	5.1304	8.9063
362	924.9	248.8	18.15	−0.15	945.2	250.3	14.30	4.05	5.2727	0.3571	3.5726
363	961.1	248.5	18.10	−0.15	973.0	259.4	13.90	4.55	2.8333	2.3191	2.5602
364	997.4	248.5	18.15	0.00	1002.0	269.8	14.50	5.20	1.2603	4.0962	3.4300
365	1033.0	248.4	17.80	−0.05	1030.0	283.0	14.00	6.60	0.7895	5.2030	4.5344
366	1069.0	248.4	18.00	0.00	1056.0	296.6	13.00	6.80	2.6000	7.0882	5.9147

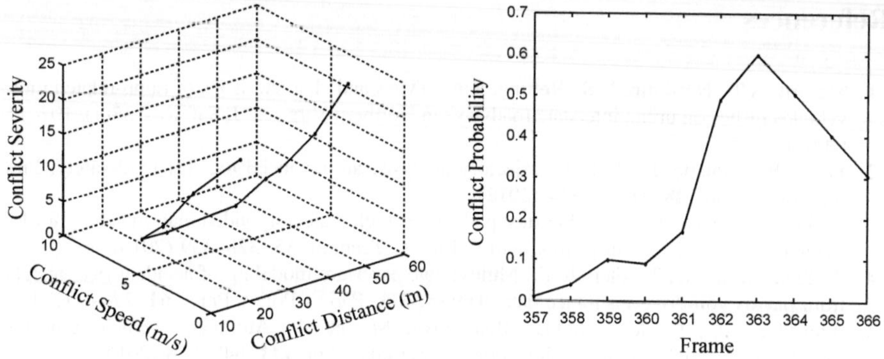

Fig. 8 The relationship between conflict probability, conflict speed and distance

4 Conclusions

The image processing technology and the intelligent algorithms facilitate the extraction and prediction of vehicle trajectory. The trajectory data includes the kinestate variables like time, distance, speed, and position. These trajectory data further traffic safety studies. The commonly used safety surrogate indicators like TTC and PET are calculated for a period, having deficiency in describing abreast driving, or distinguishing the severity levels for approximately equal TTC (PET) cases. To remedy these pitfalls, a surrogate indicator Kj is proposed and an exponential model is developed to predict the conflict probability. Kj and the conflict probability are both time frame based, illustrating the changing process at each time frame during a conflict phase.

Furthermore, the model of conflict probability provides an option to quantify the risk of a conflict. The levels of conflict severity can also be determined according to the value of probability. The value closer to 1 indicates higher risk, while smaller than 0.5 lower risk. The thresholds to classify serious conflict, slight conflict and none conflict needs later research. Also, field data is required to estimate the parameter c in the exponential probability prediction model.

Acknowledgments This research is financially supported by the National Natural Science Foundation of China (51208261, 51308192), the Science Foundation of Ministry of Education of China (12YJCZH062), the Fundamental Research Funds for the Central Universities of China (30920140132033), and China's Post-doctoral Science Fund (4438).

References

1. Mccartt, A.T., Northrup, V.S., Retting, R.A.: Types and characteristics of ramp-related motor vehicle crashes on urban interstate roadways in Northern Virginia. J. Saf. Res. **35**(1), 107–114 (2004)
2. Jacob, B., Violette, E.: Vehicle trajectory analysis: an advanced tool for road safety. Proc. Soc. Behav. Sci. **48**, 1805–1814 (2012)
3. Quddus, M., Washington, S.: Shortest path and vehicle trajectory aided map-matching for low frequency GPS data. Transp. Res. Part C Emerg. Technol. **55**, 328–339 (2015)
4. Paefgen, J., Staake, T., Fleisch, E.: Multivariate exposure modeling of accident risk: insights from pay-as-you-drive insurance data. Transp. Res. Part A Policy Pract. **61**, 27–40 (2014)
5. Azevedo, C.L., Cardoso, J.O.L., Ben-Akiva, M., et al.: Automatic vehicle trajectory extraction by aerial remote sensing. Proc. Soc. Behav. Sci. **111**, 849–858 (2014)
6. So, J.J., Dedes, G., Park, B.B., HosseinyAlamdary, S., Grejner-Brzezinsk, D.: Development and evaluation of an enhanced surrogate safety assessment framework. Transp. Res. Part C Emerg. Technol. **50**, 51–67 (2015)
7. Aoude, G.S., Desaraju, V.R., Stephens, L.H., et al.: Behaviour classification algorithms at intersections and validation using naturalistic data. In: 2011 IEEE intelligent vehicles symposium, pp. 601–606 (2011)
8. Bhattacharya, S., Idrees, H., Saleemi, I., et al.: Moving object detection and tracking in forward looking infra-red aerial imagery. In: Machine vision beyond visible spectrum, pp. 221–252 (2011)
9. Azevedo, C.L., Cardoso, J.O.L., Ben-Akiva, M.: Vehicle tracking using the k-shortest paths algorithm and dual graphs. Transp. Res. Proc. **1**(1), 3–11 (2014)
10. Ossen, S., Hoogendoorn, S.P.: Validity of trajectory-based calibration approach of car-following models in presence of measurement errors. Transp. Res. Rec. J. Transp. Res. Board **44**(2088), 117–125 (2008)
11. Peng, L., Wu, C., Huang, Z., et al.: Novel vehicle motion model considering driver behavior for trajectory prediction and driving risk detection. Transp. Res. Rec. J. Transp. Res. Board (2014)
12. Toledo, T., Koutsopoulos, H.N., Ben-Akiva, M.: Integrated driving behavior modeling. Transp. Res. Part C Emerg. Technol. **15**(2), 96–112 (2007)
13. Wu, S.J., Chiang, H.H., Perng, J.W., et al.: The heterogeneous systems integration design and implementation for lane keeping on a vehicle. IEEE Trans. Intel. Transp. Syst. **9**(2), 246–263 (2008)
14. Sun, Z., Hao, P., Ban, X.J., et al.: Trajectory-based vehicle energy/emissions estimation for signalized arterials using mobile sensing data. Transp. Res. Part D Transp. Environ. **34**, 27–40 (2015)
15. Zhang, Y., Lv, J., Wang, W.: Evaluation of vehicle acceleration models for emission estimation at an intersection. Transp. Res. Part D Transp. Environ. **18**, 46–50 (2013)
16. Oh, C., Kim, T.: Estimation of rear-end crash potential using vehicle trajectory data. Accid. Anal. Prev. **42**(6), 1888–1893 (2010)
17. Simons-Morton, B.G., Cheon, K., Guo, F., et al.: Trajectories of kinematic risky driving among novice teenagers. Accid. Anal. Prev. **51**, 27–32 (2013)
18. Souleyrette, R., Hochstein, J.: Development of a conflict analysis methodology using SSAM. Highw. Saf. (2012)
19. Gettman, D.: Surrogate safety measures from traffic simulation models. Transp. Res. Rec. J. Transp. Res. Board **1840** (2003)
20. Minderhoud, M.M., Bovy, P.H.L.: Extended time-to-collision measures for road traffic safety assessment. Accid. Anal. Prev. **33**(1), 89–97 (2001)
21. Koita, A., Daucher, D., Fogli, M.: New probabilistic approach to estimate vehicle failure trajectories in curve driving. Probab. Eng. Mech. **34**, 73–82 (2013)

22. Alhajyaseen, W.K.M., Asano, M., Nakamura, H., et al.: Stochastic approach for modeling the effects of intersection geometry on turning vehicle paths. Transp. Res. Part C Emerg. Technol. **32**, 179–192 (2013)
23. Rosey, F., Auberlet, J.: Trajectory variability: road geometry difficulty indicator. Saf. Sci. **50** (9), 1818–1828 (2012)
24. Johnson, N.S., Thomson, R., Gabler, H.C.: Improved method for roadside barrier length of need modeling using real-world trajectories. Accid. Anal. Prev. **80**, 162–171 (2015)
25. Soltani, M., Moghaddam, T.B., Karim, M.R., et al.: Analysis of developed transition road safety barrier systems. Accid. Anal. Prev. **59**, 240–252 (2013)

A More Practical Traffic Lights Cellular Automata Model for Traffic Flow Simulation

Xinlu Ma, Bei Li, Shijian Zhang and Xiaodong Guan

Abstract Based on real collected data to analyze the effects to vehicle speed near the line of parking area under different remaining of green time with or without signal countdown device on urban roads. Then we can analysis of drivers' characteristics and its spatio-temporal characteristics. Conclusion is drawn as follows: The influence of signal lights on vehicle speed, which mainly concentrate in the 70 m area near the stop line; During the green light period, at the same time point, the operational characteristics of vehicles on different positions have significant difference, which changes with the passing of green time; Compared with no signal countdown case, the vehicle speed fluctuation is larger and the average speed is obviously lower in signal countdown case; Under the traffic without blocking, most drivers want to increase speed through the stop-line. Eventually, by performing the analysis of pilot behaviors in with or without signal countdown device, an attempt is made to provide the urban road entrance lanes of more realistic CA model, which describes the operational characteristics of urban road entrance lanes with the signal controlling has obvious advantages through simulation results.

Keywords Signal countdown · No signal countdown · Measured data · Cellular automata · Simulate

X. Ma (✉) · B. Li · X. Guan
Transportation College, Chongqing Jiaotong University, Chongqing 40074, China
e-mail: paperfiles@163.com

S. Zhang
China Transport Telecommunications and Information Center, Chongqing Co.Ltd., Chongqing 40074, China
e-mail: bbl0723@foxmail.com

© Springer Science+Business Media Singapore 2017
H. Lu (ed.), *Proceedings of the Second International Conference on Intelligent Transportation*, Smart Innovation, Systems and Technologies 53,
DOI 10.1007/978-981-10-2398-9_7

1 Introduction

With the increase of the vehicles in the city, the intersection more to regulate traffic order by installing traffic signal control equipment. Through the survey we found that different cities have different preferences about use the countdown signals or the no countdown signals and different countries haven't make a clear regulation. However the differences driving behavior caused by the different signal control model has been mentioned in the previous studies. Chiou (2010) analyzed the speed of the vehicle in the parking lane 20 m and found that the average speed of green light at the entrance of the intersection can be reduced when the green light is being counted down [1]. Through video detection, Ma Tianyu (2008) compared with the speed of the vehicle in the last 3 s green light time and the yellow light time when the countdown signals was set before and after. Then he found that compared with the countdown, when there was no countdown, the car through the parking line at the speed of 5–10 m/s increased by 5 %, at the speed of 10–15 m/s increased by 20 % and the accelerated vehicles increased 18 % [2]. Also, Qian Hongbo (2011) through the questionnaire and data analysis found that 52 % of the drivers will choose to slow down at the end of the green time when the vehicles close to the stop line, however, the measured probability is only 7 % [3]. Li Zhen Long (2013) through the investigation found that the red light countdown can impact the drivers' start time and the average start time in the countdown is faster 0.5 s than without the countdown [4]. Above the research is mainly for the vehicle near the stop line. But some scholars through the study found that the difference between the driver's behavior caused by the signal lamp is not only reflected in the area near the stop line, but also to all of the vehicles that can observe the signal lamp. According to the extraction of vehicle dynamic parameters, Jiang Ling (2008) found that not all the vehicle is completely stopped at the stoplights, but a part of the vehicle decelerate at the red light and when the light is turned into the green light the vehicle which have not stopped can accelerate again [5]. Zhu Tong (2012) collected the speed and the passing time of the free stream when the vehicles through the road. And he found that the green light countdown lead a part of vehicles to decelerate, and a part of vehicles to accelerate. But the study was not clear the vehicle speed change when the signal lights at different stages [6]. In the paper, based on the measured data to analysis the speed of vehicle under different time and space. According to the analysis result, the traffic flow model can be built to describe the traffic flow of the vehicle running characteristics under the condition of the countdown time and the no countdown time. Finally, the model is verified by simulation.

2 Investigation and Analysis

This paper mainly focuses on the effect of signal lamp on the vehicle speed under different control model. Due to the saturation flow, the vehicle speed is limited by the speed of the former car, it is difficult to reflect the impact of signal lights on the vehicle speed. In this paper, the speed of the vehicle is investigated by using the hand held radar speed measurement equipment, which is focused on the free flow of the entrance of the intersection where there is have signal countdown or no signal countdown. The selection of the signal lamp countdown and no signal lamp countdown intersection need to meet the following conditions: (1) The intersection entrance lanes is flat, open field of vision, no illegal parking and other factors. (2) The road conditions in the investigation place are similar, and the road capacity is not saturated. Finally, a section of the road which has different traffic lights on the same road is selected as the investigation site. The intersection signal lights and the road speed limit is shown in Table 1.

The survey section is within 100 m from the intersection stop line, then set up a monitoring section from the stop line at the start of each interval 14 m, and each monitoring section collect fifteen signal cycle. The distance between the traffic signal equipment and the stopping line is 50 m. The monitoring point velocity distribution scatter plot of the traffic lights countdown and no traffic lights countdown as shown in Figs. 1 and 2.

Through the Figs. 1 and 2 we can conclude that in the countdown case and the no countdown case different monitoring points velocity distribution exist obvious difference: (1) In the case of countdown, within a certain monitoring range, vehicle speed is influenced larger by the lights; As the green time goes by, vehicle speed increased first and then decreased and show "∧" type distribution, but with increasing distance from the intersection, this phenomenon gradually weakened. (2) In the case of no countdown, under different monitoring ranges there is no obvious difference in the speed of the vehicle.

To further analyze the operation characteristics of the vehicle under the countdown and no countdown cases, Each cycle time is divided into early green time (Former 15 s), middle green time and late green time (Last 5 s). Speed is divided into low speed vehicle (0–18 km/h), moderate speed vehicles (18–36 km/h) and high-speed vehicles (36–60 km/h). Under the countdown and no countdown cases to analysis the average speed and the speed ratio of each monitoring section.

Table 1 Traffic signal time and speed limits

	Green time (s)	Yellow time (s)	Red time (s)	Cycle time (s)	Lanes	Width (m)	Average hourly traffic (vehicle/h)	Max speed (km/h)
Countdown	56	3	36	95	2	3.5	947	60
No countdown	67	3	15	85	2	3.5	931	60

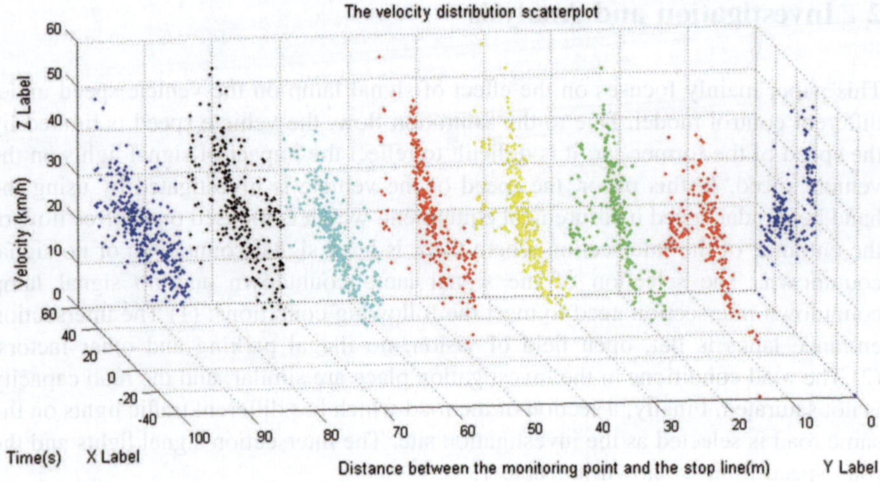

Fig. 1 The monitoring point velocity distribution scatter plot of the traffic lights countdown

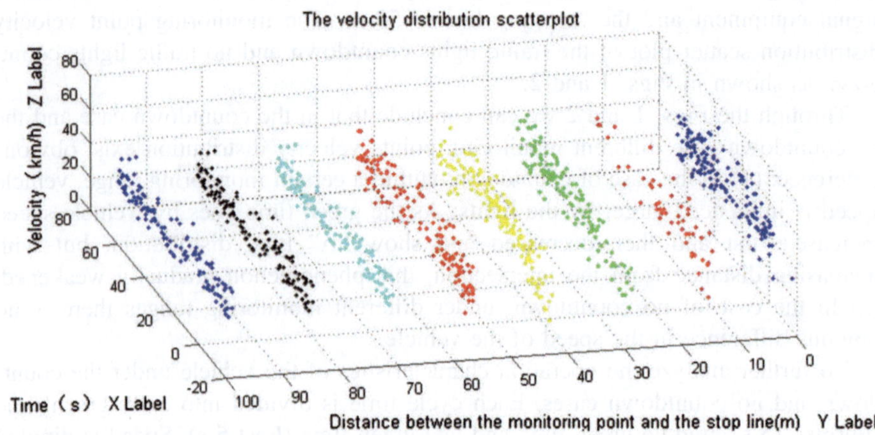

Fig. 2 The monitoring point velocity distribution scatter plot of the no traffic lights countdown

To analysis the average speed of each monitoring section in different time under the countdown, as shown in Table 2. In the early green time, the vehicle speed increases with the distance from the distance. The velocity variance of the 70, 84 and 98 was significantly smaller than that of the previous observations, that is, the speed fluctuation is small. Average velocity distribution as shown in Fig. 3.

In different periods, the average speed of the vehicle monitoring section can reflect the operating characteristics of the vehicle to a certain extent, but can't reflect the velocity distribution of the monitoring section. In order to further verify the influence range of the signal lamp on the vehicle speed and the operating

Table 2 The average velocity in the green signal countdown

Monitoring section (m)	0	14	28	42	56	70	84	98
Early green time (km/h)	15	16	21	23	28	29	30	31
Middle green time (km/h)	28	28	33	33	39	31	31	30
Late green time (km/h)	27	27	32	32	32	30	30	29
Variance	34.89	29.56	29.56	20.22	20.67	0.67	0.22	0.67

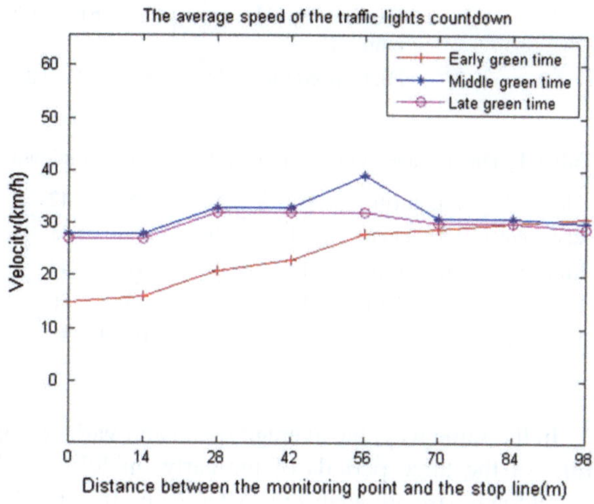

Fig. 3 The average velocity of the monitoring points during the green time in the countdown

characteristics of the vehicle at different stages, we carry out statistics on the velocity distribution of each monitoring section in different green light times, as shown in Table 3. The analysis found that the difference of the velocity distribution among the 70, 84 and monitoring sections in early green time, middle green time and late green time was small. In the early green time, with the increase of the distance from the monitoring section to the stop line, the proportion of the low speed vehicles is gradually reduced, the proportion of medium speed driving vehicles increased and the proportion of high speed running vehicles first increases and then decreases. In the middle green time, low speed driving vehicles accounted for a small proportion (below 8 %), the proportion of medium speed driving vehicles is above 50 % and the proportion of high speed running vehicles first increases and then decreases. In the late green time, each monitoring section substantially no low speed vehicle, and with the increase of the distance of the monitoring section, the proportion of the moderate speed vehicles is increasing, the proportion of high speed driving vehicles is gradually reduced, And the proportion of high speed driving vehicles in the stop line as high as 58 %.

Table 3 The proportion of the velocity in the green signal countdown

Monitoring section (m)		0	14	28	42	56	70	84	98
Early green time	Low speed ratio (%)	68	70	63	9	11	15	17	17
	Moderate speed ratio (%)	30	27	18	54	47	60	61	62
	High-speed ratio (%)	2	3	19	37	42	25	22	21
Middle green time	Low speed ratio (%)	3	8	4	0	1	0	0	1
	Moderate speed ratio (%)	70	83	53	63	50	85	86	86
	High-speed ratio (%)	27	9	43	37	49	15	14	13
Late green time	Low speed ratio (%)	0	0	0	1	0	0	0	0
	Moderate speed ratio (%)	42	61	62	73	77	81	84	85
	High-speed ratio (%)	58	39	38	26	23	19	16	15

Table 4 The average velocity in no green signal countdown

Monitoring section (m)	0	14	28	42	56	70	84	98
Early green time (km/h)	33	46	43	31	36	36	35	36
Middle green time (km/h)	40	42	41	44	43	37	37	36
Late green time (km/h)	41	47	43	36	35	35	35	35
Variance	12. 67	4. 67	0. 89	28. 67	12. 67	0. 67	0. 67	0. 22

In the same way, the average speed and variance of the monitoring section at the time of the three periods of the early, middle and late green time under the no countdown case are analyzed. After 70 m, the speed variance of the different green time of the monitoring section is small, as shown in Table 4. The average velocity of each monitoring section is above 30 km/h, as shown in Fig. 4.

Carry out statistics on the velocity distribution of each monitoring section in different green light times under the no countdown case: In the early green time, the ratio of each monitoring section of the low speed line is below 8 %, with decreasing distance from the stop line, the proportion of high speed running vehicles first increases and then decreases; In the middle green time, sections almost no low speed vehicles, many vehicles with above average speed, with decreasing distance from the stop line, the proportion of high-speed driving of the vehicle showed increasing trend, compared to the early green time, the high-speed vehicle increased significantly; In the late green time, with the distance of the distance, the speed of the vehicle is increased, and the proportion of high speed driving vehicles is increasing, as shown in Table 5.

Through the analysis of the average speed and the velocity distribution of the intersection imports under the condition of no countdown:

Fig. 4 The average velocity of the monitoring points during the green time in no countdown

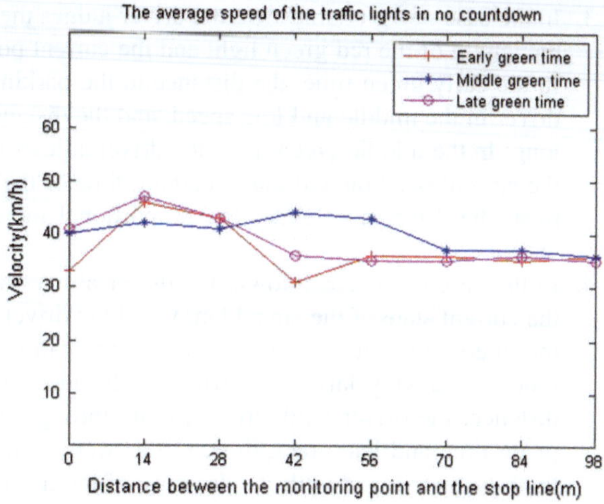

Table 5 The proportion of the velocity in no green signal countdown

Monitoring section (m)		0	14	28	42	56	70	84	98
Early green time	Low speed ratio (%)	3	8	0	6	0	0	0	0
	Moderate speed ratio (%)	57	16	26	73	53	70	70	70
	High-speed ratio (%)	40	76	74	21	47	30	30	30
Middle green time	Low speed ratio (%)	0	0	1	0	0	0	0	0
	Moderate speed ratio (%)	40	38	23	20	28	62	62	62
	High-speed ratio (%)	60	62	76	80	72	38	38	38
Late green time	Low speed ratio (%)	0	0	0	0	0	0	0	0
	Moderate speed ratio (%)	6	0	8	10	29	41	41	41
	High-speed ratio (%)	94	100	92	90	71	59	59	59

1. The impact of signal light on the speed of the vehicle is mainly concentrated in the range of 70 m (as shown in Fig. 1).
2. In the case of no countdown, vehicles traveling speed is significantly higher than the countdown, and there is a signal countdown to the speed of the impact is greater than the no signal countdown, which is consistent with the literature [1] conclusions.

3. In the case of the countdown, the driver judges the current situation according to the length of the red green light and the current position of the current situation. In the early green time, the distance to the parking line is relatively near to the driver in the middle and low speed, and the distance to the stop line is relatively long; In the middle green time, the driver adjusts the vehicle speed according to the current position and the remaining time of the signal lamp; In the late green time, the driver accelerates the proportional increase with the decrease of the distance.
4. In the case of no countdown, the driver always does not know the duration of the current state of the signal lamp, and the driver always wants to pass through the intersection in the signal cycle. In the early green time, the pilot start loss time is relatively large, and with the distance of the distance of the stop line distance, the driver tends to accelerate through the intersection; In the middle green time and late green time, in the traffic without block, the driver tends to the high speed, and with the decrease of the distance to the stop line, the driver accelerated ratio increased.
5. In the middle and late green time, with the decreasing distance from the stop line, the high-speed driving vehicles are both increased under the countdown case or no countdown case, and it prove that most of the drivers always want to increase the speed in the case of no traffic.

3 Model

Since the Nagel [7] and BML [8] Model has been introduced in 1992, in order to improve the reality of the model, many scholars amend the model rules according to the different applicable conditions and significant results have been achieved [9–17]. In recent years, CA model has been widely used in building the model in urban traffic signal control intersections. Brockfeld (2001) studied on the influence of signal cycle on the traffic capacity of cross intersections through modifications to the Nasch and BML coupling model rules [18]; Ding Zhongjun (2009) established the cellular automation model about T-shaped intersection system under signal controlling and compared with the no lights model [19]; But the above established signalized intersection cellular automata model only treat the distance to the stop line as the only limiting condition to analysis the effect of traffic light signal on vehicle acceleration and deceleration, based on this, Zhang Jian (2011) built a cellular automata model which consider the traffic lights have an impact on all vehicles on the road and taking into account the expected movement of the vehicle in front. And treat reduce the vehicle exhaust emissions as the starting point to analysis the acceleration, deceleration and idle behavior of the vehicle under different cycle time [20]; Peng Chuan (2012) improved this model as a two-lane model and build the signal control model, urban Intersections control CA model has been further improved [21].

On the basis of Refs. [7, 12, 16, 20] and based on the analysis of Chapter "Steering Angle Balance Control Method for Rider-less Bicycle Based on ADAMS", the paper proposed a more realistic model—Realistic Traffic Light Cellular Automaton (RTL). This model considers the velocity changing with the lights and the position of the vehicle. The model was used to simulate the one-lane traffic flow which was under the control of traffic lights. And in the model, traffic light states (containing the countdown) as well as the position of vehicle are taken into account to making sure the driver according to the current location, current speed, and signal time to adjust the speed. According to the above principles, the model can simulation more realistic driving characteristics to ensure the vehicles safely through the intersections in the green state and reduce waiting time in the red state.

3.1 Consideration of the Model

The RTL model which is proposed in this article shows that the driver according to the vehicle location, signal status and the space with the front car to adjust the speed before the vehicle entered the intersection road. In the case of countdown and no countdown, the model is considered from the following aspects:

1. In the countdown, the vehicle start time can be ignored. When there is no countdown, losing start time is Δt.
2. During movement of the vehicle, the vehicle speed considerations of the desired speed of the front car in the next update period.
3. Under the green light, drivers can always expect smoothly through the parking line in current cycle with the current speed. In the case of countdown, the vehicle will accelerate when determining that the current time is not enough to pass the stop line; In the case of no countdown, drivers always want to reach the maximum speed in order to ensure to through the stop line in the current green time.
4. Under the green light, in the case of countdown, the first vehicle adjusts its speed basing on the current position and the remaining green time, other vehicles adjust speed according to the current position, remaining green time and the distances with the front car.
5. Under the red state, drivers are always expect a minimum wait time to pass the stop line and adjust speed with minimal acceleration.
6. Under the red state, in the case of countdown,when the front vehicles are parking with line, driver based on the red remaining time and the current position adjust speed; On the contrary, the driver considering the speed and the distance of the front vehicle to adjust speed when the front vehicles are moving. When there is no signal countdown, no speed adjustment rules.

3.2 Definition of the Model

At the t moment, the n car's current location, speed and the distance with the front car is denoted as $x_n(t)$, $v_n(t)$ and $d_n(t)$, and the parameters of the front vehicle denoted as $x_{n+1}(t)$, $v_{n+1}(t)$ and $d_{n+1}(t)$; $T_g(t)$ is the remainder green time at t, $T_r(t)$ is the remainder red time at t, $d_l(t) = x_l - x_n(t)$ means the distance between the first car and the stop line, x_l is the position of the stop line. At the t moment, one vehicle expected to its front car's speed at next time is:

$$v_{n+1}(t+1)_{anti} = \min\left[v_l(t)_{poss}, v_l(t)_{con}, v_n(t)_{adj}, d_{n+1}(t), v_{n+1}(t)\right] \quad (1)$$

$v_l(t)_{poss} = ceil\big(d_l(t)/T_g(t)\big)$ denoted as the minimum speed if a car want to safety through the stop line under green time, *ceil* is rounding up function; $v_l(t)_{con} = ceil(d_l(t)/T_r(t))$ denoted as the head vehicle control speed under red time, $v_n(t)_{adj} = ceil\big(d_{eff\,n}(t)/T_r(t)\big)$ denoted as the vehicle control speed under red time.

Effective distance is the vehicle can increased maximum distance when considered the front car moving in next time step, formula is defined as:

$$d_{eff\,n}(t) = \mathrm{d}_n(t) + \max\left[v_{n+1}(t+1)_{anti} - d_{\sec}, 0\right] \quad (2)$$

$d_{\sec}$ is the minimum safety distance, generally the value is 1.

3.2.1 Have a Countdown

Step 0 Random slow probability

$$p_n = \mathrm{p} \quad (3)$$

Step 1 Green time

if: The first car is denoted as n, $T_g(t) > 0$
if: $v_n(t) < v_{poss}$, the first car based on the characteristics of the behavior

$$v_n(t+1) = \begin{cases} v_n(t), & \text{if } v_n(t) < v_{poss} - \dfrac{T_g(t)+1}{2} \\ \min[v_n(t)+1, v_{\max}], & else \end{cases} \quad (4)$$

Under the current speed, if the car is still can't through the intersection when accelerate, the vehicle will moving with current speed; On the contrary, if the car can through the intersection when accelerate, the car will accelerate.
else: $v_n(t) \geq v_{poss}$

$$v_n(t+1) = \begin{cases} \min[v_n(t)+1, v_{\max}], & \text{if } rand() < p'_n \\ v_n(t), & else \end{cases} \tag{5}$$

$p'_n = \frac{v_{\max} - v_n(t)}{v_{\max}}$, this representative the first car can pass the stop line at current speed, but the car is still likely to accelerate.

else:

if: $v_n(t) < v_l(t)_{poss} < d_{eff\,n}(t)$

$$v_n(t+1) = \min[v_n(t)+1, v_{\max}] \tag{6}$$

If: $v_l(t)_{poss} \le v_n(t) < d_{eff\,n}(t)$

$$v_n(t+1) = \begin{cases} \min\left[v_n(t)+1, \min\left(v_{\max}, d_{eff\,n}(t)\right)\right], & if : rand() < p''_n \\ v_n(t), & else \end{cases} \tag{7}$$

$p''_n = \frac{d_{eff\,n}(t) - v_n(t)}{\max\left(v_{\max}, d_{eff\,n}(t)\right)}$, this representative other vehicles can through the stop line at current speeds, but the car is still likely to accelerate.

if: $v_n(t) \le deff_n(t) < v_l(t)_{poss}$

$$v_n(t+1) = \min\left[v_n(t)+1, \min\left(v_{\max}, d_{eff\,n}(t)\right)\right] \tag{8}$$

if: $v_n(t) \ge deff_n(t)$

$$v_n(t+1) = \max\left[v_n(t) - 1, d_{eff\,n}(t)\right] \tag{9}$$

Step 2 Red time, $T_r(t) > 0$

If: The first car is denoted as n:

$$v_n(t+1) = \min\left[v_l(t)_{con}, v_{\max}\right] \tag{10}$$

else:

if: $v_n(t) < v_n(t)_{adj}$

$$v_n(t+1) = \min\left[v_n(t)+1, \min\left(v_n(t)_{adj}, v_{\max}\right)\right] \tag{11}$$

if: $v_n(t)_{adj} \le d_{eff\,n}(t) < v_n$

$$v_n(t+1) = \min\left[v_n(t) - 1, d_{eff\,n}(t)\right] \tag{12}$$

if: $v_n(t)_{adj} \le v_n < d_{eff\,n}(t)$

$$v_n(t+1) = \begin{cases} \min\left[v_n(t)+1, \min\left(d_{eff\,n}(t), v_{\max}\right)\right], & if: rand() < p''' \\ v_n(t)_{adj}, & else \end{cases} \quad (13)$$

$p''_n = \frac{d_{eff\,n}(t) - v_n(t)}{\max\left(v_{\max}, d_{eff\,n}(t)\right)}$, this representative other vehicles need not stop at current speeds, but the car is still likely to accelerate.

Step 3 Random slow probability

if $\{rand() < p\}$ then

$$v_n(t+1) = \max[v_n(t) - 1, 0] \quad (14)$$

Step 4 Location update

$$x_n(t+1) = x_n(t) + v_n(t+1) \quad (15)$$

3.2.2 No Countdown

Step 0 Determination of the probability of random slow

$$p_n = \mathrm{p}_0 \quad (16)$$

Step 1 Loss of green start time

$$v_{n,n-1,\ldots,n-m}(t) = 0,\ \mathrm{t} \in [t_a,\ t_b] \quad (17)$$

t_a means green start time corresponding to the moment, t_b means start loss Δt corresponding to the time, $\Delta t = 2\mathrm{s}$ [2].

Step 2 Green time

if: n is the first vehicle *and* $t \ge t_b$

$$v_n(t+1) = \min[v_n(t) + 1, v_{\max}] \quad (18)$$

else if:

$$v_n(t+1) = \min\left[v_n(t) + 1,\ \min\left(d_{eff\,n}(t), v_{\max}\right)\right] \quad (19)$$

Step 3 Red time

if: n is the first vehicle

This rule should be selected deceleration based on your current speed and distance, preferences at lower speeds.

$$v_n(t+1) = \begin{cases} \min[\max(v_n(t) - 1, 1), d_l(t)], & rand() < p'_n \\ \min(v_n(t), d_l(t)), & else \end{cases} \tag{20}$$

$p'_n = v_n(t)/v_{\max}$ denoted as that vehicles are not certain when the red time is, and drivers for non-stop waiting deceleration probability.

else

if: $v_n(t) \le d_{eff\,n}(t)$

$$v_n(t+1) = \begin{cases} \min\left[v_n(t) + 1, \min\left(d_{eff\,n}(t), v_{\max}\right)\right], & if : rand() < p'' \\ \max(v_n(t), 1), & else \end{cases} \tag{21}$$

$p''_n = v_n(t)/v_{\max}$—meaning vehicles under the current speed accelerate.

if: $v_n(t) > d_{eff\,n}(t)$

$$v_n(t+1) = \min\left[v_n(t) - 1, d_{eff\,n}(t)\right] \tag{22}$$

Step 4 Random slow probability

if $\{rand() < p_0\}$ then

$$v_n(t+1) = \max[v_n(t) - 1, 0] \tag{23}$$

Step 5 Location update

$$x_n(t+1) = x_n(t) + v_n(t+1) \tag{24}$$

4 Simulation and Discussion

The operating characteristics of the urban road vehicles are obviously different from the urban expressway and the highway. Therefore, the simulation needs to be more refined. each cell in the model length is 1.4 m, five cellular representative for a car, each time step is 1 s, units of acceleration correspond better acceleration in real is 1.4 m/s^2. Set for this model from the stop line 140 m road, vehicles travelling from left to right and do not allow overtaking, using a deterministic random deceleration probability and open boundary condition. In urban road, vehicles maximum speed limit below 60 km/h, so the maximum speed for the model is $v_{\max} = 60$ km/h = 12 cell/s. To facilitate the simulation, assuming that each cycle made up of alternating green and red and T is a signal cycle length, $T_{green} = T_{red} = T/2$, T is 60 s in the simulation. For the countdown and no countdown simulation all using the above parameters, the difference between the two is that under countdown to signal cases, drivers can directly observe the traffic light status and count down the remaining time, and when no signal countdown, drivers can only observe traffic lights. Based on the above premise of two established models in Chapter "The management of slight traffic accident based on the Internet" which makes simulation and comprehensive comparison.

In the model simulations, starting at the stop line every 14 m monitoring points are selected separately for each monitoring point without the countdown speed analysis, and analyze the origin of the speed difference under different control strategies as is shown Figs. 5 and 6.

Compared with the three Stages—a, b, c in the Fig. 5, found in the countdown case, the speed which the RTL model shown in the early, middle and late green time was consistent with the actual situation.

1. Early green time (a stage), distance from the stop line near location, Low speed vehicles' proportion is larger, with the decrease of position from the stop line, medium speed vehicles is decrease, the proportion of vehicle at high speed is first increase and then decrease.
2. Middle green time (b stage), small proportion of low speed vehicles, vehicles traveling at a moderate speed or over the speed.
3. Late green time (c stage), the proportion of low speed vehicles is small in each monitoring point, with the decrease of the distance from the stop line, medium-speed vehicles proportion decreases gradually, high-speed vehicles proportion increases gradually.

Compared with the three Stages—a, b, c in the Fig. 6, found in the no countdown case, the speed which the RTL model shown in the early, middle and late green time was consistent with the actual situation.

1. Early green time (a stage), with the decrease of the distance from the stop line, low speed vehicles increased in proportion, moderate and high speed vehicles' proportion are decrease.
2. Middle green time (b stage) and end of green light (c stage), with the decrease of the distance from the stop line, high speed vehicles increased in proportion.

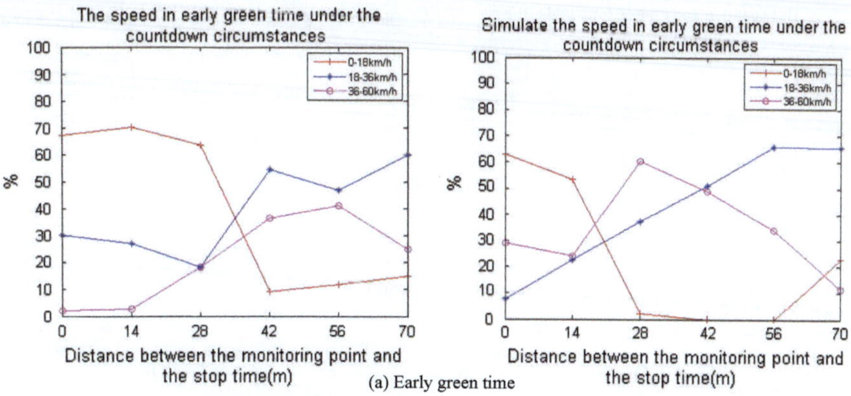

(a) Early green time

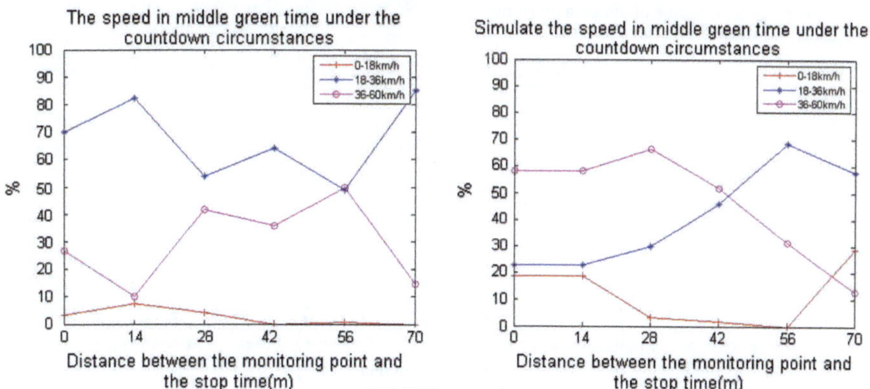

(b) Middle green time

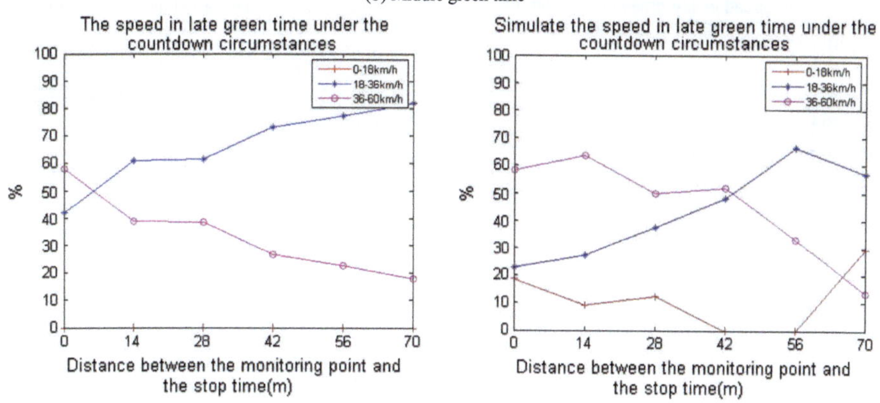

(c) Late green time

Fig. 5 The velocity and the simulate speed of the monitoring points during the green time in the countdown

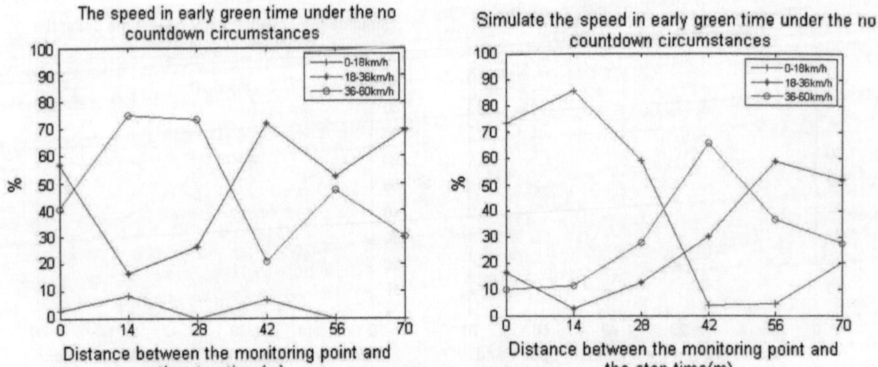

(a) Early green time

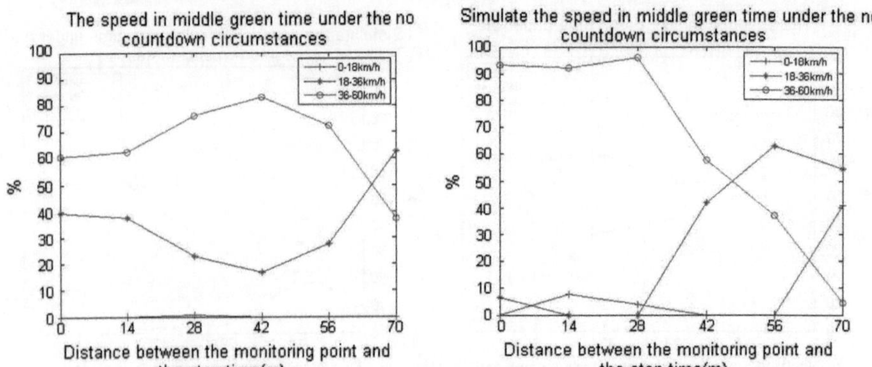

(b) Middle green time

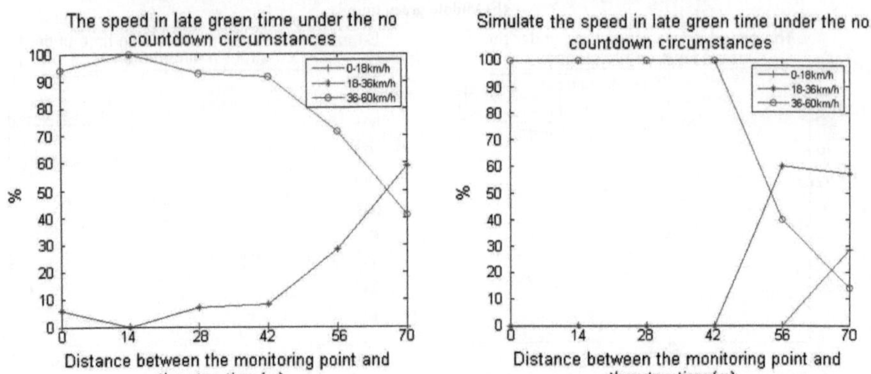

(c) Late green time

Fig. 6 The velocity and the simulate speed of the monitoring points during the green time in the no countdown

5 Conclusion

Based on the measured data, this paper analysis the speed of the intersection in the city under the signal countdown and no signal countdown. Through the analysis of the vehicle speed at different green time and position we obtained drivers' driving behavior under different conditions and obtained drivers' light-sensitive area mainly focus on distance within the stop line 70 m. Based on this, the paper presents a more realistic signal control cellular automata model (short for RTL) which considering the speed of vehicles change with signal time and vehicle position. Based on driver characteristics, under the single countdown and no single countdown, vehicle update rule are established, through simulation found that the patio-temporal speed of vehicles under this model agree well with the experimental data, can be reflected more realistic vehicle control performance under the signal control. Under the saturated flow, due to the large of traffic density on the intersection imported road, the vehicle distance is smaller, the vehicle drive with following, the sensitivity of drivers to signal light is decrease, so it can't reflect the advantages of the model. In the free stream, the driver can adjust the speed according to the current position and the signal state, so the model proposed in this paper has obvious advantages in the description of the vehicle operating characteristics under the free flow.

Acknowledgments Project supported by the National Natural Science Foundation of China (Grant No. 61403052).

References

1. Chiou, Y.C., Chang, C.H.: Driver responses to green and red vehicular signal countdown displays; safety and efficiency aspects. Accid. Anal. Prev. **42**(4), 1057–1065 (2010)
2. Ma, T.: Study on the effect of countdown. Jilin University, Changchun (2008)
3. Qian, H.: Influence of countdown of green signal on traffic safety at crossing. China Saf. Sci. J. **20**(3), 9–13 (2010)
4. Li, Z., Zhang, J.: Effect of traffic signal countdown on driver's start-reaction time. Sch. Electron. Inf. Control Eng. **5**(31), 77–81 (2013)
5. Ling, J.: Study on dynamic characteristic parameters of signal control in urban intersection. Beijing University of Technology (2008)
6. Tong, Z., Pei, X., Chun-Lin, G.: Effects of green signal countdown and other factors on vehicle speed at intersection. J. Chang'an Univ. Nat. Sci. Ed. **32** (4), 70–75 (2012)
7. Nagel, K., Schreckenberg, M.: A cellular automaton model for freeway traffic. Phys. I. (France) **2**, 2221–2229 (1992)
8. Biham, O., Middleon, A., Levine, D.: Phys. Rev. A. **46**, 6124 (1992)
9. Takasu, M., Takyasu, H.: 1/f noise in a traffic model. Fractals **1**, 860–866 (1993)
10. Benjamin, S.C., Johnson, N.F. Hui, P.M.: Cellular automaton models of traffic flow along a highway containing a junction. J. Phys. A. **29**, 3119–312 (1996)
11. Barlovie, R., Santen, L., Schreckenburg, A.: Metastable states in cellular automata for traffic flow. Eur. Phys. J. B. **5**, 793–780 (1998)

12. Li, X.B., Wu, Q.S., Jiang, R.: Cellular automaton models considering the velocity effect of a car on the successive car. Phys. Rev. E. **64**, 066128 (2001)
13. Knospe, W., Santen, L., Schadschneider, A., et al.: Towards a realistic microscopic description of highway traffic. Phys. A. **33**, L477–L485 (2000)
14. Jiang, R., Wu, Q.S.: Cellular automata models for synchronized traffic flow. Phys. A. **36**, 381–390 (2003)
15. Lee, H.Y., Barlovic, R., Schreckenberg, M., Kim, D.: Mechanical restriction versus human overreaction triggering congested traffic states. Phys. Rev. Lett. **92**, 238702 (2004)
16. Zhu, H.B., Ge, H.X., Dong, L.Y., Dai, S.Q.: A modified NaSch model with density-dependent randomization for traffic flow. Eur. Phys. J. B. **57**, 103–108 (2007)
17. Larraga, M.E., Alvarez-Icaza, L.: Cellular automata model for traffic flow with safe driving conditions. Chin. Phys. B. **23** (5), 050701 (2014)
18. Brockfeld, E., Barlovic, R., Schadschneider, A., Schreckenberg, M.: Optimizing traffic lights in a cellular autromatom model for city traffic. Phys. Rev. E. **64**, 056132 (2001)
19. Ding, Z.J., Wang, B.H.: Traffic behavior of T-shaped intersection system under signal controlling compared with unsignalized model. J. Jilin Univ. Eng. Technol. Ed. **39**(2), (2009)
20. Zhang, J., Dong, L.: A cellular automaton model for urban traffic considering anticipation effect in the presence of traffic lights. J. Shanghai Univ. Nat. Sci. **17**(5), 642–647 (2011)
21. Peng, C.: Research on signal light of intersection base on cellular automaton. East China Jiaotong University (2012)

Robot Driver's Motion Analysis and Simulation Based on Vehicle Speed Control

Kunming Liu, Guoyan Xu and Guizhen Yu

Abstract In order to improve the control performance of the robot driver, the dynamic analysis of its key parts such as throttle and brake mechanical legs are carried out, furthermore, a co-simulation platform of the robot based on speed control is built, and the movement of the robot in the process is analyzed. The mechanical leg's dynamic simulation model is built by ADAMS, and then the vehicle model is built by CarSim while the electromechanical co-simulation model "Robot Driver—Vehicle" is built based on the closed-loop speed control in MATLAB/Simulink. Simulation results show that the dynamic model of mechanical leg has good dynamic response; In addition, the electromechanical co-simulation model is able to complete the basic speed tracking simulation experiment, so as to provide a virtual simulation platform to improve the mechanical structure and control strategy.

Keywords Robot driver · Electromechanical simulation · Collaborative simulation · Dynamic analysis · Virtual prototype

1 Introduction

The development of intelligent transportation system together with the problem of high cost, low efficiency and potential safety hazard in the automobile road test have greatly increased the demand of the intelligent robot driver in automobile industry [1]. Today only a few countries have key technology of robot driver [2–4],

K. Liu · G. Xu · G. Yu (✉)
School of Transportation Science and Engineering, Beijing University of Aeronautics and Astronautics, Beijing 100191, China
e-mail: yugz@buaa.edu.cn

K. Liu
e-mail: ldkccss@163.com

G. Xu
e-mail: xuguoyan@buaa.edu.cn

© Springer Science+Business Media Singapore 2017
H. Lu (ed.), *Proceedings of the Second International Conference on Intelligent Transportation*, Smart Innovation, Systems and Technologies 53,
DOI 10.1007/978-981-10-2398-9_8

including Germany, America, Britain, Japan etc. Chinese own robot driver is primarily the DNC series developed by Southeast University and Nanjing Automotive Research Institute [5].

During the research and development of the robot driver, simulation is an important way to conserve funds and shorten the research period [6]. Chinese Ma and Niu [7] achieved the dynamic simulation of the clutch of the driving robot using software ADAMS while neglected the great influence of the drive motor. Jiang et al. [8] accomplished the simulation analysis of the position servo system of the robot in MATLAB, but there was no dynamic model of the mechanical device. Moreover, the existing simulation analysis methods do not involve the vehicle model, which brings great difficulty to distinguish the robot driver's ability to control vehicles. In this paper, a co-simulation simulation is proposed to solve these problems, which is based on ADAMS, MATLAB/Simulink and CarSim.

2 Dynamics Analysis of the Robot

The robot is composed of the driving motors and mechanical structures, as shown in Fig. 1. The robot is driven by AC servomotor. As human driver's movements on the accelerator and brake pedal are basically the same, the robot's throttle mechanical leg and brake mechanical leg are designed to have the same structure, as is shown in Fig. 2.

The robot's mechanical leg includes the mechanical arm, pedal splint, ball screw, screw bearing, slider, etc. The structure is a slider-rocker mechanism, the servomotor drives the ball screw with the help of the coupling, and the slider fixed on the screw nut makes linear motion along the axis of the screw to drive the pedal

Fig. 1 Robot's mechanical structure

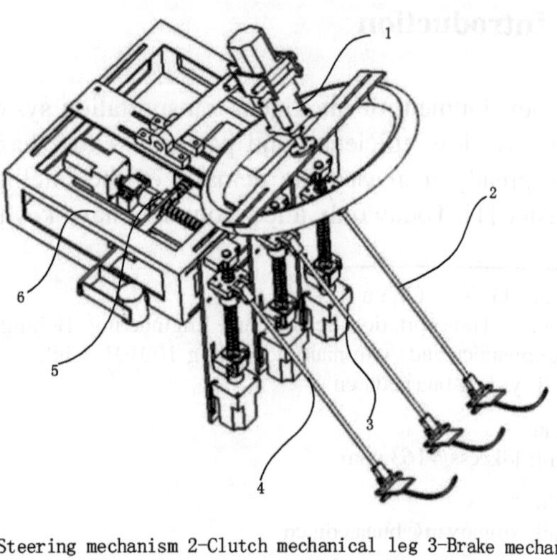

1-Steering mechanism 2-Clutch mechanical leg 3-Brake mechanical leg 4-Throttle mechanical leg 5-Shift machine leg 6-Base

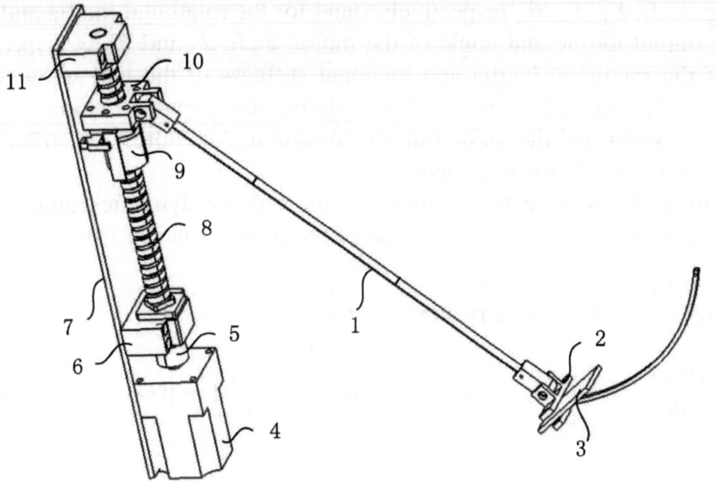

1-Arm 2-Pedal splint 3-Brake pedal 4-Servomotor 5-Joint 6-Screw bearing1 7-Mounting plate 8-Ball screw 9-Nut 10-Slider 11-Screw bearing2

Fig. 2 3D model of robot's leg

Fig. 3 Diagram of structural mechanics of the leg

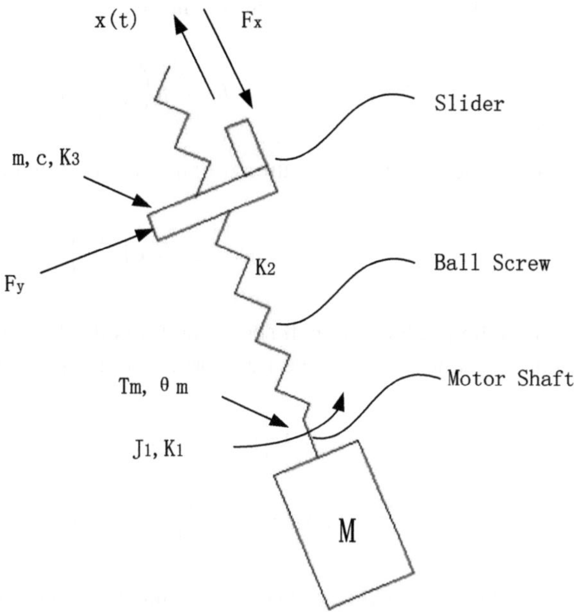

through the arm and splint. As the mechanism is driven by the servo motor, its action has good elasticity and compliance like human muscle, which is closer to the reality. The diagram of structural mechanics of the leg is shown in Fig. 3.

In Fig. 3, J_1, K_1, T_m, θ_m respectively stand for the rotational inertia and torsional stiffness, output torque and angle of the motor shaft; J_2 and K_2 is respectively on behalf of the rotational inertia and torsional stiffness of the ball screw; m, c, K_3, x represent the equivalent mass of the slider, the moving damping coefficient between the slider and the guide rail, the tension and compression stiffness of ball screw and the slider's displacement.

Transform this system to the motor shaft, and the dynamic equations of this system are obtained according to the principle of equivalence [9].

$$\begin{aligned} J_1 \frac{d^2\theta_m}{dt^2} &= T_m(t) - K\left[\theta_m(t) - x(t)\frac{2\pi}{L}\right] \\ J_0 \frac{2\pi}{L}\frac{d^2x(t)}{dt^2} &= K\left[\theta_m(t) - x(t)\frac{2\pi}{L}\right] - c_0\left(\frac{2\pi}{L}\frac{dx(t)}{dt}\right) - [(F_Y + mg)u + F_x]\frac{L}{2\pi} \end{aligned} \tag{2.1}$$

In the equation, K is the equivalent torsional stiffness of the system:

$$K = \frac{1}{\frac{1}{K_1} + \frac{1}{K_2} + \frac{1}{K_3/(L/2\pi)^2}} \tag{2.2}$$

J_O is the equivalent moment of inertia:

$$J_0 = m\left(\frac{L}{2\pi}\right)^2 + J_2 \tag{2.3}$$

c_O is the equivalent rotational damping coefficient:

$$c_0 = c\left(\frac{L}{2\pi}\right)^2 \tag{2.4}$$

L is the pitch of the ball screw, and u is the friction coefficient of the ball screw and the guide rail. Laplace transform of the Eq. (2.1) is as follows:

$$\begin{aligned} J_1 s^2\theta_m(s) &= T_m(s) - K\left[\theta_m(s) - x(s)\frac{2\pi}{L}\right] \\ J_0\frac{2\pi}{L}s^2x(s) &= K\left[\theta_m(s) - x(s)\frac{2\pi}{L}\right] - c_0\frac{2\pi}{L}sx(s) - [(F_Y + mg)u + F_x]\frac{L}{2\pi} \end{aligned} \tag{2.5}$$

So the transfer function of the mechanical structure is a second order oscillation element. Then convert it into standard form:

$$G(s) = \frac{L}{2\pi} \frac{w_n^2}{s^2 + 2\xi w_n s + w_n^2} \tag{2.6}$$

In the equation, $w_n = \sqrt{\frac{K}{J_0}}$ and it stands for the natural frequency of mechanical system.

3 Simulation Model

3.1 Dynamic Simulation Model of Mechanical Leg

Simulation model of the robot's mechanical leg is built up in ADAMS as shown in Fig. 4 and the main movement relationships are shown in Table 1.

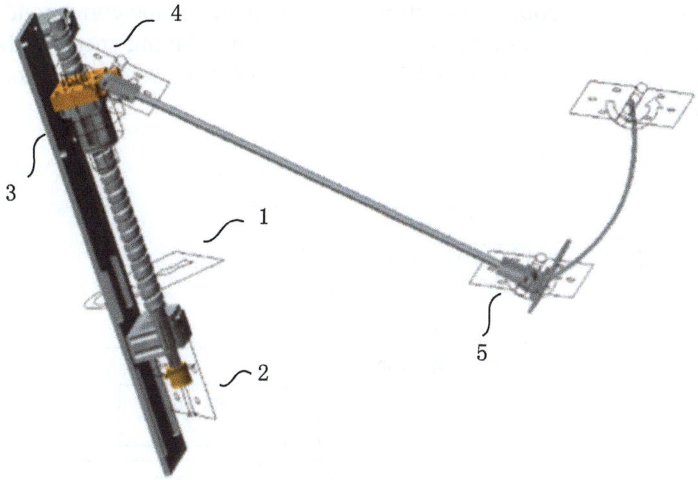

1-Joint_1 2-Joint_2 3-Joint_3 4-Joint_4 5-Joint_5

Fig. 4 ADAMS simulation model of mechanical leg

Table 1 Main movement of the mechanical leg

Name	Attribute	Member1	Member2
JOINT_1	Fixed joint	Mounting plate	Ground
JOINT_2	Revolute joint	Mounting plate	Ball screw
JOINT_3	Screw pair	Nut	Ball screw
JOINT_4	Revolute joint	Slider	Arm
JOINT_5	Revolute joint	Arm	Brake paddle

3.2 Co-simulation Model of the Robot Driver and Vehicle

In order to explore the robot driver's control process, the vehicle model is introduced into the model by CarSim, and the simulation platform is established based on closed-loop speed control. The key techniques of the simulation platform include dynamic analysis, vehicle modeling, control system and software interface.

The dynamic simulation model established in ADAMS/View accomplishes the data exchange with MATLAB/Simulink by interface module in ADAMS/Control. Set rotation angle of the servo motor as input variable of the dynamic model. The output variables include the angle of the ball screw and pedal angle. The vehicle model established in CarSim, and the input variables include rotation angle of accelerator pedal and pressure in master brake cylinder, the output variable is speed. Figure 5 reveals the schematic diagram of the software interface of the collaborative simulation model.

The mechanical and electrical collaborative simulation model based on closed-loop speed control is shown in Fig. 6. The first step in this process is comparing the output speed of the vehicle model with the target vehicle speed and generates error. Then control algorithm is used to generate the corresponding action of the servomotor according to the error to control the driving speed of the vehicle, finally the speed of the vehicle is fed back to the control system to accomplish the closed-loop control.

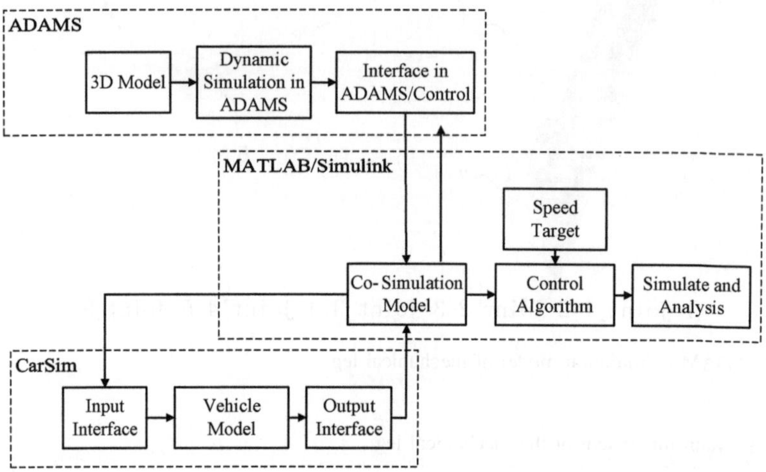

Fig. 5 Schematic diagram of software interface

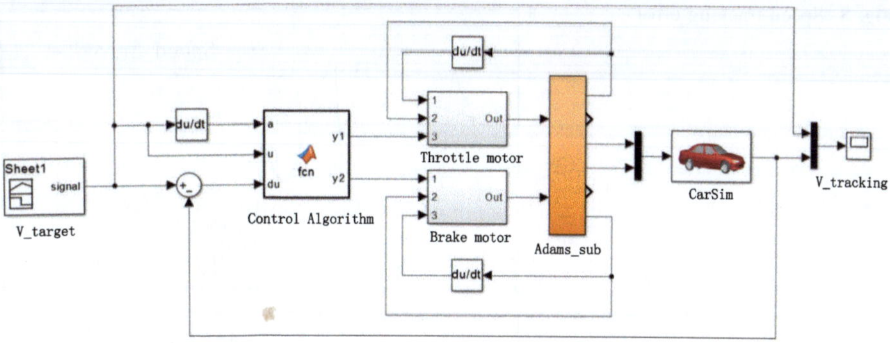

Fig. 6 Collaborative simulation model of "Robot Driver-Vehicle"

4 Simulation Experiment and Analysis

The vehicle speed curve is part of the road test condition of the Chinese limits and measurement methods for emissions from light-duty vehicles [10], the speed tracking results and speed tracking error are shown in Figs. 7 and 8. In this process, the change of throttle opening and pressure in brake pipe are as Fig. 9.

It can be determined from these figures that the model is able to complete the tracking task. Given the control method, the robot driver successfully takes proper action of the accelerator pedal with its throttle mechanical leg to speed up in the acceleration stage (2–5 s, and 10–15 s), while the maximal value of speed tracking error is 3 km/h owing to the hysteresis and nonlinearity of the vehicle; In the uniform phase (5–10 s, and 15–20 s), robot driver can maintain the relative stability of the speed and the speed tracking error is less than 2 km/h; In the deceleration stage, the throttle mechanical leg is able to quickly return to zero while the brake mechanical leg brakes.

Fig. 7 Speed tracking process

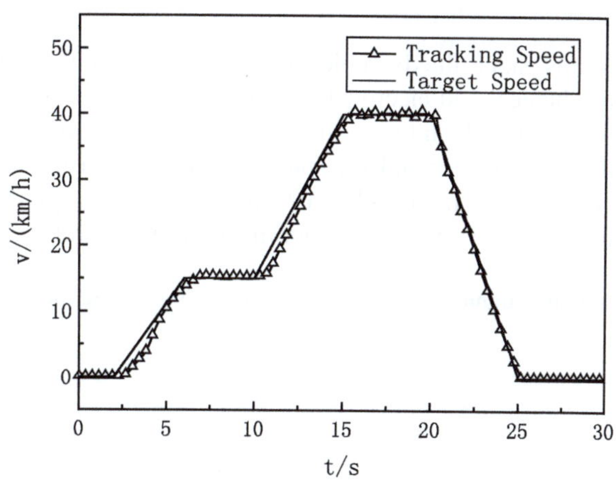

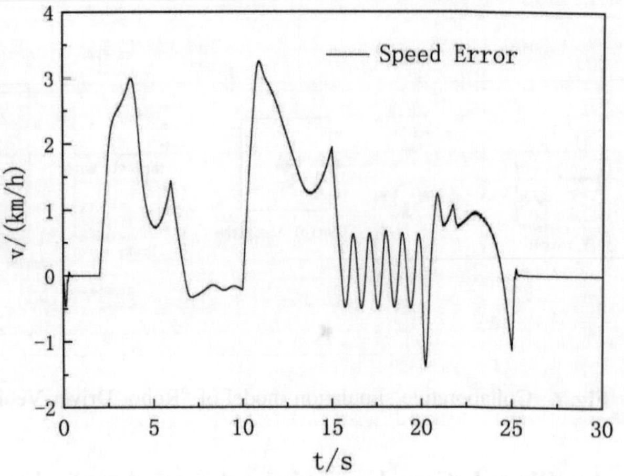

Fig. 8 Speed tracking error

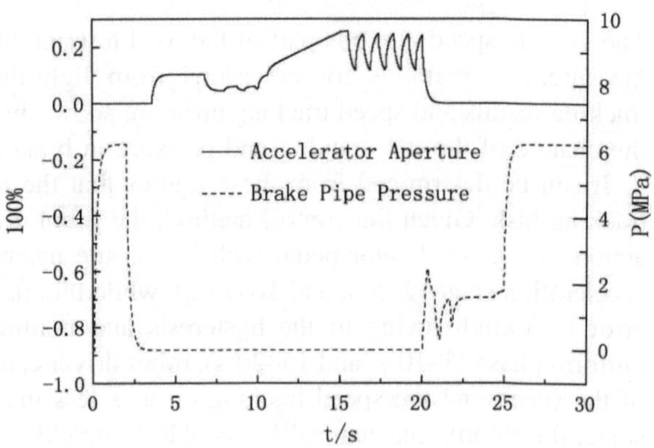

Fig. 9 Diagram of throttle and braking parameters

5 Results

In this paper, the dynamic simulation model of the robot driver is built and then the mechanical and electrical collaborative simulation platform based on speed control is established, and the control strategy is designed. The simulation results show that the robot driver and the integrated simulation platform meet the requirements of engineering. It provides a virtual prototyping platform for the improvement of mechanical structure and optimization of control strategy.

Acknowledgments Thanks to National Natural Science Foundation of China for funding (No.61371076, No.51105021).

References

1. Lu, H.P., Li, R.M.: Developing trend of ITS and strategy suggestions. J. Eng. Stud. **3**(1), 6–19 (2014)
2. Thiel, W., Grof, S., Hohenberg, G., et al.: Investigations on robot driver for vehicle exhaust emission measurements in comparison to the driving strategies of human drivers. SAE Paper, 982642 (1998)
3. Namik, H., Inamura, T., Stol, K.: Development of a robotic driver for vehicle dynamometer testing. In: Proceedings of 2006 Australasian Conference on Robotics and Automation, Auckland, New Zealand, pp. 1–9 (2006)
4. Holschuh, N., Winckler, J., Probst, H., et al.: Performance of a mechanized driver for measurement of automobile exhaust gas emissions and fuel economy on chassis dynamometer. SAE paper No. 910037, pp. 1–5 (1991)
5. Zhang, W.G., Chen, X.B.: Key technologies of vehicle robot driver. J. Jiangsu Univ. (Nat. Sci. Ed.) **26**(1), 20–23 (2005)
6. Wang, X.R.: Development and application of modeling and simulation technology. Mach. Build. Autom. (01), 1–45 (2010)
7. Ma, Y.H., Niu, Z.G.: Dynamic simulation of clutch robot leg of robot driver based on ADAMS. Mech. Manage. Dev. **24**(3), 152–153 (2009)
8. Jiang, H., Yue, J.G., Hu, L.D.: Fuzzy control strategy study of servo system in robotic driver. Syst. Simul. Technol. **9**(1), 61–65 (2013)
9. Li, Q.X.: Servo System and Electrical Control of Machine Tool. China Machine Press, Beijing, pp. 240–242 (1994)
10. Ministry of Environmental Protection of the People's Republic of China, GB 18352.3-2005. Limits and Measurement Methods for Emissions from Light-duty Vehicles (CHINA 5). Standards Press of China, Beijing, pp. 44–48 (2005)

Comparative Analysis the Rectifier Part of Electric Locomotive Auxiliary Converter

Quanzhu Zhang, Xiaohui Lu, Min Lei and Yonghong Deng

Abstract Based on the development of the auxiliary converter of SS4B electric locomotive, the key research is focused on the front stage of the rectifier. Comparative analysis the two widely used in the auxiliary converter circuit topology (PWM rectifying circuit and phase-controlled rectifier), and finally choose the phase controlled rectifier circuit for auxiliary converter front stage the rectification scheme.

Keywords Auxiliary converter · PWM rectifier · Phase controlled rectifier

1 Introduction

At present, China's electric locomotive is developing rapidly. The transmission technology is gradually transformed from AC-DC to AC-DC-AC, and it has realized the full promotion of the level of our country's locomotive equipment. But the domestic electric locomotive with SS series as the representative of the railway line still plays an important role in passenger and freight transport, the AC-DC transmission technology can't be completely replaced. SS4B electric locomotive uses split-phase motor as auxiliary circuit power system. However, since the unbalanced and unstable output voltage, it is necessary to replace the split-phase motor with the auxiliary converter as the power supply. In this paper, a comparative analysis is made mainly aimed on the front stage rectification scheme of auxiliary converter. We transform the auxiliary converter of two SS4 types electric locomotive, it is based on the development of the SS4 type electric locomotive auxiliary converter

Q. Zhang · X. Lu (✉) · M. Lei · Y. Deng
North China Institute of Science and Technology, Institute of Information and Control Technology, Sanhe, Yanjiao 065201, Hebei, China
e-mail: luxen2014@foxmail.com

Q. Zhang · X. Lu
Anhui University of Science and Technology, Huainan 232001, Anhui, China

© Springer Science+Business Media Singapore 2017

H. Lu (ed.), *Proceedings of the Second International Conference on Intelligent Transportation*, Smart Innovation, Systems and Technologies 53,
DOI 10.1007/978-981-10-2398-9_9

system witch developed by the China Shenhua Group at the Beijing Jiaotong University.

2 Introduction of Auxiliary Power Supply System [1]

2.1 *Auxiliary Power Supply System Based on Split-Phase Motor*

The origin auxiliary power supply system of SS4B electric locomotive relayed on split-phase motor supplying power for auxiliary unit. The split-phase motor can be regarded as the combination of a single-phase motor and three-phase generator. It accepts the auxiliary winding of the main transformer single-phase power and clefts into three-phase power to supply the auxiliary unit [2].

As shown in Fig. 1, the split-phase motor itself is only outputting one of the three-phase voltages, the other two phases are supplying by transformer auxiliary winding directly. And we can also think that the transformer auxiliary winding single-phase voltage as the line voltage of three-phase voltage between A and B phase. From Fig. 1, we can see that the stator winding W1 and W2 of split-phase motor not only act as an electric winding but also act as a power generator winding. Though embedded stator winding in the split-phase motor stator side as the asymmetric way, the split-phase motor output voltage finally will be unbalanced with the change of the electric locomotive operating conditions.

Second, the two phase voltages of the auxiliary power supply system based on split-phase motor are provided by the auxiliary unit of main transformer directly. Because of the grid voltage is unstable and the fluctuation range is −24 to +20 % (19–30 kV), so the output voltage of the split-phase motor is unstable.

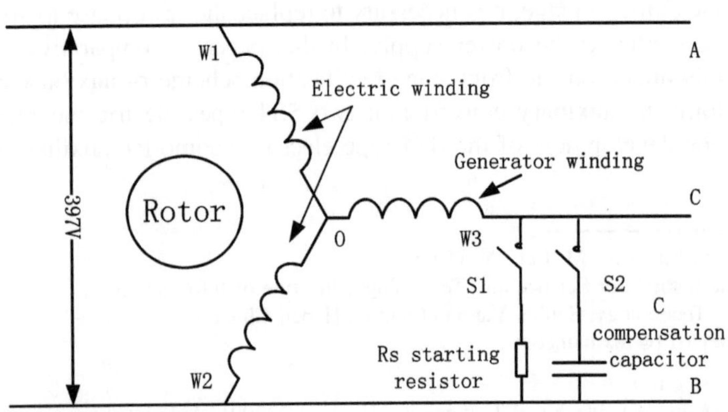

Fig. 1 Internal structure and principle diagram of the split-phase motor

2.2 *Auxiliary Power Supply System Based on Static Converter*

Static converter is using the current mainstream AC-DC-AC power conversion. The output of the auxiliary winding of the main transformer is supplied to the auxiliary system. Auxiliary converter introduces modern power electronic power transformation. It has a higher automation and its output is more controllable and more stable comparing with the power supply scheme of split-phase motor. The auxiliary converter is composed of a front stage rectifier and a back stage inverter, which is shown in Fig. 2.

The main idea of auxiliary converter is rectified the output AC of the main transformer auxiliary winding to DC signal, and then convert it to AC to supply for auxiliary system. It makes the whole auxiliary system controllable and stable. The auxiliary converter can assignment and management load according to different auxiliary winding types. It avoids the current impact on auxiliary power supply system with sudden load in soft start mode.

The auxiliary unit of the inverter scheme is three-phase 50 Hz, 380 V power supply. We choose the voltage type inverter that the inverse coefficient can reach 0.78 in theory. But in the actual engineering, considering the loss and the efficiency, the inverter coefficient is calculated according to 0.7. According to the three-phase output voltage effective value 380 V, inverter coefficient 0.7, we determine that the DC bus voltage is 540 V, which is a classic engineering value.

The task of the rectifier is to rectify the AC output of the auxiliary windings with the power electronic device to a stable 540 V DC voltage. At present, there are two main types of rectifier. One is phase-controlled rectifier that uses thyristor as a switch device. The other is the PWM rectifier scheme which relies on IGBT as a switching device. In general, these two schemes have advantages and disadvantages

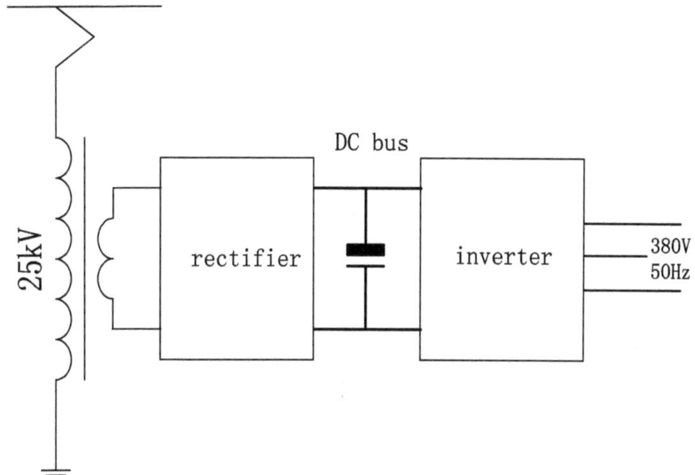

Fig. 2 Auxiliary converter structure block diagram

in the practical application. The following will be a comparative analysis of these two schemes.

3 Phase-Controlled Rectifier Scheme

3.1 Basic Principle of Phase-Controlled Rectifier Circuit

The phase-controlled rectifier scheme uses a single phase bridge half controlled rectifier circuit topology, as shown in Fig. 3. The rectifier circuit contains two thyristor (K1, K2), and two diode (D1, D2), square wave reactor (L), and the DC side support capacity (C_d). It can control the output DC voltage by controlling the turn-on angle of a. We will not introduce the circuit principle since it is uncomplicated.

In the practical application, we give K1, K2 suitable trigger pulse to achieve the control of the main circuit of the whole phase-controlled system. The trigger pulse has the following two characteristics. First, the frequency of the trigger pulse is the same as the input voltage frequency of the system. Second, the zero phase reference point of the trigger pulse is the zero crossing of the input voltage. At present, the most popular use of integrated trigger chip KJ009 for pulse transmission.

In the ideal situation, the average value of the system is:

$$U_o = \frac{1}{\pi}\int_{\alpha}^{\pi} \sqrt{2}U_{in}\sin(\omega t)d(\omega t) = 0.9U_{in}\frac{1+\cos\alpha}{2} \tag{1.1}$$

The formula (1.1) shows that, the output voltage of the system with control angle is nonlinear, the variation range of the control angle is 0∼PI. If we want output

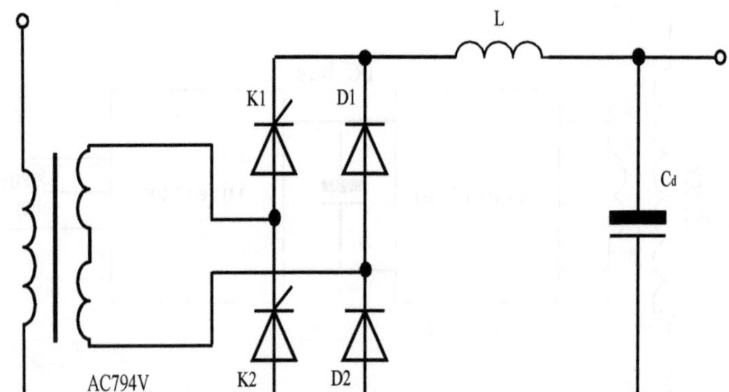

Fig. 3 Principle of phase-controlled rectifier scheme

voltage is 540 V DC, the input voltage effective value shall not be less than 600 V. Considering the net voltage fluctuation, the output of the auxiliary winding is not less than 600 V, that is, the rated output of the auxiliary winding shall not be less than 600/74 % = 790 V.

3.2 *Phase-Controlled Rectifier Scheme Control Strategy*

In order to make the output voltage stable and controllable, the double closed-loop control is introduced. The control block diagram is shown in Fig. 4. The voltage loop is for stabilizing the output voltage, and the current of control system does not exceed the current threshold. The difference ΔU between reference signal U_{ref} and the actual system output voltage U_o is regulated by the PI regulator formation of the phase angle reference signal a. It comes into being phase shift angle a' after the current inner loop. At the same time, the phase locked (PLL) circuit samples and follows the output synchronization signal of frequency and phase position of the input voltage. Finally, it will be sent into PWM generator with a'. Then the PWM generator will output driving pulse. After controlled by the phase-controlled rectifier scheme control strategy, the output voltage of the system can fluctuate around the set value in a certain range.

3.3 *Simulation of Single-Phase Controlled Rectification Circuit*

We designs the phase controlled rectifier circuit of the SS4B type auxiliary converter according to the theoretical analysis and practical requirement. And we

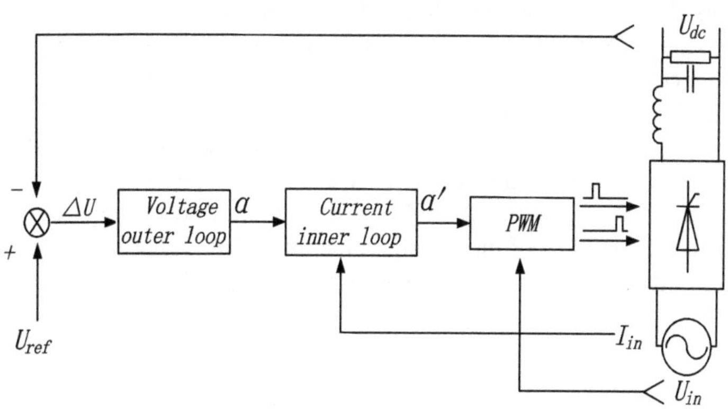

Fig. 4 Phase-controlled rectifier scheme control strategy

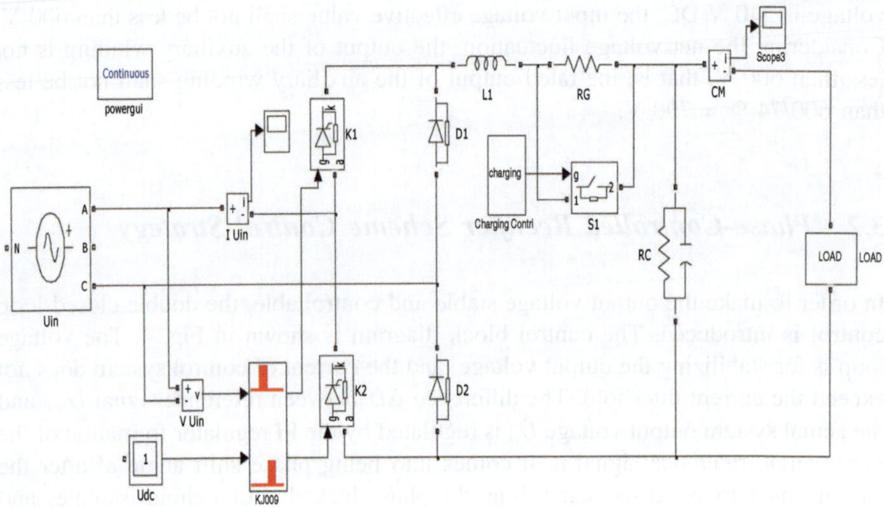

Fig. 5 Phase controlled rectifier Simulink simulation model

simulate this circuit with simulink. The simulink model is shown in Fig. 5 which encapsulate some of the sub modules.

The model of KJ009, which is used to control the thyristor control, is built up by us. The main task of KJ009 in the system is to control the generation of the thyristor pulse and the synchronization with the input voltage. It is the core device of rectifying system, and it has achieved the control of the output voltage. The internal structure of KJ009 is shown in Fig. 6.

When the system starts, the instantaneous charge current of the supporting capacity at both ends of the DC bus will exceed current limit without any measure of hard start. The large current will instantly breakdown of the main circuit components. To avoid the components being broken down, we have taken some measures. First, series connect a charging resistor (RG) for pre charge for bus capacitor in the end of the bus. Second, the soft start mode is adopted for the rectifying system. At the beginning of the start, the setting value of DC voltage is raised with a certain rake ratio, and then it achieves the expected set value within a certain period of time and maintain.

In this model, the function of the control charging sub module is to do the access control of the charging resistance. When we start the system, the charging resistor L1 and the flat wave reactor RG will be series connected to the circuit. When the bus voltage exceeds 70 % of the setting value, the control charging module controls the short circuit S1 action, and the charging resistance RG is shorted.

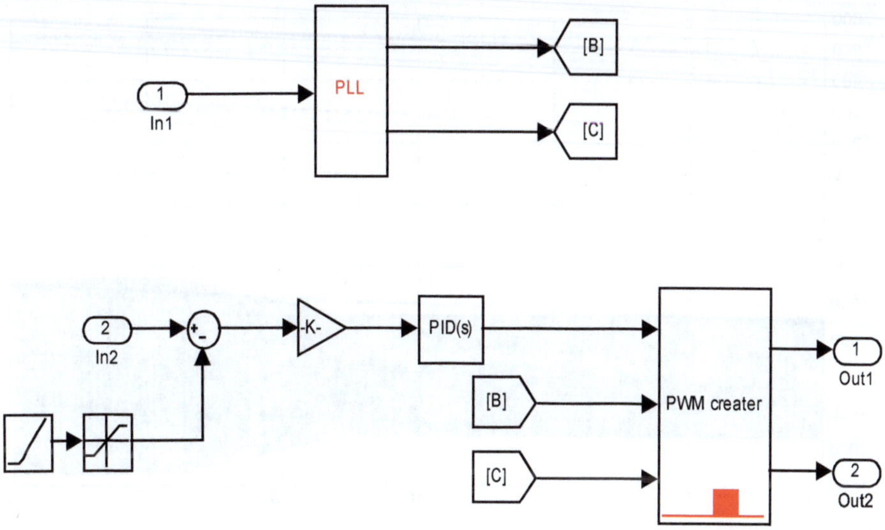

Fig. 6 Internal structure of KJ009 simulation model

Four seconds after the system start, when the DC voltage being stable, will be loading the load with soft start mode. It will load the load with soft start at the moment of 5th second and finish loading until 13th second, while the load is 30 A. The model uses programmable power supply considering the network voltage fluctuation in the range of 19–30 kV. In 0–18 s, the source voltage is 604 V, while the corresponding grid voltage is 19 kV. From 18th second, the voltage amplitude of the grid is increased by 0.1 times of the positive slope. Until 24th second, the grid voltage will reach at 30 kV, and it will hold this value till 30th second. Rectifier output and input voltage waveforms are shown in Fig. 7. It can be seen that when the pre charge resistance is short, the DC voltage will be fluctuations, but it does not constitute a voltage impact on the system. The output voltage of the rectifier system is stable and reliable despite of the fluctuation of load and grid voltage, and the ripple voltage is less than 10 V.

Because the modulation frequency of phase controlled rectifier has the same frequency with the input voltage, the input current waveform after loading and harmonic analysis are shown in Fig. 8.

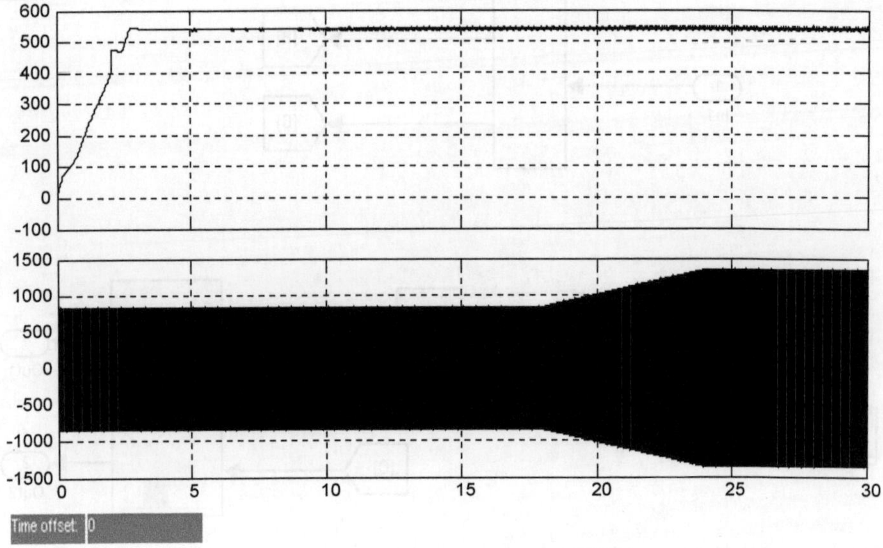

Fig. 7 Phase-controlled rectifier output and input voltage waveform

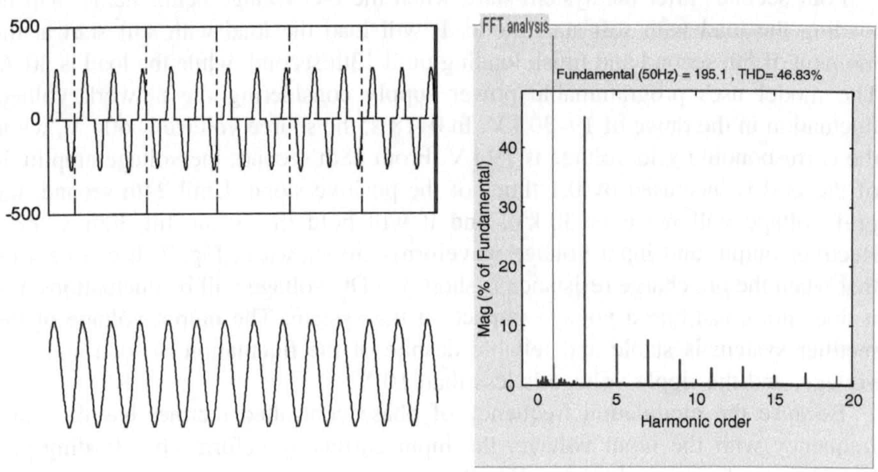

Fig. 8 Phase-controlled circuit input current and voltage waveform contrast and current harmonic analysis

4 PWM Rectifier

4.1 Basic Principle of PWM Rectifier Circuit [3]

Recent years, PWM rectifier circuit has developed a high level of technology with a input and high speed, etc. It also can transmit the energy bidirectionally. We use IGBT as the switching device. It has a higher switching frequency, lower loss and higher power integration compared with the thyristor.

The main function of the PWM rectifying circuit and the phase-controlled rectifying circuit are the same, they both can achieve stable DC output under the condition of net voltage fluctuation, and ensure the normal operation of auxiliary inverter. Compared with the phase-controlled rectifier circuit constituted by Semiconductor Controlled Rectifier (SCR), the PWM rectifier has some outstanding advantages. The net side current contains very few harmonics, and the input side current is approximately sinusoidal. And it is achievable that the input power factor close to 1 by certain control. The two obviously superior characteristics of the PWM rectifier circuit comparing with the phase-controlled rectifier make it to be the main trend of the current AC/DC circuit.

Circuit structure and equivalent circuit diagram of single-phase PWM rectifier circuit is shown in Fig. 9. The conversion of the electric energy is accomplished by the H-bridge which is composed by four IGBT. From the structure, it can be seen as a single phase inverter and an AC side reactor L. The main role of L is filtering and storing energy. We detect the voltage real-time across the L to achieve the control of PWM rectifier. It needs to parallel a support capacitor on the DC side of voltage type PWM rectifier to ensure the DC bus voltage stable. If the current transmit from left to right, the circuit is rectifier circuit. On the other hand, it's inverter circuit, so this circuit topology can achieve two-way transmission of electric energy. In fact, from the relationship between the input current and the input voltage of PWM rectifier, the PWM rectifier can work in any quadrant, and therefore PWM rectifier is also called as the four quadrant converter.

Neglecting the input side resistance, the input vector Eq. (3.1) of the PWM rectifier is established.

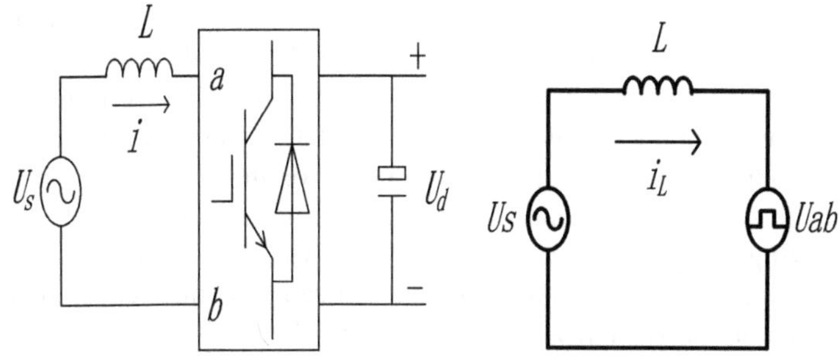

Fig. 9 Circuit structure and equivalent circuit diagram of PWM rectifier circuit

$$\overrightarrow{U_s} = j\omega L\overrightarrow{I_L} + \overrightarrow{U_{ab}} \tag{3.1}$$

According to the equivalent circuit and vector equation set, when the input side AC power supply US and AC side inductor are certain, the size and phase of i_L can be controlled by controlling the U_{ab}. And then it can control the transmission power and power factor of the converter are controlled. The basic mathematical model and its analysis have been described in detail in the relevant literature. It is needed to explain that the PWM rectifier is a boost circuit, and the DC side voltage cannot be too low. If the input voltage is lower than the peak voltage of AC side, the U_{ab} will not be able to get enough fundamental voltage and cannot achieve effective control. So the limit control of US is shown in Eq. (3.2).

$$|U_S| \leq \frac{U_{dc}}{\sqrt{2}} \tag{3.2}$$

The output voltage of the DC side is 540 V, and the input side voltage shall not exceed 380 V. Considering the fluctuation of the grid voltage, the input voltage of PWM rectifier will not exceed 380 V. If we use the PWM rectifier, the rated output voltage of the auxiliary winding shall be 320 V.

4.2 The Control Strategy of PWM Rectifier Circuit

We identify two control objectives by analyzing the principle of the PWM rectifier. One is the output stable adjustable DC voltage, the other is to ensure that the input current harmonic content is small and the input high [4, 5].

Control mode of PWM rectifier circuit is divided into direct and indirect current control which determined by whether the input current is directly involved in the control [6]. Method in this paper since it has the advantages of fast system dynamic response and strong current following to the insensitive change of system parameters. The control strategy is shown in Fig. 10.

This is a typical double closed loop control strategy with stable and realizable which inner loop is the voltage loop and outer loop is current loop. We detect the DC voltage to compare with the reference voltage to get voltage error signal e_v. It will be the input signal of outer loop controller G1, then this output signal as the scalar given signal I_L^* of AC side current. AC side input voltage minus U_L^* can be obtained the modulation signal U_{ab}^*. The modulation signal input to the SPWM generation module to generate a driving signal for the inverter. It is the control strategy of the whole system and the physical meaning of the signal.

By analyzing the principle above, as long as we control the size and the fundamental phase of the U_{ab}, the input side current can be controlled, so that the input side power factor can be adjusted between 0 and 1.

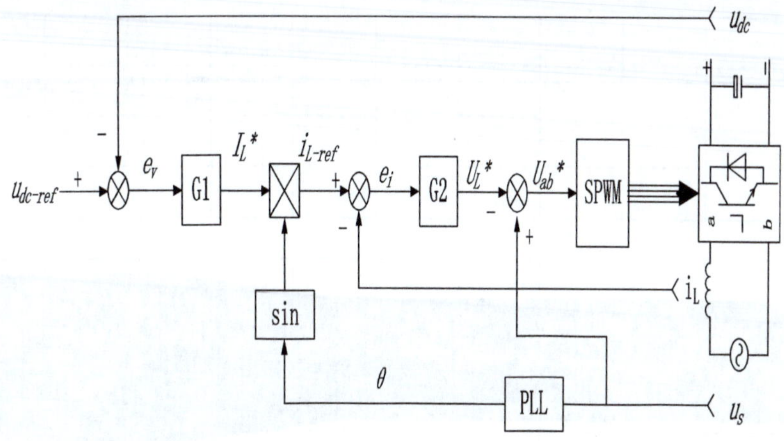

Fig. 10 The control strategy of PWM rectifier

4.3 Simulation of Single-Phase PWM Rectifier Circuit

Based on the previous analysis, a single-phase PWM rectifier circuit is built under the simulink module of the MATLAB software. It is shown in Fig. 11.

The sub function modules have been packaged. The external parameters and the charging method are basically the same as the principle of the phase control circuit, which is not repeated. PWM rectifier input and output voltage is shown in Fig. 12.

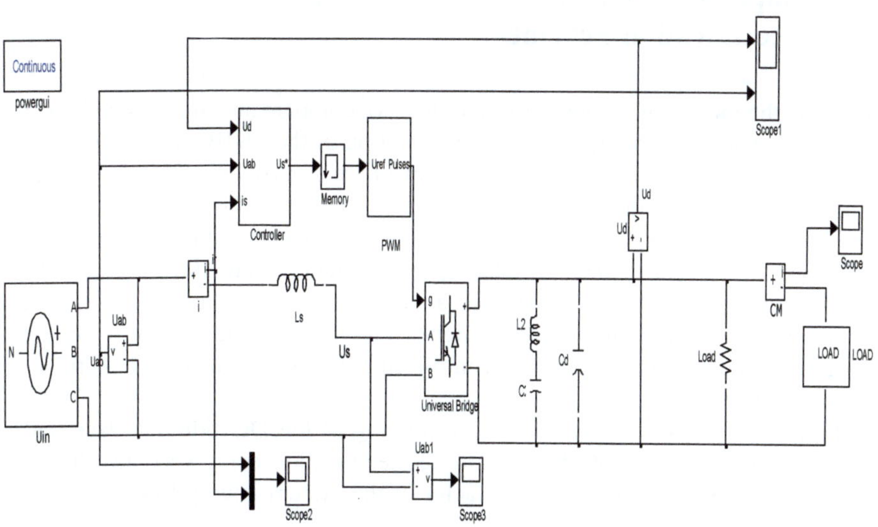

Fig. 11 Simulink simulation model of PWM rectifier

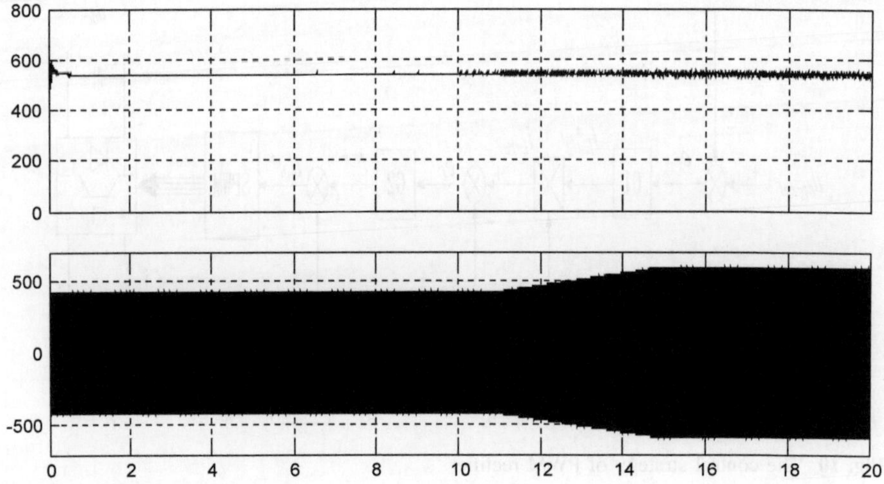

Fig. 12 The input and output voltage waveform of PWM rectifier circuit

Under the situation of the grid voltage change and load, the voltages are able to achieve stable output, the ripple is less than 10 V.

The harmonic analysis of the network side current of the PWM rectifier is shown in Fig. 13. It can be seen that the harmonic content of the PWM rectifier is significantly decreased compared with the phase-controlled rectifier.

5 Comparison and Analysis of Two Sets of Rectification Schemes

Through the modeling of two sets of rectifier, we compare and analysis with the two rectification schemes based on the actual situation of the SS4B electric locomotive.

At the technical level, the technology of phase-controlled rectifier is relatively simpler and more mature. The control system with KJ009 as the core is simpler and has a lower cost. While the control system of PWM rectifier scheme is more complex, it must use DSP or other programmable logic device to meet the requirements. And PWM rectifier is a boost rectifier device, once controlled improperly, DC side voltage will be higher than the AC side voltage many times. Therefore, the technical cost and control complexity of the PWM rectifier are relatively higher.

From the view of energy, the forced filtering of the DC side flat wave reactor makes the power factor of phase-controlled rectification being very low due to its own characteristics. And the power factor of the system is very low.

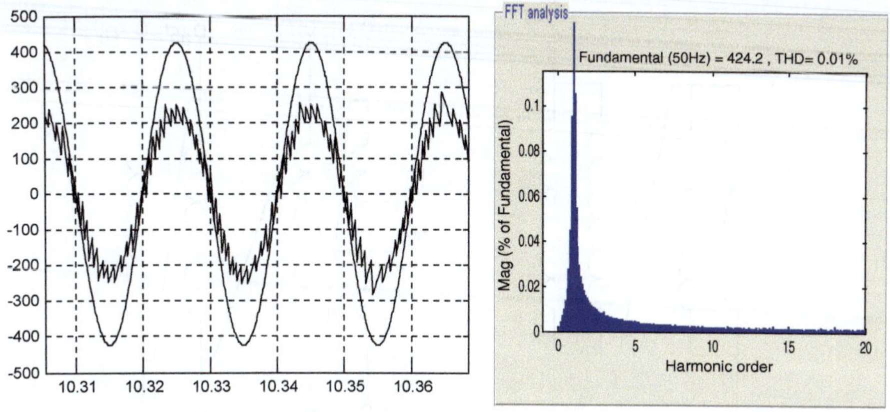

Fig. 13 Input current, voltage waveform and current harmonic analysis of PWM rectifier

The power utilization efficiency of PWM rectifier is higher. The power factor is close to 1. It can be automatically adjusted according to different operating conditions, which is obviously better than the phase-controlled rectifier.

The input current of the PWM is nearly sine wave with lower harmonic content and less affected on grid through the impact on power grid and the whole transmission system. The phase-controlled rectifier scheme injects a lot odd harmonics to the power grid, and there is a large reactive current. Although it is effectual to improve the current quality by using a certain compensation device, but if use too much, it will make the grid disorders [7].

The actual situation is that the two programs are used for the auxiliary power supply system of SS4B type electric locomotive. They need the auxiliary winding to supply power. But the rated power supply voltage of the original auxiliary power supply winding is 397 V. And it is not suitable for the phase control and PWM rectification scheme to use double pancake winding parallel winding. So it is necessary to re-design the winding for PWM rectifier. For phase-controlled rectifier scheme, the rated output voltage shall not be less than 790 V by changing the parallel connection to series connection of the double pancake winding of original auxiliary winding. The rated voltage of the auxiliary winding is doubled while the rated current is halved, so the output power of the whole winding is constant. The wiring is shown in Fig. 14.

According to the transformed diagram, we can see that this kind of transformation is of less work and easy to implement. And it is still feasible to resume it as a split-phase motor for the auxiliary system again. All we should do is just change the external winding connection mode.

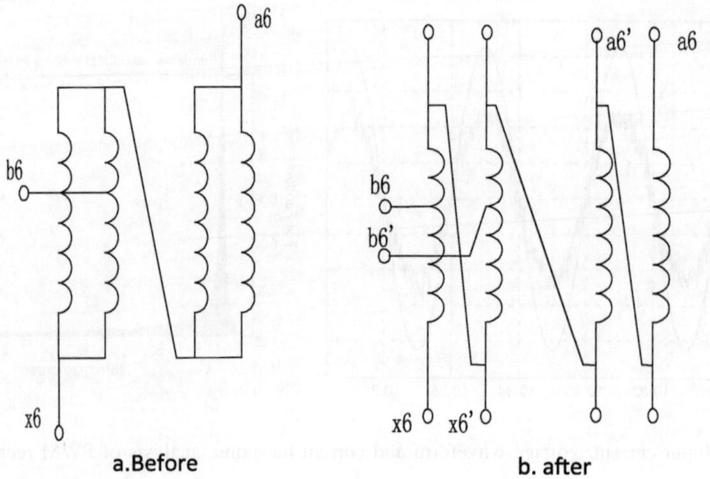

Fig. 14 Schematic diagram of auxiliary winding of phase-controlled rectifying scheme

6 Conclusion

By comparison with two kinds of rectification schemes, we choose the phase-controlled rectifier as the first stage rectifier of the SS4B electric locomotive auxiliary converter. Because the number of SS4B type electric locomotive used in the transformation is very small. Therefore, the inverter can improve the benignant effect on the power grid by using a certain compensation device. Although the PWM rectifier circuit is used as the final rectifier, the excellent performance of the PWM rectifier circuit is very considerable. It has become the current electric locomotive traction and auxiliary power supply converter mainstream program.

References

1. Wang, Y., Liu, W., Yang, Z., Ge, X., Liu, X.: Research on design evaluation of high-speed train auxiliary power supply system based on the AHP. In: 2014 IEEE Conference and Expo Transportation Electrification Asia-Pacific (ITEC Asia-Pacific) (2014)
2. Zhang, Q.: Research and design of power supply system for electric locomotive auxiliary motor. Power Supply Technol. **34**(4), 391–394 (2010) (in Chinese)
3. He, B.: Research on Single-Phase PWM Rectifier. Huazhong University of Science and Technology, Wuhan (2012) (in Chinese)
4. Zhang, H.: Research on the Control of PWM Rectifier for Electric Locomotive Auxiliary Converter. Beijing Jiaotong University, Beijing (2008) (in Chinese)
5. Zhang, C., Zhang, X.: PWM Rectifier and its Control. Machinery Industry Press, Beijing (2003) (in Chinese)
6. Zhao, G., Chao, S.: The simulation of new arc welding power based on Matlab. Adv. Mater. Res. (2011)
7. Wang, Z., Huang, J.: Power Electronic Technology. Machinery Industry Press, Beijing (2000) (in Chinese)

An Improved Multi-Kernel Estimation Method for Vehicle Localization

Wenjie Lu, Liangwei Jiang, Min Wang and Qing Ma

Abstract Vehicle localization is one of the key functions in Intelligent Vehicles (IVs). Localization result is usually provided by the combination of GPS data and additional sensors, which are able to improve the localization precision. In this paper, a monocular camera and map database help vehicle localization, aiming to enhancing the localization performance. To this end, map-based road lane markings are constructed according to open source map. Then, vision-based markings and map-based markings are fused to obtain the improved vehicle fix, using an improved Multi-Kernel Estimation (MKE) method. The results using real data show that our method leads to an obvious improvement in vehicle localization accuracy.

Keywords Vehicle localization · Multi-Kernel estimation (MKE) · Map database · Road lane marking

W. Lu (✉) · L. Jiang · M. Wang · Q. Ma
Traffic Management Research Institute of Ministry of Public Security, Wuxi, China
e-mail: luwenjie0122@msn.cn

L. Jiang
e-mail: jlw18@126.com

M. Wang
e-mail: wangmin_tmri@126.com

Q. Ma
e-mail: 13338110260@126.com

W. Lu · L. Jiang · M. Wang
Key Laboratory of Ministry of Public Security for Road Traffic Safety, Wuxi, China

© Springer Science+Business Media Singapore 2017
H. Lu (ed.), *Proceedings of the Second International Conference on Intelligent Transportation*, Smart Innovation, Systems and Technologies 53,
DOI 10.1007/978-981-10-2398-9_10

1 Introduction

Vehicle localization plays a critical role in Intelligent Transportation System (ITS) because it is the prior task for higher level operations, such as traffic flow control [1] and vehicle network communication [2]. A state-of-the-art on vehicle localization methods is summarized in [3].

Map-matching algorithm is a frequently used method in localization with the development of digital maps. Normally, the trajectories of vehicles are restricted in the road area. Hence, a digital map of the road network is used to constraint the navigation solution of the in-car navigation system. This process is known as map matching [4], and is detailed in the following subsections: geometric approaches [5], Topological approaches [6], probabilistic methods [6] and advanced algorithms.

Map-matching algorithms associate the positioning data from GPS sensors with Geographic Information Systems (GIS). By means of this process, it is possible to identify the current road section on which the host vehicle is traveling, and redefine the position of the host vehicle on the identified road section. The purpose of a map-matching method is to enhance the localization performance, and in advance to improve navigation function of ITS. In the following the state-of-the-art related to map-matching approaches are classified into the categories: geometric approaches, topological techniques, probabilistic methods, and advanced algorithms [7].

One branch of advanced method is road lane marking based methods. Road marking based localization approaches Publications as [8] on this direction utilize high-accuracy digital maps, including the precise positions of the markings. In [8], a transformation vector between vision space and map space are derived according to the features of traffic signs (i.e. markings and pedestrian crossings) captured via camera, which helps to adjust vehicle location. High-accuracy digital maps help to improve localization results. However, such maps are specially customized, which is time-consuming to fabricate and range-limited. In [9], a map of lane marking features is built. Vehicle localization is implemented using this map and vision based marking features.

A lane marking based vehicle localization method is proposed in this paper, using map database, low-cost GPS and camera to adjust rough position signals. The thinking is to extract and fit vision-based and map-based markings to optimize vehicle position in real time. The road segments are estimated using multi-criterion estimation, the map-based lane markings of current state are then determined from an open source map database. After that, vehicle location is estimated according to integrating both vision-based lane markings and map-based lane markings, using an improved multi-kernel estimation method.

2 Road Matching

In this section, the map database is at first reconfigured from a "node-way" topology to a "lane marking" topology. Then the current lane is selected according to multi-criterion. Thus the map-based lane markings are derived.

A. *Map reconfiguration*

A "lane marking" map topology is constructed. Different from the construction of lane marking based maps using External Kalman Filter (EKF) in [10], an open source database, called Open Street Maps (OSM) is introduced to create map topology. At first, the map organization composed of "nodes" and "ways" is derived from map data. Then the "node-way" topology is reconfigured to a serial of "road segments". The "segments" are transformed to "cells". A cell represents a direct road lane in highway environment, and thus the left and right boundaries are road markings of the lane. Therefore, a "lane marking" topology is achieved. The transmissions from "way" and "node" to "segment" and "cell" are depicted as Fig. 1.

As shown in Fig. 1, a "segment" is a line segment of a poly-line "way", and a "cell" is a single lane in a "segment". The left and right cell boundaries of a cell (red lines in Fig. 1) are exactly lane markings.

Table 1 enumerates all the specific conditions met in the transmission from way to segment. Specifically, the width of link road (e.g. trunk link road and motorway link road in Table 1) is unable to be determined by only the OSM map. Because the straight link roads own a width of 3.5 m, while the width of roundabout link roads is 5 m. Therefore, a strategy to judge a relative straight link road or a roundabout is required. To this end, an average derivative cell slope $\bar{k}_i^{way}$ of the ith way is defined, which represents the slope variation degree of a way. A threshold Th_k is set to differentiate these two link roads, and thus the road width is determined.

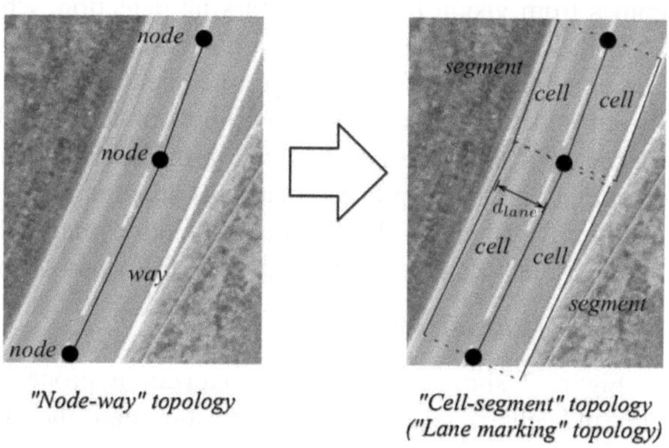

Fig. 1 Map reconfiguration to "lane marking" topology

Table 1 Relationship from OSM road type to lane number and lane width

"Node-way" topology			"Marking" topology	
Highway value	Lane value	Oneway value	Lane number	Lane width (m)
Trunk_link	1	Yes	1	3.5/5
Motorway_link	1	Yes	1	3.5/5
Motorway_link	2	Yes	2	3.5
Motorway	3	Yes	3	3.5
Motorway	4	Yes	4	3.5
Trunk	1	Yes	1	3.5
Trunk	2	Yes	2	3.5
Trunk	3	Yes	3	3.5
Primary	2	No/yes	1	3.5
Primary	2	Yes	2	3.5

B. *Lane selection*

The lane markings need to be selected in the constructed lane marking topology. It is possible that lane selection algorithm based on multi-criterion is implemented. At first, the current segment is estimated according to a criterion of the linear distance in ENU space between the rover and segment candidates. Then the current cell is determined by several criteria derived from vision based lane detection and filtered GPS positions. Considering that the left and right boundaries of current cell are exactly the lane markings of current lane, thus when the current cell is determined, the map-based lane markings are derived.

3 Marking Based Vehicle Localization

The method is to match the two different kinds of lane markings in Bird's Eye View (BEV) space, in order to help improve vehicle localization. One marking source comes from vision based lane marking detection, which has been accomplished in [10], the other comes from map database. At first, map based markings in BEV space are detected. Two parts are included in the marking of BEV space, one part includes the left and right lane markings of current cell, and the other owns the lane markings of the corresponding cell in the next segment. Then both the vision based markings and the map based markings are compared and fitted in BEV space, using an improved MKE. The 2D transformation matrix between these two sources is able to adjust vehicle position filtered from raw GPS.

A. *Vehicle localization*

Figure 2 shows the vehicle localization procedure. On one hand, when the current "cell" is selected in lane selection part, the markings of current cell and the cell in front are projected to BEV space, as shown the black lines in Fig. 2. Markings of a cell are a pair of paralleled lines in map topology, so the map-based

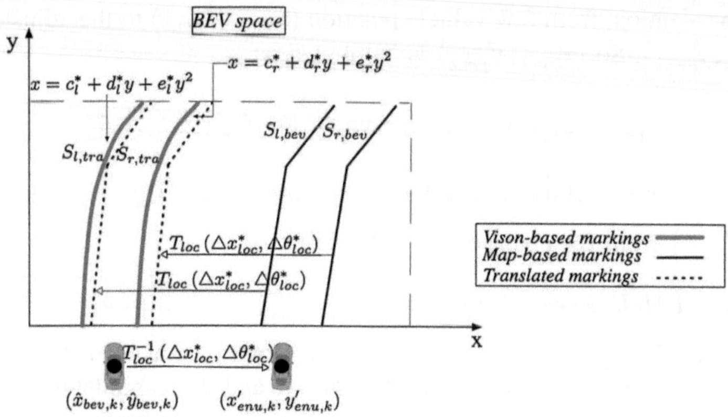

Fig. 2 Vehicle localization procedure

markings are combination of several straight lines. The sets of left and right marking pixels are denoted as $S_{l,ipm}$ and $S_{r,ipm}$ respectively. On the other hand, vision-based lane markings are represented as the form of quadratic model $(c_i^*, d_i^*, e_i^*), i = l, r$ from previous work, as shown the gray curves in Fig. 2.

The translation from $(c_i^*, d_i^*, e_i^*), i = l, r$ to $S_{l,ipm}$ and $S_{r,ipm}$ can be regarded as the relative rigid transformation from rough GPS positioning to positions according to map information.

However, it is difficult to compute a translation matrix from a model (vision-based markings) to a set of pixels (map-based markings). The idea is to at first estimate the translation from the set of pixels to the marking model. The required matrix is then derived using inverse matrix transform.

A translation matrix from map-based markings to vision-based lane markings is estimated according to an improved MKE method. The translation matrix is defined as $T_{loc}(\Delta x_{loc}, \Delta\theta_{loc})$, where Δx_{loc} is lateral displacement, and $\Delta\theta_{loc}$ is vehicle rotation. A multi-kernel based descriptor $G_p(c_i^*, d_i^*, e_i^*, x, y)$, is introduced to describe the "distance" between a single pixel on a map-based marking and a vision-based marking model. The optimized lateral displacement Δx_{loc}^* and rotation $\Delta\theta_{loc}^*$ are optimized according to $G_p(c_i^*, d_i^*, e_i^*, x, y)$ as

$$(\Delta x_{loc}^*, \Delta\theta_{loc}^*) = \operatorname{argmax}_{\Delta x_{loc}, \Delta\theta_{loc}} \left[\sum_{i=l,r} \sum_{(x,y)\in S_{i,tra}} G_p(c_i^*, d_i^*, e_i^*, x, y) \right],$$

where

$$S_{i,tra} = \{(x,y) | (x, y, 1) = (x', y', 1) \cdot T_{loc}, (x', y') \in S_{i,ipm}\}.$$

Translated markings through $T_{loc}(\Delta x_{loc}^*, \Delta\theta_{loc}^*)$ are marked as dashed lines in Fig. 2. The inverse matrix $T_{loc}^{-1}(\Delta x_{loc}^*, \Delta\theta_{loc}^*)$ is the translation of lane marking based vehicle localization, from raw vehicle position $(\hat{x}_{ipm,k}, \hat{y}_{ipm,k})$ to the adjusted vehicle fix $(x'_{ipm,k}, y'_{ipm,k})$. So $(x'_{ipm,k}, y'_{ipm,k})$ is computed as

$$(x'_{ipm,k}, y'_{ipm,k}, 1) = (\hat{x}_{ipm,k}, \hat{y}_{ipm,k}, 1) \cdot T_{loc}^{-1}(\Delta x^*_{loc}, \Delta\theta^*_{loc}).$$

Therefore, the marking based localized position is $(x'_{enu,k}, y'_{enu,k})$, translating $\left(x'_{ipm,k}, y'_{ipm,k}\right)$ from BEV space to ENU space.

B. *Improved Multi-Kernel Estimation*

The original MKE process is called Statistical Hough Transform (SHT) [11]. The parameters are modeled as multi-kernel density, and the candidates are found by comparing the probability of different line parameters. The basic descriptor of this algorithm is the similarity between an image pixel (x_i^{bev}, y_i^{bev}) and the model (c, d, e), presented as

$$G_{pi}\left(c, d, e, x_i^{bev}, y_i^{bev}\right) = \int_{-\infty}^{+\infty} K_x^{ori}\left(x_i^{bev}\right) K_y^{ori}\left(y_i^{bev}\right) dy, \tag{1}$$

where,

$$K_y^{ori} = \frac{1}{\sqrt{2\pi\left(\sigma_y^{ori}\right)^2}} exp\left(-\frac{\left(y - y_i^{bev}\right)^2}{2\left(\sigma_y^{ori}\right)^2}\right), \tag{2}$$

$$K_x^{ori} = \frac{1}{\sqrt{2\pi\left(\sigma_x^{ori}\right)^2}} exp\left(-\frac{\left(c + dy + ey^2 - x_i^{bev}\right)^2}{2\left(\sigma_x^{ori}\right)^2}\right). \tag{3}$$

In Eq. (1), Gauss-Hermite quadrature method [12] is employed to compute the numerical solution of G_{pi}. However, the computation time of this method can be improved. So the original method is modified to represent better real-time performance.

At first, the kernels K_y^{ori} and K_x^{ori} in Eqs. (2) and Eq. (3) are simplified using triangle model, represented as

$$K_x^{tri} = \begin{cases} 0, \quad x \in \left(-\infty, -2\sigma_x^{ori} + x_i\right] \\ \frac{1}{2\sigma_x^{ori}\sqrt{2\pi\sigma_x^{ori}}} \cdot (x - x_i) + \frac{1}{\sqrt{2\pi\sigma_x^{ori}}}, \quad x \in \left(-2\sigma_x^{ori} + x_i, x_i\right] \\ -\frac{1}{2\sigma_x^{ori}\sqrt{2\pi\sigma_x^{ori}}} \cdot (x - x_i) + \frac{1}{\sqrt{2\pi\sigma_x^{ori}}}, \quad x \in \left(x_i, 2\sigma_x^{ori} + x_i\right] \\ 0, \quad x \in \left(2\sigma_x^{ori} + x_i, +\infty\right] \end{cases},$$

$$K_y^{tri} = \begin{cases} 0, \quad y \in (-\infty, -2\sigma_y^{ori} + y_i] \\ \frac{1}{2\sigma_y^{ori}\sqrt{2\pi\sigma_y^{ori}}} \cdot (y - y_i) + \frac{1}{\sqrt{2\pi\sigma_y^{ori}}}, \quad y \in (-2\sigma_y^{ori} + y_i, y_i] \\ -\frac{1}{2\sigma_y^{ori}\sqrt{2\pi\sigma_y^{ori}}} \cdot (y - y_i) + \frac{1}{\sqrt{2\pi\sigma_y^{ori}}}, \quad y \in (y_i, 2\sigma_y^{ori} + y_i] \\ 0, \quad y \in (2\sigma_y^{ori} + y_i, +\infty] \end{cases},$$

where σ_x^{ori} and σ_y^{ori} are the variance of original Gaussian kernel, $x = c + dy + ey^2$. This modification is able to simplify the calculation of $G_{pi}(x_i^{bev}, y_i^{bev})$.

In summary, the original MKE is enhanced in order to purchase a real-time ability. Three modifications are implemented: looking-up table to approximate kernels, discretizing $G_{pi}(x_i^{bev}, y_i^{bev})$ to simplify computation, and limiting the range of $p_{mke}(d, e)$ to reduce calculation time.

4 Results

The proposed method was evaluated with real data obtained by an experimental vehicle on highway roads. The experimental vehicle is equipped with a monocular pinch hole camera, a low-cost GPS receiver, and a high-cost RTK-GPS receiver. The camera is mounted in front and at the central part of the vehicle, capturing the road features ahead of the rover. The camera height is 1.2 m with respect to the road surface. The GPSs are located on top of the vehicle roof, the low-cost GPS is used to produce rough vehicle positions, while the RTK-GPS one is considered as the ground truth vehicle position for result verifying. The two GPSs are mounted outside the vehicle, to ensure strong connection with satellites.

The first step in vehicle localization is lane selection stage, which determines which lane the vehicle is in within multi-lane road sections. Figure 3 shows the result of lane selection in a stretch of the road. In our algorithm, 6 different states are considered in a maximum 3-lane road section, which are 1 lane, right lane of 2-lane, left lane of 2-lane, right lane of 3-lane, middle lane of 3-lane and left lane of 3-lane. These states are depicted on the y-axis of Fig. 3. The original state, represented as blue lines, is the lane section state by the original GPS positions. Red lines are the filtered lane states after our lane selection stage. The green lines are the benchmark in this scenario, which are defined manually. The accuracy of original state is 54.1 %, while the accuracy of our algorithm is 87.7 % in this scenario.

The error curve of the localization result is shown in Fig. 4. Blue curve is the error of rough GPS position, and the localization result using MKE method is the red curve. In 95.6 % of the frames, the localization result has smaller error than rough coordinates. A localization result with more than 3 m's error is caused by several reasons. For instance, at frames around 2700 and frames around 2950, a bad lane marking is detected because of the heavy traffic, which affects the MKE localization. During these wrong lane detections, the marking confidences support the results, because a number of traffic vehicle bodies are regarded as lane marking

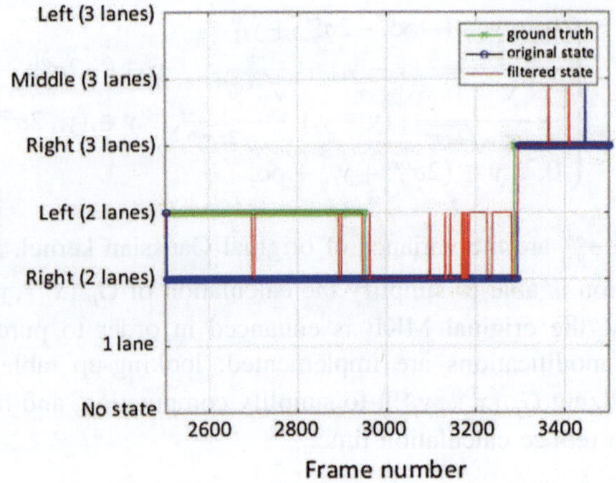

Fig. 3 Lane selection results

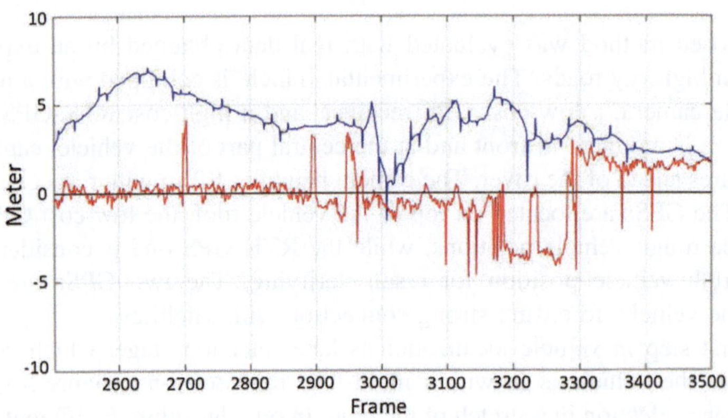

Fig. 4 Error of localization results

"signatures" in image processing. At frame 3200–3300, a false lane state is produced in the lane selection stage, the vision based lane markings are fitted with the lane marking of a wrong road lane, so the localization errors increase. At frame 3300–3500, the MKE based vehicle localization method works without problems, but the error at this part is around 2 m. This is caused by inaccurate road location of OSM.

The statistics are listed in Table 2. The MAE is reduced from 3.96 m to 1.10 m after the MKE localization. The errors are greatly reduced by our algorithm, which represents that our localization method is able to adjust the rough vehicle position, using open source database and camera vision.

Table 2 Error statistics

	Low-cost GPS (m)	MKE localization (m)
Average error	0.536	−0.848
Standard deviation	3.72	1.87
Max value	7.37	7.07
95th ‰	6.65	4.52
MAE	2.95	1.40

5 Conclusion

This paper proposes a marking feature based vehicle localization using low-cost GPS, monocular camera and open source map. Our method makes use of multi-criterion confidences to infer potential errors, and in advance, to enhance the vehicle localization. At first, the vision-based lane marking models are obtained. Meanwhile, the map-based lane markings of current state are derived from map databases. Both lane marking sources are fused together to implement vehicle localization, using an improved MKE method.The results show that a clear improvement in localization accuracy is achieved by our method.

Acknowledgments The work of the authors was supported inpart by the National Natural Science Foundation of China under Grant 61303174, the NationalKey Technology Program of China under Grant 2014BAG01B03, and the Chinese PostdoctoralScience Foundation under Grant 2015M571648.

References

1. De Wit, C.C.: Best-effort highway traffic congestion control via variable speed limits. In: Proceedings of 50th IEEE conference on decision and control and european control conference pp. 5959–5964. IEEE (2011)
2. Wang, L.Y., Syed, A., Yin, G., Pandya, A., Zhang, H.: Coordinated vehicle platoon control: Weighted and constrained consensus and communication network topologies. In Proceedings of IEEE 51st IEEE conference on decision and control (CDC) pp. 4057–4062. IEEE (2012)
3. Skog, I., Handel, P.: In-car positioning and navigation technologies—a survey. IEEE Trans. Intell. Transp. Syst. **10**(1), 4–21 (2009)
4. Bernstein, D., Kornhauser, A.: An introduction to map matching for personal navigation assistant. Transp. Res. Part C: Emerg. Technolo. **1**, 91–108 (2000)
5. Bentley, J.L., Maurer, H.A.: Efficient worst-case data structures for range searching. Acta. Informatica. **13**, 155–168 (1980)
6. Ochieng, W., Quddus, M., Noland, R.: Map-matching in complex urban road networks. Brazilian Society of Cartography, Geodesy, Photogrammetry and Remote Sensing (SBC) (2003)
7. Barth, A., Siegemund, J., Schwehr, J.: Fast and precise localization at stop intersections. In: Proceedings of IEEE intelligent vehicles symposium workshops (IV Workshops), pp. 75–80 (2013)

8. Wu, T., Ranganathan, A.: Vehicle localization using road markings. In: Proceedings of IEEE intelligent vehicles symposium (IV), pp. 1185–1190 (2013)
9. Tao,Z., Bonnifait, Ph., Frémont, V.: Lane marking aided vehicle localization. In: Proceedings of IEEE conference on intelligent transportation systems (ITSC), pp. 1509–1515 (2013)
10. Lu, W., Rodriguez, F.S.R., Seignez, E., Reynaud, R.: Monocular multi-kernel based lane marking detection. In: Proceedings of IEEE conference on cyber technology in automation, control, and intelligent systems (CYBER), pp. 123–128 (2014)
11. Dahyot, R.: Statistical hough transform. IEEE Trans. Pattern Anal. Mach. Intell. **31**(8), 1502–1509 (2009)
12. Liu, X., Song, Q., Li, P.: A Parabolic Detection Algorithm Based on Kernel Density Estimation. Emerging Intell. Comput. Technol. Appl. **5754**, 405–412 (2009)

The Calculation Method with Grubbs Test for Real-Time Saturation Flow Rate at Signalized Intersection

Weiming Luo, Yang Wu, Jianhua Yuan and Wenjie Lu

Abstract In order to provide accurate and real time method for the calculation of saturation flow rate at signalized intersection, by the analysis of the motion law of traffic and the composition of the queue, this paper proposes that it should to eliminate the data which more probability contains outlier data and exclude the outliers in the rest of the measured headway. Finally, the saturated headway and saturation flow rate are calculated. The effectiveness of the proposed approach is verified through an example.

Keywords Signalized intersection · Real-time saturation flow rate · Saturated headway and Grubbs test method

1 Introduction

After the United States, the number of vehicles in china is second in the world by the statistics of the Ministry of Public Security. The growing number of cars leads to plenty of stress to urban transport system, people makes for enhanced demand on the reliability and controllability of the city traffic system. As an important parameter of the traffic signal control, Traffic capacity measures the traffic dispersion capability. The capacity of different urban road is different [1]. Yang et al. reviewed the relative results of signalized intersection capacity [2] and proposed the saturation flow rate method is the main method in calculating signal intersection traffic capacity.

W. Luo · Y. Wu · J. Yuan (✉) · W. Lu
Traffic Management Research Institute of Ministry of Public Security Ministry, Wuxi 214151, Jiangsu, China
e-mail: Yjh1308@163.com

Y. Wu · J. Yuan
Jiangsu Province Collaborative Innovation Center of Modern Urban Traffic Technologies, Nanjing 210096, Jiangsu, China

© Springer Science+Business Media Singapore 2017

H. Lu (ed.), *Proceedings of the Second International Conference on Intelligent Transportation*, Smart Innovation, Systems and Technologies 53,
DOI 10.1007/978-981-10-2398-9_11

Saturation flow rate at signalized intersection is an important parameter of the calculation of traffic capacity with the saturation flow rate method [3]. Saturation flow rate can usually be calculating by modeling or measured. The calculation methods of the model need to be corrected by the various factors that influence saturation flow rate, these factors are hard to be quantified or measured [4]. Therefore, in application of actual engineering, saturation flow rate is calculated by actual measurements of the headway. The method measures the headway of the standard car, which is usually defined as the time that elapses between consecutive vehicles when vehicles in a queue start crossing the stop line or any other reference line at a signalized intersection. To avoid the influence of green loss time, after the light turns green, it should delete the data of forward cars of queue. The saturation flow rate is the reciprocal of the average queue discharge headway times 3600 [5]. The coil detector is one of the widely used traffic detector. It is sensitive, tunable, adaptable and low price. The single coil detector can only measure the vehicles arrival and the time of cars through a coil. But this data cannot distinguish vehicle types without the speed.

In this paper, a headway preprocessing method based Grubbs test is proposed, preprocessing the measured data can eliminate the influence of median and large vehicles. The method provides support for the calculation of the traffic capacity and the design of signal timing scheme at signalized intersection.

2 The Motion Law of Traffic and the Composition of Queue at Signalized Intersection

The cars in China are classified into three types: small (passenger car), median and big vehicle [6], Table 1 lists the criteria of the classification. In this study, the big vehicle refers primarily to buses.

The movement law of traffic flow at the intersection is analyzed under the circumstances of unknown vehicle type. When the light turns green, with the first five cars of queue to cross the stop line, the rest of the queue form saturated traffic flow with a relatively constant speed. The difference of headway with the same type is relatively small. With different vehicle types, the difference of headway is relatively big.

On the basis of classic statistics, the probability of the three vehicle types in the queue is equal. It means that when the number of cars tends to infinity, the probability of each type is 1/3. Therefore, the number of passenger car accounted for one-third.

Table 1 Vehicle classification at signalized intersections

Parameters	Vehicle type		
	Small	Median	Big
Length (m)	<6.0	6.0–9.0	6.0–12.0
Weight (t)	<4.0	4.0–11.0	11.0–16.0

The composition of the queue in each period has three cases. First of all, traffic flow composed by passenger cars. The velocity of saturated traffic flow is a constant value, so the difference of headways is small. Secondly, three types of vehicles form the stream of cars, the difference of headways is big, because the length of cars is different. Finally, traffic is composed of median and big vehicle. The difference between measured headways and the saturation headway is too large, so all of the data should be rejected. Therefore, the Accuracy of real-time saturation flow rate is hardly assured by calculating with samples of one single cycle. Suppose the number of cycles tends to infinity, the probability of the three cases is equal to 1/3. In this paper, the calculation of real-time saturation flow rate based on actual observations of headways of multiple periods.

3 The Motion Law of Traffic and the Composition of Queue at Signalized Intersection

In order to estimate the real-time saturation flow rate at signalized intersection, the traditional saturation flow rate is determined by the queue discharge headway as follows

$$S = \frac{3600}{t_{\mathrm{s}}} \tag{1}$$

where, S is the saturation flow rate (veh/h); t_{s} is the queue discharge headways (s) which are defined as the time intervals between two successive vehicles passing a stop line or any predetermined reference line at the signalized intersections.

Akçelik et al. [7] defined the queue discharge headway t_{s} as a function of the response of drivers during the queue

$$t_{\mathrm{s}} = t_{\mathrm{d}} + \frac{l}{v} \tag{2}$$

where, t_{d} is the response of driver; l is the space length of a car in the queue; v is the speed of the queue. Because of complex environments of traffic and roads, the driver response t_{d} is difficult to be measured. The difference of vehicle length caused the discrepancy of the space length of the cars in queue discharge. Equation (2) is difficult to apply to practical engineering. The measured headway is a random variable under the influence of driver response, road conditions, or to other factors like weather. Therefore, the average of measured queue discharge headways is used as saturation flow rate in engineering applied fields

$$\bar{t}_{\mathrm{s}} = \frac{1}{n}\sum_{i=1}^{n} t_{\mathrm{s}i} \tag{3}$$

where, $\bar{t}_s$ is the average of measured queue discharge headways; $t_{si}, i = 1, 2\ldots, n$ is the measured queue discharge headway of the *i*th car in the queue. When $n \to \infty$, $\bar{t}_s$ is equal to the saturated headway t_s.

Thus, Eq. (1) can be rewritten as

$$S = \frac{3600}{\bar{t}_s} \tag{4}$$

4 The Calculation of the Saturated Headway with Grubbs Test

In this paper, Grubbs test is used to exclude outlier, because of the method is suit to small sample size problem. Suppose the actual saturated headway, the observation target, is a constant t_s. The measured data include the random noise such as traffic, road conditions, weather, driver psychological, etc. [8]. Suppose the statistical distribution of the random noise is normal distribution. Noise time series can be written as $w_i \sim N(0, \sigma^2)$. The measured data model equal the observation target added random noise time series [9].

$$t_{si} = t_s + w_i, w_i \sim N\left(0, \sigma^2\right) \tag{5}$$

where, t_{si} is the *i*th measured queue discharge headway; t_s is the actual saturated headway; w_i is white noise with mean 0 and variance 1, $i = 1, 2\ldots, n$.

The mean and variance of the measured data can be expressed as

$$\bar{t}_s = \frac{1}{n} \sum_{i=1}^{n} t_{si} \tag{6}$$

$$\bar{s} = \sqrt{\frac{1}{n-1} \sum_{i=1}^{n} (t_{si} - \bar{t}_s)^2} \tag{7}$$

Constructed the discrimination statistic G_i

$$G_i = \frac{\max_{1 \le i \le n} |t_{si} - \bar{t}_s|}{\bar{s}} \tag{8}$$

Equation (8) can be transformed into t-distribution with the freedom (n − 2)

$$\frac{G_i}{\sqrt{(n - 1 - G_i^2)/(n-2)}} \sim t_{\alpha i}(n-2) \tag{9}$$

where, α is the test levels. Then, the critical value $G_{\alpha i}$ which is used for determining the outlier can be expressed as

$$G_{\alpha i} = \frac{t_{\alpha i}\sqrt{n-1}}{\sqrt{n-2-t_{\alpha i}^2}} \tag{10}$$

If $G_i > G_{\alpha i}$, in other words, when $P(G_i > G_{\alpha i}) = \alpha$, the sample t_{si} is an abnormal value. Then, t_{si} is rejected, the number of samples minus 1. Repeat above-mentioned process until $G_i < G_{\alpha i}$, $i = 1, 2..., n_1, n_1 \leq n$.

Chen et al. [10] proposed that Grubbs test was the best in various test methods of the abnormal value when the ratio of the outlier under 10 %. However, the delay time of vehicle starting and the different of the vehicle type infects the calculation of headway by the actual measurement, these causes the queue discharge headways are bigger than the actual saturated headway. In the queue, the number of median and big vehicle is likely to above 10 % according to above-mentioned analysis results. The radio of outlier in the measured headways is too large which can lead a high result about the estimate of the variance. But, the critical value G is underestimate which can cause the false negative or even the test is failure. To solve this problem, it is supposed to reduce the influence of the outlier on the estimate of the variance by rejecting a portion of the data which contain a significant number of abnormal values.

The probability of small vehicle type in the queue discharge can be expressed as

$$P_{\text{small}} = \frac{1}{3}, \quad n \to \infty \tag{11}$$

where, P_{small} is the probability of small vehicle type in the queue. n is the number of cars in the queue.

When the light turns green, it is reasonable to suppose the velocity of traffic flow is a constant according to the motion law of the queue discharge at signalized intersection. Because of the queue discharge headways of small cars is relatively small, the measured samples of one period can be sorted from small to large. Then, the first third of the sorted samples are selected for the specified outlier test by Grubbs test.

When the number of period tends to infinity, the probability of traffic flow at signalized intersection composed by passenger cars is equal to 1/3. The probability can be expressed as

$$P_1 = \frac{1}{3}, \quad m \to \infty \tag{12}$$

where, P_1 is the probability of traffic flow at signalized intersection composed by passenger cars. m is the number of period. The processed data of m periods should be merged for sorting from small to large. After data preprocessing, the first third of the samples are selected for calculate the mean which is the saturated headway at signalized intersection.

The calculate process of the method can be expressed as:

1. $t_s^j = \{t_s^j(1), t_s^j(2), \ldots, t_s^j(n)\}$ is the measured headway of the jth period, $j = 1, 2\ldots, m$.
2. t_s^j is sorted from small to large. Then, the first third of the sorted samples are selected into $\bar{t}_s^j$, $j = 1, 2\ldots, m$.
3. Grubbs test is used for excluding the outlier of $\bar{t}_s^j$, the processed data is stored in $\boldsymbol{X}$.
4. Repeat from step 1 to step 3 with m period, the processed data of each period is stored in X.
5. Repeat from step 2 to step 3 with the data in X.
6. Calculating the saturated headway and real-time saturation flow rate.

5 Comparison of Saturation Flow Rate Estimations

To verify the effectiveness of the proposed method, comparison on the calculation of the saturation flow rate using the traditional way and the new one. The measured data is surveyed at the signalized intersections of Dachi Road and Qianrong Road in Wuxi. Dachi Road and Qianrong Road are important section of urban road in Wuxi. There is much more traffic on the roads between 4:30 p.m. and 5:30 p.m. Traffic data are collected by video cameras and remote analyzed in the laboratory for the accuracy and recycling. Table 2 lists the saturation flow rates based on different methods by using the measured data of 75 periods. Where, the m is 5.

The saturation flow rates are computed using traditional method and the proposed method. It can be found that the computed saturation flow rates based on the new method are higher than the values computed by Eq. (4). The interval of the saturated time headways are computed using the method based Grubbs test is [1.80, 2.10] s. The upper limit of the interval is less than the minimum value of headway computed by the traditional method. The saturation flow rates are computed by the proposed method, which are bigger than the saturation flow rates are computed using traditional method, are between 1714 and 2000 pcu/h. The standard deviations indicate that the pretreated headways are distributed evenly. The calculated results show that the traditional method of the calculation of the saturation flow rate is influenced by vehicle types and underestimate the saturation flow rate. The new method based Grubbs test, which is more accurate, can exclude the impacts of median and big vehicles.

Table 2 Vehicle classification at signalized intersections

Number	The samples size	Headway (s)		Standard deviation		Saturation flow rates (pcu/h)	
		Traditional	Grubbs	Traditional	Grubbs	Traditional	Grubbs
1	70	2.44	1.96	0.61	0.14	1477	1840
2	61	2.51	2.03	0.45	0.14	1437	1777
3	56	2.41	1.99	0.37	0.17	1496	1807
4	65	2.34	1.87	0.52	0.15	1536	1928
5	59	2.40	1.89	0.48	0.13	1500	1903
6	71	2.43	1.99	0.38	0.11	1484	1807
7	74	2.29	1.89	0.62	0.06	1575	1901
8	55	2.65	2.10	0.59	0.24	1359	1714
9	78	2.21	1.84	0.32	0.13	1632	1951
10	73	2.30	1.80	0.56	0.05	1562	2000
11	63	2.49	1.98	0.69	0.13	1447	1812
12	60	2.56	1.96	0.74	0.14	1403	1836
13	66	2.42	1.81	0.71	0.21	1487	1993
14	59	2.89	1.97	1.48	0.23	1242	1824
15	72	2.86	1.89	0.47	0.09	1574	1903

6 Conclusion

A new calculation method based Grubbs test of the saturation flow rate at signalized Intersection has been demonstrated and experimentally validated. On the basis of the analysis of the motion law of traffic and the composition of the queue, pre-processing the measured data of multiple periods and eliminating the data which is more probability contains outlier data. Grubbs test is used for excluding the outlier of pre-processed data. The proposed method solved the problem which because of the radio of outlier is too large which lead a high result about the estimate of the variance and the critical value G was underestimate which can cause the false negative or even the test is failure. The method can eliminate the influence of median and large vehicles.

Acknowledgments The work of the authors was supported by the National Key Technology Program of China under Grant 2014BAG01B04.

References

1. Leutzbach, W.: Introduction to the Theory of Traffic Flow. Spriner, Berlin (1988)
2. Yang, X., Zhao, J., Ma, W., et al.: Review on calculation method for signalized intersection capacity. China J. Highw Transport **05**, 148–157 (2014)

3. Manual, H.C.: Transportation Research Board, p. 113. National Research Council, Washington, DC (2000)
4. Tarko, A.P., Tracz, M.: Uncertainty in saturation flow predictions. Red **1**, P2 (2000)
5. Jin, X., Zhang, Y., Wang, F., et al.: Departure headways at signalized intersections: a log-normal distribution model approach. Transp. Res. Part C: Emerg. Technol. **17**(3), 318–327 (2009)
6. Shao, C.Q., Rong, J., Liu, X.M.: Study on the saturation flow rate and its influence factors at signalized intersections in China. Procedia-Soc. Behav. Sci. **16**, 504–514 (2011)
7. Akçelik, R., Besley, M.: Queue discharge flow and speed models for signalised intersections. In Transportation and Traffic Theory in the 21st Century. Proceedings of the 15th International Symposium on Transportation and Traffic Theory, pp. 99–118 (2002)
8. Shao, C.Q., Rong, J., Zhao, L.: Developing adjustment factors of saturation flow rates at signalized intersections. Beijing Gongye Daxue Xuebao (J. Beijing Univ. Technol.) **37**(10), 1505–1510 (2011)
9. Wood, S.N.: Data driven statistical methods. Biometrics **54**(4), 1678 (1998)
10. Rui, C., Shumin, Z.: The application of improved Grubbs' criterion for inspecting the count of radon concentration (2009)

Traffic Optimization Design and Simulation Evaluation of Isolated Intersection

Muting Ma, Xiaodong Ma, Yichi Yang and Fengchun Han

Abstract Scientific optimization design of isolated intersection, which as the key point of city arterial road, is of great significance to improve traffic capability. The core intersection on the arterial road in Shenyang was selected as research object. Based on traffic parameter investigation and delay evaluation of intersection status, bus station settings, traffic organization and signal control were optimized and the optimization effect was simulated. The results show that intersection saturation and queue length decrease, while traffic flow average speed increase, further, vehicle average delay is reduced from 60 to 25 s, while intersection level of service is raised from E to C. The optimization effect is obvious and perfect. It suggests that the optimization design of intersection can actually not only reduce the traffic delay, improve the level of service, but also ensure city arterial road being safe and smooth.

Keywords Intersection · Traffic optimization · Traffic delay · Level of service · Traffic simulation

1 Introduction

Isolated intersection in traffic management has an important role in regulating vehicle operation, assigning road resources, easing the traffic congestion and keeping traffic safe. The traffic operation of isolated intersection is of great complexity, which involves signal control, traffic organization and road condition settings. Appropriate regulation of traffic flow and coordination of intersection operation have a huge influence on easing traffic delay status.

Given the success of research and development of traffic control systems such as TRANSYT [1], SCOOT [2] and SCATS [3], research on traffic control are

M. Ma (✉) · X. Ma · Y. Yang · F. Han
Traffic Management Engineering Department, People's Public Security
University of China, Daxing District, Beijing, China
e-mail: mmt_gongda010@163.com

© Springer Science+Business Media Singapore 2017

H. Lu (ed.), *Proceedings of the Second International Conference on Intelligent Transportation*, Smart Innovation, Systems and Technologies 53,
DOI 10.1007/978-981-10-2398-9_12

becoming more and more intensive. Yang and Yagar [4] optimized traffic volume distribution and signal timing model based on saturated traffic network. Sen and Head [5] optimized the delay, stop numbers and queue length of signal intersection on the basis of real-time purpose algorithm. Steven [6] evaluated three disadvantages of adaptive signal control algorithm in isolated intersection. Stevanovic et al. [7] optimized four basic signal timing parameter and vehicle priority control pattern VISSIM-based. Robert and Alexander [8] evaluated three conditions about disadvantage effect of over-saturated intersection condition on adaptive signal control system. Zhai [9] analyzed and designed the traffic actuated signal control method. Li [10] optimized the city traffic actuated signal control parameter. Wang et al. [11] established the non-linear model of signal timing parameter based on genetic algorithm in isolated intersection. Sun and Xu [12] Synchro-optimized the signal timing status in isolated intersection. He et al. [13] established the dynamic signal timing model under every conditions of traffic flow arriving characteristic in isolated intersection. Fu and Zhao [14] compared BRT bus station settings based on average bus delays.

The core intersection on the arterial road in Shenyang was selected as research object and its traffic parameter status was investigated. Accordingly, intersection delay status was evaluated, and then optimization scheme, which involves bus station settings, traffic organization optimization, actuated signal timing and phase design, was discussed, and subsequently the optimization scheme was simulated and contrasted by VISSIM software, and finally the feasibility of it was verified.

2 Data Collection

The data used in this study came from the intersection of Jianshe Road and Xinghua Street in Tiexi District, Shenyang. Jianshe Road is two-way eight-lane and Xinghua Street is two-way ten-lane. Both of them are arterial roads. The intersection status is shown in Fig. 1. The data were collected from 7:00 to 9:00 a.m., Monday to Sunday, August 17 to 23, 2015. The weather was fine. Intersection signal timing parameters, traffic volume, speed, queue length, bus route and vehicle delay data were collected and extracted by video cameras set in every direction corner of the intersection.

3 Intersection Status Analysis and Evaluation

3.1 Status Analysis

The intersection of Jianshe Road and Xinghua Street lies in business and recreational area. The volume of traffic and people flow through the intersection, with

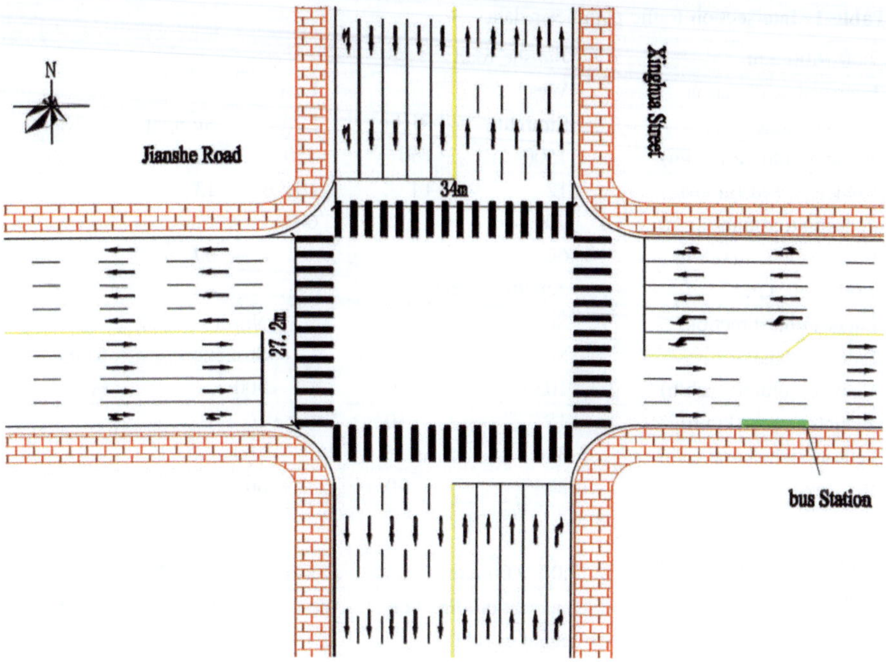

Fig. 1 The intersection status

frequent traffic congestion, is huge and the delay status is severe. The causes of delay states as follows:

1. A huge amount of buses with high departure frequency stop at the bus station, which is non-bus-bay aside the road, being 50 m away from East exit lane. Due to the long waiting time and occupancy on lanes of buses and passengers, social vehicles cannot go through smoothly and have to change their lanes at low speed.
2. Due to the huge amount of vehicles that come straight from West entrance lane and turn right from South entrance lane flowing into the narrow East exit lane together, vehicles at low speed often congest in the intersection.
3. The huge number of vehicles that turn left from East entrance lane, without exclusive left turn phase, stop at the intersection waiting for chance to turn left.
4. Vehicles in the South-North phase, hindered from non-evacuated vehicles in the East-West phase, cannot flow through the intersection. A large amount of vehicles delay in the intersection, status of which is in a mess.

Table 1 Intersection traffic parameter data

7:00–9:00 a.m.	Jianshe road				
Intersection direction	West		East		
Entrance lane	Straight	Right	Left	Straight	Right
Vehicle volume (veh/h)	1500	150	450	1200	150
Vehicle speed (km/h)	12	14	10	13	15
Pedestrian volume (p/h)	900		600		
Bus frequency (veh/h)	60	–	–	60	–
7:00–9:00 a.m.	Xinghua street				
Intersection direction	South		North		
Entrance lane	Straight	Right	Straight	Right	
Vehicle volume (veh/h)	1050	300	900	90	
Vehicle speed (km/h)	15	10	16	18	
Pedestrian volume (p/h)	300		450		
Bus frequency (veh/h)	30	90	60	–	

Table 2 Intersection signal timing data

7:00–9:00 a.m.	Jianshe road		Xinghua street	
Intersection direction	West	East	South	North
Signal cycle (s)	120			
Signal phase	First phase		Second phase	
Green time (s)	60		40	
Flow ratio	0.49		0.31	
Green time ratio	0.50		0.33	
Saturation	0.98		0.94	

Intersection traffic parameter status data is shown in Table 1.

Two signal phases, East-West and South-North phase, which are fixed signal timing, control the intersection. Intersection signal timing status data is shown in Table 2.

3.2 Status Evaluation

The intersection traffic operation status in the morning rush is saturated, status evaluation data is shown in Table 3.

In summary, intersection traffic organization pattern, bus station settings and signal phases are not appropriate, saturation is high, queue length and average delay is long, and the level of service is low.

Table 3 Intersection status evaluation data

7:00–9:00 a.m.	Jianshe road				
Intersection direction	West		East		
Entrance lane	Straight	Right	Left	Straight	Right
Queue length (m)	85	50	96	70	55
Average delay (s)	70	45	82	50	48
Level of service	E	D	F	D	D

7:00–9:00 a.m.	Xinghua street			
Intersection direction	South		North	
Entrance lane	Straight	Right	Straight	Right
Queue length (m)	87	108	65	45
Average delay (s)	60	88	54	40
Level of service	E	F	D	D
Total level of service	E			

4 Intersection Simulation

4.1 Intersection Simulation Scheme

4.1.1 Bus Station Adjustment

The original bus station was closed to East exit lane and inflows and outflows of passengers were time consuming, therefore, occupying the road resource and hindering vehicles flowing through the intersection. Bus-bay stations aside the road are set 100 m away from West entrance lane for buses passing through and 100 m away from South entrance lane for buses turning right.

4.1.2 Traffic Organization Optimization

East exit lane was often in congestion under the influence of itself narrow lanes and only one bus station that all the buses converge. Only one exclusive left turn lane and left turn waiting zone are set at the East and West entrance lane and exclusive left turn phase is set. Road axis is shifting one lane from West entrance lane to West exit lane to ensure normal East to West traffic. The pavements are shared by pedestrians and bicycles, purpose of which is to advocate slow traffic system. Traffic organization optimization is shown in Fig. 2.

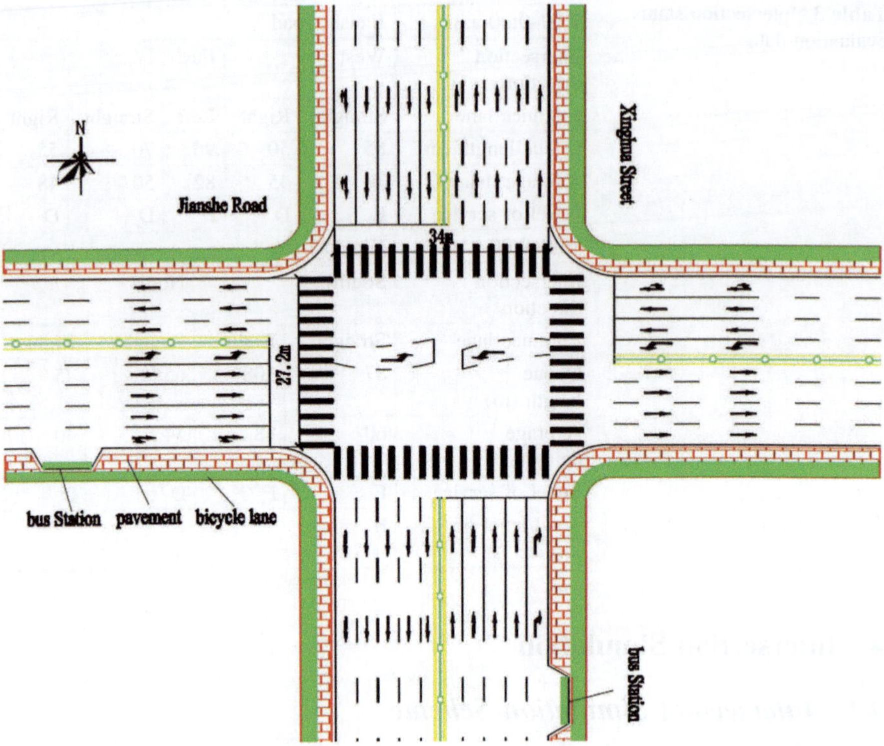

Fig. 2 Traffic organization optimization

4.1.3 Signal Control Optimization

Original inappropriate signal phase, inflexible signal timing scheme and long signal cycle intensified the intersection delay. Signal cycle and phase and timing scheme are re-designed based on intersection traffic flow status. The signal phase is shown in Fig. 3.

The optimal signal cycle [15] computes as follow:

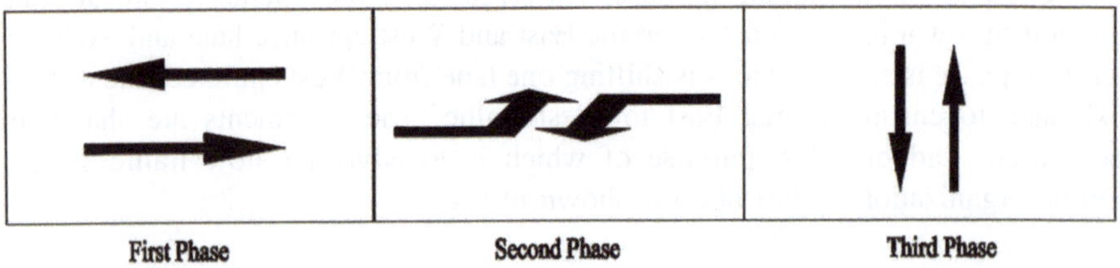

Fig. 3 The signal phase

Table 4 Actuated signal timing parameter data

Optimal signal cycle (s)	110		
Saturation	0.88		
Key phase	First phase	Second phase	Third phase
Shortest effective green time (s)	24	16	13
Optimal effective green time (s)	39	33	24
Longest effective green time (s)	50	46	32
Unit extension green time (s)	10	10	7

$$C = \frac{(1.4 + K) \cdot L + 6}{1 - Y} \tag{1}$$

where C is optimal signal cycle (s); L is phase loss time (s); Y is the sum of key phases' flow ratio; K is stop compensation factor, and K is equal to 0 in order to make delay minimum.

Actuated signal timing parameter data is shown in Table 4.

4.2 *Simulation Evaluation*

The original intersection control pattern is compared with the optimization of bus station, traffic organization and signal control by the simulation software VISSIM 5.4. Three different random seeds are selected for simulation, each of which takes 3600 s. Three phases VAP actuated signal control detectors are 40 m away from entrance lanes. The exclusive left turn lane and pavement are set for route choice. Expected speed distribution is from 5 to 30 km/h. The bus-bay stations aside the road are 100 m away from entrance lane. Average 30 passengers are carried by each bus, departure interval time of which is from 5 to 10 min and average waiting time of which is 20 s. Data collection point and queue count detector are 10 m away from each entrance lane. Data collection interval time is 120 s. Simulation optimization evaluation data is shown in Table 5.

In summary, it can be obviously found that intersection saturation is reduced from 0.98 to 0.88, average speed is raised from 14 to 19 km/h, queue length is reduced from 73 to 47 m, and average delay is reduced from 60 to 25 s. Level of service of each entrance lane is improved and intersection level of service is raised from E to C. Intersection operation status is improved obviously. It suggests that the optimization scheme is feasible.

Table 5 Simulation optimization evaluation data

	Jianshe road				
Intersection direction	West		East		
Entrance lane	Straight	Right	Left	Straight	Right
Vehicle speed (km/h)	16	20	15	15	22
Queue length (m)	51.7	29.1	86	49.8	35.8
Average delay (s)	29.2	17.4	34.9	29.7	28.5
Level of service	C	B	C	C	C
	Xinghua street				
Intersection direction	South		North		
Entrance lane	Straight	Right	Straight	Right	
Vehicle speed (km/h)	17	20	21	23	
Queue length (m)	52	44.2	45.2	25.9	
Average delay (s)	30.6	15.9	17.9	16.5	
Level of service	C	B	B	B	
Total level of service	C				

5 Conclusions

Based on the selected core intersection in Shenyang, intersection traffic operation status parameter are analyzed, delay status and its reason are simulated, and the optimization of bus station, traffic organization and signal control is designed. The simulation result shows that intersection saturation and queue length and average delay decrease, while traffic flow average speed increase and level of service of each entrance lane and the whole intersection is improved. It can be concluded the optimization scheme is appropriate and feasible.

References

1. Robertson, D.I.: TRANSYT: a traffic network study tool. In: RRL Report LR253, Road Research Laboratory (1969)
2. Stevanovic, A., Martin, P.T.: Split-cycle offset optimization technique and coordinated actuated traffic control evaluated through micro simulation. Transp. Res. Rec.: J. Transp. Res. Board **2080**, 48–56 (2008)
3. Sims, A.G., Dobinson, K.W.: The Sydney coordinated adaptive traffic (SCAT) system philosophy and benefits. IEEE Trans. Veh. Technol. **29**(2), 130–137 (1980)
4. Yang, H., Yagar, S.: Traffic assignment and signal control in saturated road networks. Transp. Res. Part A: Policy Pract. **29**(2), 125–139 (1995)
5. Sen, S., Head, K.L.: Controlled optimization of phases at an intersection. Transp. Sci. **31**(1), 5–17 (1997)

6. Steven, G.S.: Single-intersection evaluation of real-time adaptive traffic signal control algorithms. Transp. Res. Rec. **1**, 183–192 (2004)
7. Stevanovic, J., Stevanovic, A., Martin, P.T., et al.: Stochastic optimization of traffic control and transit priority settings in VISSIM. Transp. Res. Part C **16**(3), 332–349 (2008)
8. Robert, C., Alexander, S.: Issues affecting performance of adaptive traffic control systems in oversaturated conditions. Transp. Res. Rec.: J. Transp. Res. Board **2438**, 23–32 (2014)
9. Zhai, R.P.: Research on actuated signal control timing design method. J. Chin. People's Public Secur. Univ. **11**(3), 43–46 (1998)
10. Li, H.Q.: Study on the optimization methods of signal timing parameters of urban traffic control. Doctoral dissertation, Jilin University (2004)
11. Wang, Q.P., Tan, X.L., Zhang, S.R.: Signal timing optimization of urban single-point intersections. J. Traffic Transp. Eng. **6**(2), 61–64 (2006)
12. Sun, C., Xu, J.M.: Traffic signal timing optimization at city single point intersection. Comput. Commun. **26**(6), 6–10, 17 (2008)
13. He, Z.C., Yang, W.C., Liang, S., et al.: Study of real-time traffic signal optimized control and simulation for urban isolated intersection. Comput. Eng. Appl. **46**(33), 239–243 (2010)
14. Fu, J.Y., Zhao, J.: Optimal layout of bus rapid transit stops at intersections based on delay analysis. J. TONGJI Univ. (Nat. Sci.) **42**(11), 1677–1681, 1699 (2014)
15. Quan, Y.S.: City Traffic Control. China Communications Press, Beijing (1989)

Characteristics of Intelligent Transportation Development in Cities

Qiu Rui

Abstract With rapid urbanization, environmental pressure on urban resources and ecological degradation, the government's public management and public demand generated the "Smart City" concept. The Smart City: to log into the "intelligent transport cloud" platform with a mobile phone before going out, query bus arrival information, taxi location information, the location of parking and real-time traffic information and other comprehensive traffic information, to say goodbye to "traffic jam city" [1]. facilitated by intelligent transportation and satellite navigation systems integrated into the communication network, information communicated and exchanged among satellites, the ground receiving station, the terminal equipment and smart devices in people's hands. Smart city's application prospects bring opportunities for Beidou satellite navigation system which was independently developed by China. The Chief Engineer of the North Satellite Navigation System, researcher Sun Jiadong, is full of hope for the compass: "The Beidou Navigation Satellite System will not only play a role in the sky, but on the ground it will also apply more applications to benefit people." Transportation security, traffic congestion and the reduction of environmental pollution will be greatly aided by Intelligent Transportation Technology (ITS). Intelligent Transportation Technology (ITS) in the Smart City optimizes the traffic infrastructure.

Keywords Smart city · Intelligence transportation · Develop · Effectiveness

Q. Rui (✉)
Sichuan Science & Technology Vocational College, No. 96 Tuanjie Xueyuan Road, Gao Xin West-District, Chengdu 611745, Sichuan, China
e-mail: sckj5186@163.com

© Springer Science+Business Media Singapore 2017
H. Lu (ed.), *Proceedings of the Second International Conference on Intelligent Transportation*, Smart Innovation, Systems and Technologies 53,
DOI 10.1007/978-981-10-2398-9_13

1 The Current Situation of Intelligent Transportation Technology in Smart Cities in Different Countries

1.1 Current Situation of Intelligent Transportation in Smart Cities in Some Asian Countries

1. Laos—ASEAN's first big dipper CORS station went into operation, on August 22, 2013, it was the first Beidou continuous operation reference station built in an ASEAN country, which can provide a model for agriculture and forestry management, etc. It not only meets the requirements of the high precision positioning in land surveys, agriculture, forestry management and other aspects of high precision positioning, but it also provides highly accurate service capabilities to promote the use of space-time monitoring and early warning systems for agriculture in Laos. It also plays an important role in the fields of agriculture, forestry, urban planning, transportation and land management.
2. Cambodia—"Cambodia's Beidou continuous operation reference station system construction and police positioning demonstration of the application" [2] system held a successful opening ceremony in Phnom Penh, Cambodia. Kampuchea police management demonstration applications, including police patrol management based on location services, emergency police mobilization, police officers and police officers to locate evidence and comprehensive management and other content, to help the Kampuchea police to grasp real-time, accurate information on the location of police resources, reasonable dispatch, rapid deployment, improve the efficiency of local police work.
3. Thailand—through the terminal, demonstration system can track the running path of the Beidou navigation vehicle clearly, measuring the positioning accuracy from 3 to 5 cm, the speed accuracy reached 0.2 m/s, real-time navigation services to reach 0.5 m level, while the previous precision in 10 m.
4. Pakistan—Karachi City, Pakistan City, the establishment of 5 benchmark stations and a processing center, in urban planning, land survey, land management, environmental monitoring, disaster prevention and reduction, traffic monitoring and other modern information management.
5. Burma—Beidou receiver used in agriculture and land management. In 2014, Myanmar's state department of agriculture announced publicly that they want to purchase high precision global satellite navigation and positioning system receiver. Chinese company won the 520 sets of Beidou/GPS dual-mode satellite navigation receiver real-time dynamic difference [3]. These instrument will be used to collect agricultural data, valuate land and manage land, etc.
6. Brunei—based on the Beidou location services to promote urban construction, in the satellite navigation chip technology, terminal products, value-added services development and system integration and other fields of cooperation, to provide services to meet the needs of local navigation and location-based services products, foster local value-added services in Brunei to protect the long-term sustainable development based on the Beidou navigation and

location-based services, in order to bring better navigation services and more employment opportunities.

1.2 Current Situations of Smart Cities in the United States, Canada, Australia, the UK

1. The United States: in terms of the development of traffic information, the U. S. Department of transportation enacts laws and regulations, administered by the deputy director and the departments responsible for information resource management. The chief executive of the Department of transportation is responsible for the work of the entire Department of transportation related to government information. The Information Resource Management Manual DIRMM (Department), which is used to support the unified planning of information resources in the protocol of the Department of Transportation, was issued in 1989. The handbook has strong authority and legal standing and it has a positive effect on the realization of the information object.
2. Canada: Canada's government believes in the sustainable development of the national railway, the international maritime transport, the trans continental air service system, the management of the transportation infrastructure, and the business management of the transport infrastructure. In order to maintain and strengthen the transport system in Canada and enhance the quality of life of Canadian people, transport policy provides a framework for the development of the 3 elements of sustainable transport: social, economic and environmental, to provide the carrier and infrastructure to meet the business opportunities, innovation and competition.
3. Australia: Australian Transport and Regional Services (DoTRS) takes the most important information on transportation to develop an online action plan that businesses access through the network to optimize logistics. Transportation and Regional Services for Australia's mission is to provide a better transport system to help the government achieve its policy objectives in transport and regional services, and to provide policy recommendations to the cabinet ministers and meetings. It also provides administrative management, research, regulations, investigations, security, grants and service territory [4].
4. UK: in the second half of 1996, the British government launched the "Electronic Government" program in which companies can use the latest information technology to obtain government services. Electronic Government makes full use of Internet and other new electronic technology, to provide enterprises with taxes, changing business licenses, consultative policy and access to all types of government information. The British government is ready to invest 38 billion pounds in the next 10 years to build the information superhighway, encourage enterprises to vigorously carry out global electronic commerce activities and the development of online trade.

2 Intelligent Traffic in the Smart City with Chinese Characteristics

2.1 *In August 2014, the National Development and Reform Commission*

The Ministry of Industry and eight other ministries jointly promote the healthy development of the Smart City. The mission is to manage the construction of traffic guidance, travel information services, public transport, integrated passenger transport hubs, integrated coordination of command and control and other intelligence systems. The program uses the Beidou navigation satellite ground system in the development of differential traffic information and value-added services. This is actually intended to promote the Beidou navigation system in the application of intelligent traffic. "Planning for the Beidou satellite navigation system in the expansion of public applications also has a high priority: focusing on promoting the satellite navigation and smart phone terminal standard configuration as well as promoting its social services, travel, smart city and aiding vulnerable groups to promote the development of mass applications." The development of intelligent traffic in big cities is within the purview of urban intelligent transportation. Beijing and Guangzhou are leading the way. The 4 major types of ITS systems were initially built: road traffic control, public traffic command and scheduling, highway management, emergency management, and about 30 subsystems, which are scattered in various traffic management and operation departments [5]. At present, the road network of Beijing city has formed a circular and radial road network over several decades. The urban population density has maintained at a constant of about 27,000 people per square kilometer, but the urban population is growing, population flow is continuous, and the construction of intelligent transportation systems has become the only way for the sustainable development of traffic in Beijing.

1. Beijing During the 2008 Beijing Olympics: The Olympic route and Olympic venues around the traffic signal will be 120. In addition, it also has a comprehensive traffic monitoring system, which includes three subsystems, such as video surveillance and traffic flow detection. At the same time, in the Olympic Games venues and surrounding areas and associated road building, there are 80 television monitoring points, 15 sets of traffic incident detection systems and 80 sets of video systems.
2. Guangzhou, as one of the first round of intelligent transportation demonstration cities in China, has constructed intelligent transportation systems, including the Guangzhou city traffic information sharing main platform, the logistics information platform, the static traffic management system and other intelligent transportation systems. The common information platform has begun to take shape, realizing the connection of the network system, the network planning system, the taxi management platform, the network ticketing system, the 96900 call center and other subsystems. Guangzhou City's transportation infrastructure

construction has made great progress, but because of the economic conditions, geographical location and environmental constraints, the road traffic network construction will encounter difficulty to meet the needs of transportation growth in the future. The demands of the intelligent transportation system in Guangzhou city are not only to meet the requirements of Guangzhou city development and traffic development, but also to meet the requirements of the Asian Games in 2010.

2.2 The Future Development of Urban Intelligent Transportation According to the Country's Future Development Plan, the Construction of an Urban Intelligent Transportation System Will Continue to Develop

First of all, we will promote the construction of the traffic information service platform, provide a traffic information query, traffic guidance and other services. In more than 200 cities the goal is to develop an urban intelligent control signal system and the formation of intelligent traffic control systems; in more than 100 cities to promote the construction of public transport areas and the corresponding systems, increase the construction and application of electronic ticketing. In light of developing urban traffic obstacles, the integrated information platform of urban traffic, global positioning and vehicle navigation systems, the city's comprehensive emergency system will usher in greater development opportunities [6]. Overall, the trend of the urban intelligent transportation system will be an integrated, multi-sector driven development model. As the urban intelligent transportation system will involve the key stakeholders, public security traffic management, public traffic management, urban construction, communications and other related departments, the future development of the urban intelligent transportation process is necessarily related to traffic and public security as the main driver of the process.

3 Development Status and Trend of Vehicle Terminal Market

1. The development of the vehicle terminal market in Europe and the United States: the ITS application statistics, the GPS based road guidance vehicle navigation system has become the largest ITS user market, and occupied 29 % of all ITS users

At present, the domestic sales of vehicle GPS is divided into simple positioning, anti-theft and scheduling functions of basic model GPS, static navigation type GPS and real-time communication type GPS3. The growth of the basic GPS markets is slow, in contrast, the static navigation GPS market development is rapid, and navigation GPS is gradually becoming the mainstream of China's GPS market.

2. Prospects of China automotive terminal market and development trends
   (a) with the increase of the amount of private cars in China and the change of use, the resulting market capacity of the Beidou satellite vehicle system has continued to grow. According to statistics, at present, China's private car market has more than 2.5 million, even at only 20 % of the allocation ratio, there is a 5 million market capacity. In the future, when the Beidou satellite is further improved, the Beidou satellite system will have broad applications.
   (b) with the rapid development of RFID technology, the vehicle with the RFID tag device or RFID tag card will provide great convenience for a driver related to parking fees, highway fees. The further improvement of infrastructure and supporting facilities will also greatly promote the future development of the intelligent vehicle terminal. In the future, the technology and equipment of the new vehicle will be developed and applied. In the next few years, with the full realization of the Beidou satellite system, 4G technology and other wireless communication technologies, the vehicle will become a driving control system in and of itself. Vehicles will be integrated into a unified vehicle information and communication apparatus that will play an important role in the development of future intelligent transportation system.

4 The Development Trend of Road Traffic Intelligent Systems

4.1 Electronic Toll Collection (ETC) Market

The toll free system (ETC) is a kind of electronic automatic toll collection for dry roads, bridges and tunnels. According to the Bayes Advisory launched in 2007, China's Intelligent Transportation Market Research and development forecast 2008–2012 Research Report, the highway related communication monitoring and charging system investment increased from 4.1 billion yuan in 2001 to 14.6 billion yuan in 2007. In the Ministry of communications, "the long-term scientific and technological development plan (2006–2020)" [7] the highway and waterway transportation, the highway network toll and parking fees will be two important points of focus. Along with the increase of highway construction, the demand of communications, the monitoring and charging systems of highways will be expanded.

4.2 The Development Trend of Road Traffic Intelligence

The demand of the highway toll free (ETC) system will increase year to year. Although there is already a considerablc part of the highway to build up regarding the parking fee system, there is still a big gap between the comprehensive construction and application. In the future, it will be a problem to be considered in the process of further development of ETC. Another trend in the development of road traffic intelligence is the road traffic video surveillance system. It includes the electronic police law enforcement and penalty system, vehicle speed detection system, vehicle traffic monitoring system, intelligent multimedia network license plate recognition and the urban comprehensive emergency system, etc. From a single GPS application to multi-system compatible applications, from the application of navigation and mobile communications, the internet and other applications to change; from the end of the application oriented products and services to change, the national satellite navigation industries long-term development plan (hereinafter referred to as planning) clearly pointed out this trend. The Beidou navigation system will be widely used in urban transportation, communications, power, finance, meteorology and hydrology monitoring. And in these applications is the important content of the construction of the "smart city".

The Beidou satellite navigation system aids urban traffic, logistics, unmanned and other applications in the field of application prospects. According to the Wuhan Institute of Smart Cities Vice President Wang Fan, car networking will use the Beidou navigation system and positioning function. "The so-called 'car networking', is the people, the car and the road as realized by the internet." "Information technology is used to build up a 'wisdom of the brain', that is, the traffic control center, the information on people, vehicles and the road, so as to provide people with traffic advice."

References

1. Deakin, M., Al Waer, H.: From intelligent to smart cities. J. Intell. Build. Int.: Intell. Cities Smart Cities **3**(3) (2011)
2. Wang, D.H., Peng, J., Wang, F.X.: Study of Three Frequency Ambiguity Resolution Based on CIR in Beidou Carrier System. Gnss World of China (2009)
3. Gan, X., Yu, B., Chao, L., Liu, S.: The Development, Test and Application of New Technology on Beidou/GPS Dual-Mode Pseudolites, vol. 340, pp. 353–364. Springer, Berlin (2015)
4. Baker, D., Merkert, R., Kamruzzaman, M.: J. Transp. Geogr. **43**, 140–150 (2015)
5. Chang, W.C.: The Design and Implementation of City Emergency Management System. ShangDong University Press (2008)
6. Wang, Y., Zheng, X.: Radiant City? Broadacre City? Look from the Land for the New Standard and the Future Development of China's Urban Residential Land. Urban Development Studies (2014)
7. The Ministry of Transport Research Institute: The Long-term Scientific and Technological Development Plan (2006–2020). Chinese Ministry of Transport, 2006-00103 (2006)

An Appointment Scheduling Method for China's Driving License Exam Using Network Flow Modelling

Yongqiang Bao, Huiying Xu and Chenlu Qiu

Abstract In this work, we consider the appointment scheduling problem for China's driver licenses exam with session capacity constraints. It is modelled as a maximum flow problem solved by Edmonds-Karp algorithm. A more general case with appointment priorities is also considered as a minimum cost flow problem solved by Busacker-Gowan algorithm. Experimental comparison of proposed approaches with conventional used sorting and dispatch method is given, showing the improvement of resource utilization for proposed approaches.

Keywords Bipartite · Maximum flow · Minimum cost maximum flow

1 Introduction

Automobile ownership in China has increased greatly since 2000. Consequently, the demand for driving license exam soars, especially in urban areas with large population and limited exam sites. Previously, government officers handle each driving test appointment on a walk-in basis: driver license applicant needs to make an appointment at local Department of Motor Vehicles (DMV); no examination resources exchange and sharing exists between different DMVs. The result is high appointment cost and exam site congestion. With the rapid development of internet technology, offering an online appointment service becomes the focus of construction for traffic management administrative department.

Appointment scheduling is one core component of online appointment service system. An effective scheduling scheme can go a long way toward improving

The original version of this chapter was revised: Acknowledgment has been included. The erratum to this chapter is available at http://dx.doi.org/10.1007/978-981-10-2398-9_29

Y. Bao (✉) · H. Xu · C. Qiu
Traffic Management Research Institute of the Ministry of Public Security, Wuxi 214151, China
e-mail: yongqiangbao@aliyun.com

© Springer Science+Business Media Singapore 2017

H. Lu (ed.), *Proceedings of the Second International Conference on Intelligent Transportation*, Smart Innovation, Systems and Technologies 53,
DOI 10.1007/978-981-10-2398-9_14

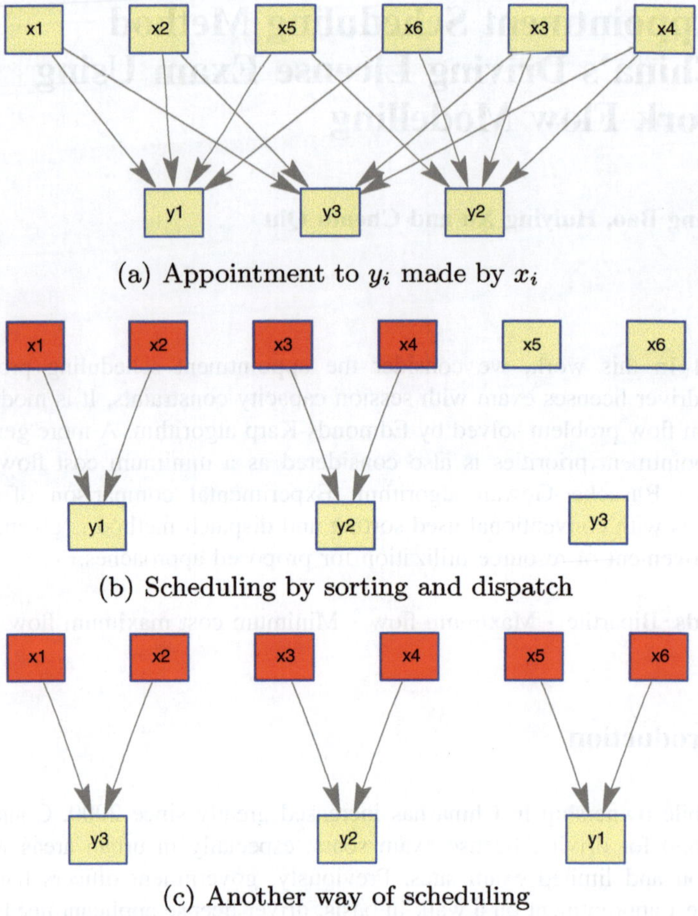

Fig. 1 Toy example of Appointment Scheduling

resource utilization and increasing applicants' satisfaction. Let us call each time slot at each exam site as an exam session. Assume that each applicants can choose multiple sessions (upper bounded by some limit specified by administrative departments), which is a valid and reasonable assumption in practice. Conventional used sorting and dispatch method simply sorts each applicant by their IDs (generated according to their appointment time or some other priority rules), and then sequentially dispatch exam session to the applicant from all available sessions. The result is the congestion of popular exam sites and the vacancy of infrequently selected exam sites (shown in Fig. 1), often in rural areas. Apparently, intelligent scheduling is needed to make the examination resources more evenly distributed, which in turn benefits people as more applicants can get allocated. We model the appointment scheduling problem as a network flow problem [1]. The main idea is modelling each applicant and exam session as a vertex in a weighted digraph and adding one source vertex and one sink vertex (as shown in Fig. 2, see details in Sect. 2). By maximizing

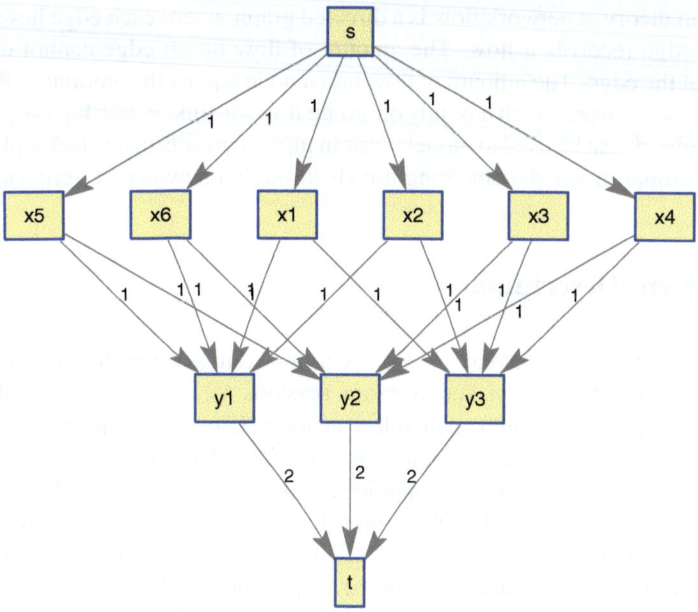

Fig. 2 Network flow modelling of Fig. 1a

the network flow from the source to the sink, we obtain a scheduling which could assign the maximum number of applicants to exam sessions under current capacity constraint. More general cases when applicants may have priorities according to administrative rules, or applicants may have own preference on their selections, are also considered as a minimum cost flow problem [1].

The paper is organized as follows. Section 1.1 gives preliminaries on graph and network flow. Section 2 gives problem description. Appointment scheduling based on network flow modelling is then developed in Sect. 3. Experimental results are presented in Sect. 4. Conclusions is given in Sect. 5.

To the limit of our knowledge, this is the first work studying appointment scheduling for China's driving license exam.

1.1 Preliminaries on Graph and Network Flow [2, 3]

A graph $G = (V, E)$ is a collection of vertices joined by edges. It comprises a set V of vertices together with a set E of edges, each of which connect two vertices. An undirected graph is a graph in which edges have no orientation. A directed graph, or digraph for abbreviation, is a graph in which edges have orientations. A weighted graph, also called network, is a graph in which a number (the weight) is assigned to each edge. Such weights represent costs, lengths, or capacities. A bipartite graph is a graph where vertices can be divided into two sets V_1 and V_2 such that no edge connects the vertices in the same group.

In graph theory, a network flow is a directed graph where each edge has a capacity and each edge receives a flow. The amount of flow on an edge cannot exceed the capacity of the edge. The amount of flow into a node equals the amount of flow out of it, unless it is a source, which has only outgoing flow, or sink, which has only incoming flow. A network can be used to model fluids in pipes, currents in an electrical circuit, or anything similar in which something travels through a network of vertices.

2 Problem Description

Let us call each time slot at each exam site as an exam session. Suppose we have m applicants $x_i, i = 1, 2, \ldots, m$ and n exam sessions $y_j, j = 1, 2, \ldots, n$ with session capacity c_j. Each applicant x_i can select at most K exam sessions, where K is a parameter specified by DMV or other administrative departments. Figure 1a shows a toy example where we have 6 applicants and 3 exam session. Each applicant selects 2 exam sessions and each session has a capacity 2. Figure 1b shows the appointment scheduling given by conventionally used sorting and dispatch method. Applicant 1 and 2 is allocated to session 1, applicant 3 and 4 is allocated to session 2, while applicant 5 and 6 can not get allocated because the congestion of session 1 and 2. Session 3 is vacant even though is has a capacity 2. Figure 1c shows another scheduling which can successfully assign exam sessions to all applicants, noticing that applicant 1 and 2 also select session 3. For appointment scheduling with small amount applicants and sessions like the toy example shown in Fig. 1a, intelligent scheduling can be found by human brain through carefully observation and calculation. However, for the situation with large amount applicants and sessions, which is the case in practice, some scheduling algorithm is needed.

3 Appointment Scheduling Using Network Flow Modelling

Note that Fig. 1a constitutes a bipartite graph where $V_1 = \{x_1, x_2, \ldots, x_6\}$ and $V_2 = \{y_1, y_2, y_3\}$. Inspire by Maximum Bipartite Matching (MBP), the appointment scheduling problem in Fig. 1a can be modelled as a network flow problem with vertex capacities as shown in Fig. 2 [1–3]. Each applicant and exam session is modeled as a vertex in a weighted digraph. One source vertex s and one sink vertex t are added. The edges pointing from the applicant to the exam session represents the choice made by the applicants, each of which has weight 1. We add edges from source to all applicants, and from all exam sessions to sink. The weight for the edges directing from the source to each applicant is 1. The weight for the edges directing from exam session to the sink equals session capacity. We should mention that it is different from MBP because the capacity of each session is not 1. Instead of sorting and dispatch which is straightforward but poor performing, we model it as maximum flow problem that maximizes the network flow from the source s and the sink t, which in turn, maximizes the number of applicants who can get assigned to exam.

The Ford–Fulkerson method [3] is an algorithm that computes the maximum flow in a flow network. The idea is as follows: as long as there is a path from the source to the sink, with available capacity on all edges in the path, we send flow along one of the paths. Then we find another path, and so on. A path with available capacity is called an augmenting path. The Edmonds–Karp algorithm [4] is an implementation of the Ford–Fulkerson method, where the augmenting path is defined as a shortest path that has available capacity. In this work, we apply the Edmonds–Karp algorithm to solve the maximum flow problem for driving test exam appointment scheduling. It computes the maximum flow, or equivalently the maximum number of allocated applicants, in $O(VE^2)$ time, where V is the number of total applicants and E is the number of appointments.

3.1 Extensions

In practice, applicants would have different priorities according to administrative rules. Or, they may have own preference over their selections. These can be modeled as a minimum cost maximum flow problem where each edge has capacity and cost associated. The objective is to find the maximum flow that has minimum cost. This problem combines maximum flow (getting as much flow as possible from the source to the sink) with shortest path (reaching from the source to the sink with minimum cost). For example, if there are priorities over the applicants' appointments, we could set the cost of the edge directing from the source s to each applicant proportional to its priority precedence, i.e. use small cost for applicant with high priority, and use large cost for applicant with low priority. If the applicants have preference over their selections, we could set the cost of the edge pointing from the applicant to its selection proportional to the preference precedence, i.e. use the smallest cost for the first choice, use the the second smallest cost for the second choice, and so on. The costs for the edges directing from exam sessions to the sink t are always same. In this work, we solve minimum cost maximum flow problem by Busacker-Gowan algorithm [5].

4 Experimental Results

We compare our proposed methods, appointment scheduling by maximum flow (MF) and appointment scheduling by minimum cost maximum flow (MCMF), against the sorting and dispatch method (S&D), which first sorts applicants and then sequentially dispatch exam session to the applicant from all available sessions.

In all experiments, we have m applicants, $x_1, x_2, \ldots, x_m$, and n exam sessions, $y_1, y_2, \ldots, y_n$. Every applicant, x_i, has made k_i choices with $k_i \geq 1$ and $n \geq k_i$, i.e., every applicant chose k_i sessions randomly from $y_1, y_2, \ldots, y_n$. Every session, y_i, has a capacity of $c_i \geq 1$. We use $k_i = K$, $c_i = c$ being constants for simplicity. Let $C := n \times c$ denote the total capacity of all exam sessions.

For sorting and dispatch method, we use the subscript i, as the appointment order of x_i, i.e., x_i makes an appointment before x_{i+1}.

4.1 Appointment Scheduling with No Priority: A Maximum Flow Problem

Figure 3a shows the comparison of proposed scheduling method by MF with conventionally used S&D method. In Fig. 3a, the x-axis plots the number of applicants and the y-axis plots the number the applicants who can be assigned to an

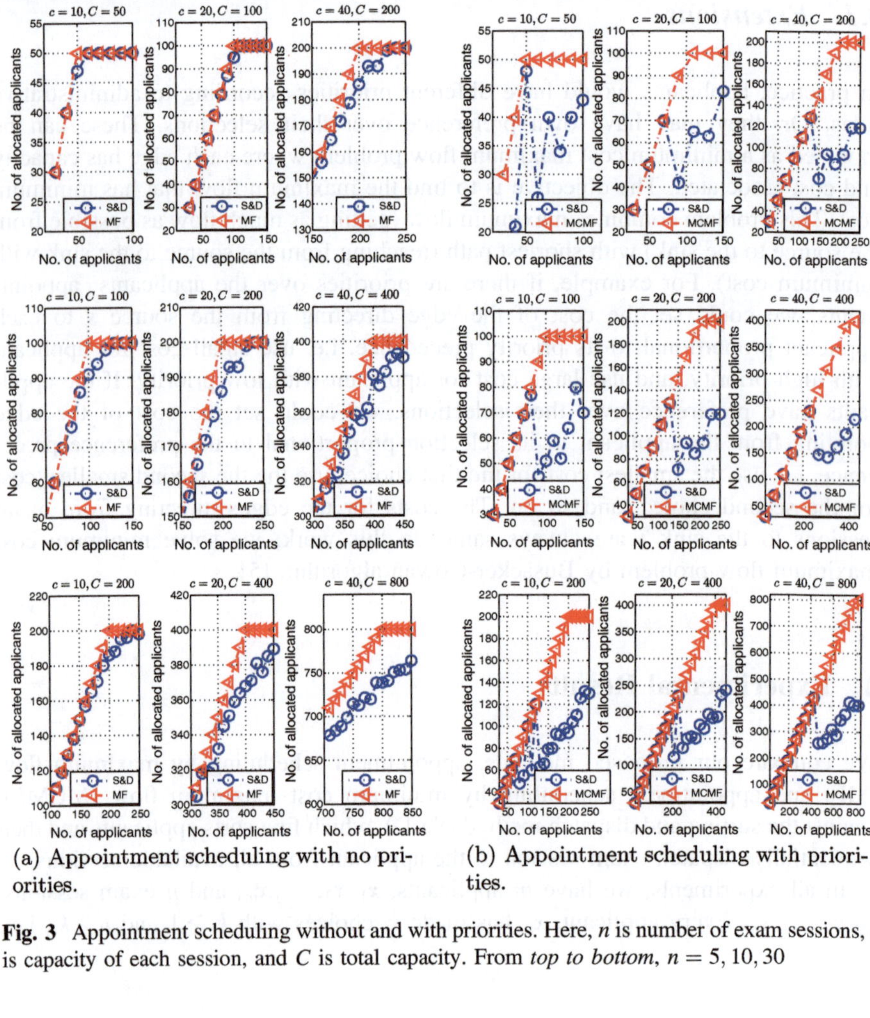

(a) Appointment scheduling with no priorities. (b) Appointment scheduling with priorities.

Fig. 3 Appointment scheduling without and with priorities. Here, n is number of exam sessions, c is capacity of each session, and C is total capacity. From *top to bottom*, $n = 5, 10, 30$

exam session. Recall that m is the number of applicants, n is the number of exam sessions and C is the capacity of all sessions. Obviously, MF scheduling generally allocates more applicants than S&D method. For example, in Fig. 3a when $m = 800$, $n = 20$ and $C = 800$, MF allocates about 60 more applicants than S&D, which gives an improvement of 7 % for resource utilization.

4.2 Appointment Scheduling with Priorities: A Minimum Cost Maximum Flow Problem

In this experiment, we consider the case where each applicant has appointment priorities. We generate applicants' appointment priorities uniformly at random from 1 to 3. Let the cost of the edge from the source to each applicant equals to the appointment priority, and let the cost of all other edges being 1. Figure 3b shows the comparison of MCMF with S&D method. As can be seen from Fig. 3b, MCMF makes significant improvement than S&D method for appointment scheduling with priorities. For example, in Fig. 3b when $m = 800$, $n = 20$ and $C = 800$, MCMF allocates about 400 more applicants than S&D, which gives an improvement of 50 % for resource utilization.

5 Conclusion

This work studies the appointment scheduling problem for China's driving license exam using network flow modeling. We model it as a maximum flow problem solved by Edmonds-Karp algorithm. A more general case with priorities is modeled as a minimum cost flow problem solved by Busacker-Gowan algorithm. Experimental results show the great improvement of proposed scheduling method on resource utilization, indicating that it could significantly increase applicants' satisfaction, reduce exam sites congestion, and improve the utilization of resources effectively.

Acknowledgments This research was supported by Central Public-interest Scientific Institution Basal Research Fund 2016SJA07 and National Key Technologies R&D Program 2014BAG01B04.

References

1. Ahuja, R.K., Magnanti, T.L., Orlin, J.B.: Network Flows: Theory, Algorithms and Applications. Prentice-Hall, Englewood Cliffs (1993)
2. Wilson, R.J.: Introduction to Graph Theory. Prentice Hall, Englewood Cliffs (1996)
3. Newman, M.: Networks: An Introduction. Oxford University Press, Oxford (2010)
4. Edmonds, J., Karp, R.M.: Theoretical improvements in the algorithmic efficiency for network flow problems. J. ACM **19**, 248–264 (1972)
5. Busaker, R.G., Gowen, P.J.: A procedure for determining a family of minimal-cost network flow patterns. Technical Report, John Hopkins University (1961)

Integration Design of Highway Traffic Safety Networked Control System

Yongqiang Bao, Huiying Xu and Chenlu Qiu

Abstract Aimed at solving China's main problems in highway traffic management and security services, we design and construct ministerial, provincial and prefectural highway traffic safety active prevention and control platform. Aggregation, analysis, exchanging and sharing of the cross regional highway traffic safety information and resources, and the real-time early warning on the effect of traffic emergencies are achieved. The construction of highway traffic safety active prevention and control platform will setup rapid and efficient traffic command system, establish new service mechanism, improve the capacity and performance level of traffic police, regulate road traffic order, and effectively prevent and reduce road traffic accidents.

Keywords Highway · Traffic safety · Networking and integrated control

1 Introduction

In recent years, with the rapid increasement of China's highway mileage, motor vehicle and driver quantities, traffic safety situation becomes increasingly severe, especially for unexpected traffic incidents, such as blizzard and snowstorm. Cross regional, cross road network, and cross sectoral emergency command are still facing many difficulties and problems. In order to prevent and reduce traffic accidents, it is urgent to strengthen the application of information technology and advanced equipment to improve the effectiveness and efficiency of highway traffic safety management.

The original version of this chapter was revised: Acknowledgment has been included. The erratum to this chapter is available at http://dx.doi.org/10.1007/978-981-10-2398-9_29

Y. Bao (✉) · H. Xu · C. Qiu
Traffic Management Research Institute of the Ministry of Public Security, Wuxi 214151, China
e-mail: yongqiangbao@aliyun.com

© Springer Science+Business Media Singapore 2017

H. Lu (ed.), *Proceedings of the Second International Conference on Intelligent Transportation*, Smart Innovation, Systems and Technologies 53,
DOI 10.1007/978-981-10-2398-9_15

Major problems of traffic management and security services in China's highway network include: (1) the lack of information exchange and resource sharing seriously affects the traffic control response speed and processing efficiency; (2) the lack of effective means to identify and evaluate overall traffic safety risk delays early warning and response to traffic emergencies; (3) the lack of intelligent control means and the intervention technology results in unsatisfactory highway network traffic flow science organization and control ability; and (4) the dispersion of management rights and confliction of obligations makes it difficult to form highway traffic safety management and control force. In order to solve these problems, a national highway traffic safety networked and control platform is constructed which effectively improves the national highway traffic safety management capability.

2 Highway Traffic Safety Active Prevention and Control Platform

In order to realize the aggregation, analysis, exchanging and sharing of cross regional highway traffic safety information and resources, to provide real-time early waring on area overall traffic safety situation and unexpected incidences, we design and construct traffic safety active prevention and control platform. It is a multi-system integrated platform, which consists of road traffic safety information integration system, multi-scale traffic safety risk assessment and alarm system, wide range traffic flow intervention and control system, police command and guard control system, traffic safety comprehensive service system and some other related systems. The platform functional architecture is shown in Fig. 1.

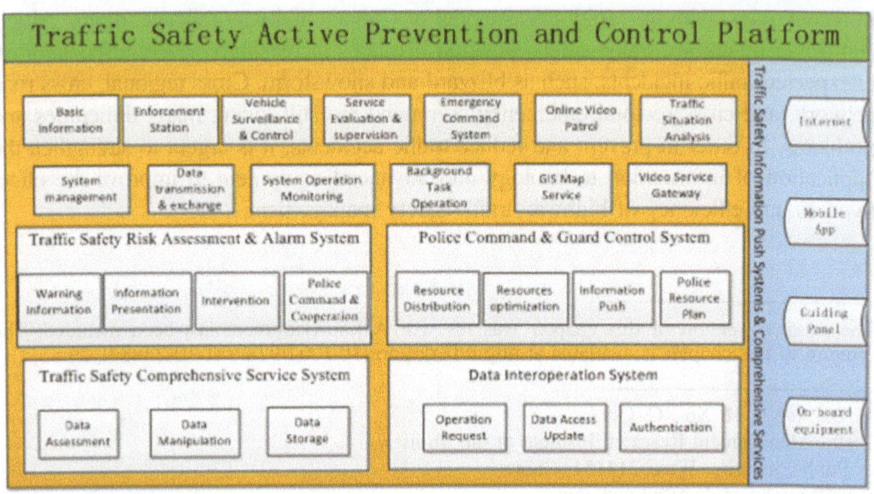

Fig. 1 Platform functional architecture

2.1 Road Traffic Safety Information Integration System

Through the establishment of the traffic safety information system based on GIS-T, the traffic safety information integration system, and a cross platform service workflow system are constructed. It consists of two subsystems: traffic safety data access and security management system and cross platform traffic safety information interoperability system.

2.2 Multi-scale Traffic Safety Risk Assessment System

Utilizing traffic safety risk analysis technology, we identify and evaluate vehicle violations risk, illegal vehicles impaction on surrounding vehicles, road accidents risk, road congestion risk and many other factors affecting high way traffic safety by applying multi-source data fusion technology and dynamic comprehensive evaluation technology. This system includes the following:

- Vehicle traffic safety risk assessment and early warning
- Road traffic safety risk assessment and early warning
- Regional road traffic safety risk assessment and early warning
- Multi-scale traffic safety risk assessment and early warning.

2.3 Wide Range Traffic Flow Intervention System

Based on highway traffic flow characteristics for national holidays, emergencies, as well as bad weather conditions, we developed road traffic flow active intervention system and speed control system to realize the scientific traffic organization and control. Intelligent control of massive cross regional traffic flow is realized. The system mainly include: massive cross regional and cross road network traffic flow induced cooperative control, speed control, wide range traffic organization and intervention evaluation scheme.

2.4 Police Command and Guard Control System

Through the study of police resources distribution and dynamic optimization, we developed police command and guard control system. It solves the problems of highway traffic management rights dispersion and obligations confliction. It helps to build unified highway traffic safety control force. Police command and control system includes police command and control service, recording service, traffic analysis and statistics, traffic safety active preventing and control, agency services, database services, etc.

2.5 *Traffic Safety Comprehensive Service System*

We developed Internet based comprehensive traffic safety service system. Information push service is provided sending announcement for traffic flow conditions, weather conditions, traffic incidents, traffic congestion, emergencies and secondary accidents, etc. Mobile application softwares are also developed for mobile phones, onboard equipment, and roadside information release terminals.

3 Key Technologies Used for Highway Traffic Safety Active Prevention and Control Platform

3.1 *Semantic Standardization of Traffic Safety Information and Interoperability Between Heterogeneous Platforms*

The data collected by difficult public security departments and transportation departments varies a lot. Semantic standardization of traffic safety information is a prerequisite to achieve cross department and cross system data exchanging and sharing. The difficulty lies in dynamic identification and format conversion of static information (i.e., road network, facilities, etc) and dynamic information (i.e., traffic flow, control strategies, etc). Key technologies include cross regional and cross network traffic information assessment using cloud architecture, Lane-GIS-T data model [2, 3], fast preprocessing of massive structured and unstructured data, concurrent write storage and efficient data retrieval, etc.

3.2 *Analysis and Prediction of Network Traffic Flow by Multiple Sources of Traffic Information*

Integrating multi-source and multi-format traffic data, we presented wide range and large scale traffic flow guidance and organization scheme. It realizes large scale traffic flow diversion for national holidays, unexpected traffic incidences, as well as bad weather conditions. Key technologies include traffic flow control and dynamic route guidance techniques [1, 4].

3.3 Prediction of Spatial and Temporal Influences of Emergency on Traffic Flow with Traffic Flow Intervention

The occurrence and evolution of traffic incident would directly affect the police resource configuration and organization for traffic flow intervention. Key technologies include identification of spatial and temporal effects of traffic emergency changes [5], traffic incidents' spatial and temporal interactions, and emergency police resource configuration.

4 Conclusions

Through the construction of highway traffic safety networked control platform, traffic management data from multiple public security departments and transportation departments, such as highway fundamental facilities information, highway video surveillance data, etc, is integrated. The construction of the platform will setup rapid and efficient traffic command system, establish new service mechanism, improve the capacity and performance of traffic police, regulate road traffic order, and effectively prevent and reduce road traffic accidents. It provides technical support for coordinated management, unified control and emergency response and rescue.

Acknowledgments This research was supported by National Key Technologies R&D Program 2014BAG01B04.

References

1. Harris, P.S., Rabone, J.A., Randall, D.S.: A simulation of dynamic Route Guidance systems. Traffic Eng. Control 327–329 (1992)
2. Michael, G.F.: Geographic information system and disaggregate transportation planning. Geogr. Syst. 19–44 (1998)
3. Shi, J., Xu, G., He, M., Song, Y.: Research progress on GIS-T data model. J. Beijing Univ. Technol. 318–322 (2004)
4. Tomkewitsch, V.R.: ALI-SCOUT, A universal guidance and information system for road traffic. Second International Conference on Road Traffic Control (1986)
5. Wang, J., Zhao, X., Cong, H.: Prediction model of freeway network traffic incidence space time effect: an introduction to topic III of Project VI of the national road traffic safety science and technology action plan. J. Transp. Inf. Saf. (2013)

Analysis of Vehicular Behavior at Bottlenecks Considering Lateral Separation

X. Shen and Z.C. He

Abstract The vehicular behaviors at bottlenecks are complex, and lateral separation makes traffic flow more unstable. Based on the lateral separation characteristic, this paper analyzes the vehicular behaviors at traffic bottlenecks and the movement is described. What's more, video recognition technology is introduced to obtain the track of the vehicle to verify the decision-making process. The lane-changing behavior and gap-following behavior can be predicted well. And in further study, more factors should also be considered.

Keywords Traffic bottleneck · Lateral separation · Vehicular behavior

1 Introduction

Lateral effect in realistic vehicular behavior exists widely especially in developing countries. However, as far as we know, a large number of existing microscope models do not take into account the effect of the lateral separation. And the traditional car-following models or lane-changing models have some inability to describe this behavior in complex driving environment.

Gunay [1] proposed that the vehicular lateral position within the current lanes obey normal distribution. Figure 1 shows the distribution of vehicular lateral position. The vehicle trajectory data was collected by quadrotor in Guangzhou, China. And the video recognition technology was introduced to obtain the track of the vehicle. On the shoulder lane, the lateral separation exists because this is a bottleneck which caused by the bus stop.

X. Shen · Z.C. He (✉)
School of Engineering, Sun Yat-Sen University,
Guangzhou, People's Republic of China
e-mail: hezhch@mail.sysu.edu.cn

X. Shen · Z.C. He
Guangdong Provincial Key Laboratory of Intelligent
Transportation System, Guangzhou, People's Republic of China

© Springer Science+Business Media Singapore 2017
H. Lu (ed.), *Proceedings of the Second International Conference on Intelligent Transportation*, Smart Innovation, Systems and Technologies 53,
DOI 10.1007/978-981-10-2398-9_16

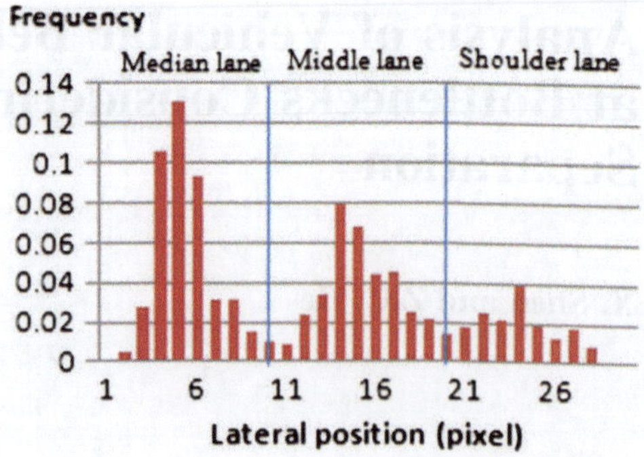

Fig. 1 Distribution of vehicular lateral position

To study the vehicular behavior, the car-following model and lane-changing model should be reviewed. Car-following and lane-changing behaviors are vital components of traffic flow theories. Car following and lane changing models describe vehicular longitudinal and lateral interactions on the road, respectively. The microscope models describing vehicular longitudinal interactions in the single lane were proposed by Reuschel [2] and Pipes [3]. The model was based on the stimulus-response framework, and the relative velocity between the preceding vehicle and following vehicle was regarded as the stimulus. While, Newell [4] regarded the distance headway as the stimulus instead of the velocity difference. And then Bando et al. [5] proposed the OV model. Soon afterwards, according to the actual data, Helbing and Tilch [6] calibrated the OV model, and then proposed a generalized force model. What's more, Jiang et al. [7] developed a FVD model and then on the basis of Jiang's work, Wang et al. [8] proposed a multiple velocity difference model which considering the velocity differences of multiple vehicle.

As for models considering the lateral separation, the first one was proposed by Gunay [1] which considering the lateral discomfort. Then based on FVD model, Jin et al. [9] proposed a non-lane-based car following model. In these models the follower has the chance to overtake the leading vehicle. However, it was hard for driver to perceive minor lateral separation, speed and distance directly. To solve this problem, Jin et al. [10] introduced visual angle information to develop a new non-lane-based car following model. The Time-To-Collision (TTC) was introduced to explain how the following drivers perceive the change of lateral separation between the leader and the follower.

Although the staggered car following model proposed by Jin takes lateral separation into account, the description of motion is ignored. To overcome the shortcomings, He et al. [11] build up the equations of motion based on target steering angle (TSA) and then establish a series of rules for determining TSA.

Note that the vast majority of car following models focus on the movement in a single lane, and the models can't reflect the real situation. And the driving behaviors are disorder especially in China, so the models lack the general applicability. Therefore the new model should be proposed to focus on the lateral and longitudinal movement in multiple lanes.

Although car following model has been widely studied for many years, lane-changing did not receive much attention until recently. Comparing with the car-following model, lane-changing model takes more factors into account, such as the more neighboring vehicles, road conditions and so on. The decision-making process is more complex, and the movement is hard to describe. What's more, large number of vehicle trajectory data is needed to describe the lane-changing behavior.

As for the lane-changing models, there are two kinds of these, one is the mandatory lane-changing model and the other is discretionary. The primary purpose of a discretionary lane changing is to gain a speed advantage or a better driving environment, whereas the primary motivation of a mandatory lane changing is to reach the planned destination. And the process of lane changing includes the generation of lane-changing intention; feasibility analysis and implementation of lane changing. As for the generation of the lane-changing intention, the classical models are utility theory model, probability of lane changing (PLC) model and speed perception model. With reference to feasibility analysis, the classical models are gap acceptance theory model, safety factor evaluation model, and fuzzy logic model.

Note that the lane-changing decision-making process is dynamic, whereas it is ignored by the nowadays lane-changing model. What's more, the lateral separation isn't taken into account. In addition, for the lack of large number of vehicle trajectory data, the model can't be calibrated. Therefore the lane-changing model does not apply well.

This paper focuses on the effect of the lateral separation on vehicular behavior and the target steering angle is learned from the reference materials to describe the vehicular movements. Based on the lateral separation characteristic, this paper analyzes the vehicular behaviors at the traffic bottleneck. And the vehicular behavior decision–making process is set to describe vehicular longitudinal and lateral interactions on the road. At last, the video recognition technology is introduced to obtain the vehicle trajectory to evaluate the decision-making process.

2 Analysis of Vehicular Behavior

To our knowledge, many factors have effect on vehicular behaviors. According to the observation of video, daily experience and related materials, the velocity and position of the neighboring vehicle are the main factors. And in the real world, not only the preceding car 2 but also the next-preceding car 1 has impact on the vehicular behavior decision [11, 12] (Fig. 2).

Therefore when considering the vehicular behavior in the case of multi-lane, the neighboring vehicles are shown in Fig. 2. As shown, the neighboring vehicles include: the next-preceding vehicle 1, preceding vehicle 2, left-preceding vehicle 3, left-following vehicle 4, right-preceding vehicle 5, the right-following vehicle 6, left-next-preceding vehicle 7, and right-next-preceding vehicle 8.

In this paper, the vehicle position is the vehicle's centroid position. The x-axis and y-axis means the lateral and longitudinal position respectively. Based on the Fig. 2, the involved parameters and meaning are shown as follows:

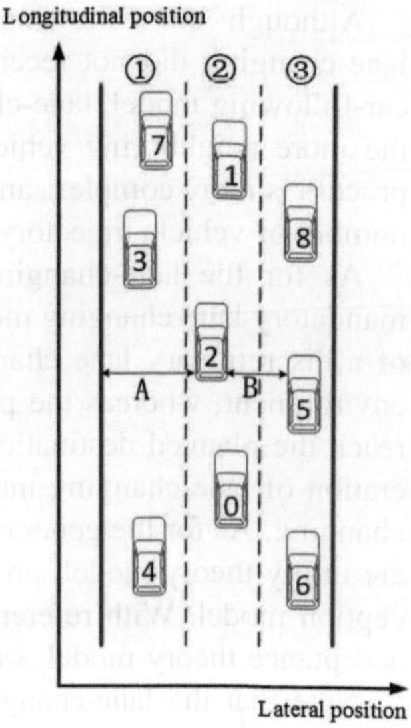

Fig. 2 Position relationship between the subject vehicle and the neighboring vehicles

w_c	width of the vehicle (m), w_c = 1.8 m;
l_c	length of the vehicle (m), l_c = 5 m;
v_i	velocity of vehicle i (m/s)
x_i	lateral position of vehicle i (m)
y_i	longitudinal position of vehicle i (m)
$\Delta v_{ij} = v_i - v_j$	velocity difference between vehicle i and vehicle j (m/s)
$\Delta x_{ij} = x_i - x_j$	lateral distance between vehicle i and vehicle j (m)
$\Delta y_{ij} = y_i - y_j$	longitudinal distance between vehicle i and vehicle j (m)

where $i, j = 0, 1, 2, 3, 4, 5, 6, 7, 8$;

Then the classification of vehicular behavior is discussed. In this paper the vehicular behaviors are classified into 4 kinds according to current driving conditions. There are lane-changing behavior, gap-following behavior, centerline-following behavior, and preceding-vehicular-following behavior. The major difference between the lane-changing behavior and gap-following behavior is shown in Fig. 2. If the car follows the gap A, it is the lane-changing behavior. If it follows the gap B, it is the gap-following behavior.

The main difference is the longitudinal distance between the preceding vehicle 2 and the lateral-preceding vehicle 3, 5. And the equation is

$$|\Delta y_{2j}| = |y_2 - y_j| \le l_c, \quad \text{where } j = 3, 5 \tag{1}$$

The target steering angle [2] is introduced in this paper to describe the vehicular movements. Regardless of the vehicle skid, steering and acceleration are regarded as two independent procedures. There are the two assumptions of the equations of motion based on target steering angle.

One is that turning the wheel only changes the direction of the vehicle velocity, and the magnitude of the velocity won't change. And the change of the direction happens instantaneously.

The other one is when the vehicle moves forward, the longitudinal motion is always in car-following condition. And the acceleration will be affected by the following target, so the longitudinal motion of the vehicle could be described by OV model.

The equation of traditional OV model is shown as follows.

$$\frac{dv_n(t)}{dt} = \alpha[V(\Delta y_n(t)) - v_n(t)] \tag{2}$$

where $V()$ is the optimal velocity function, $v_n(t)$ is the velocity of car n at time t, $\alpha = 1/\tau$ is the sensitivity coefficient.

Based on the assumptions mentioned above, Eq. (2) is rewritten as follows

$$\frac{dv_{n,y}(t)}{dt} = \alpha\left[V(\Delta y_n(t)) - v_{n,y}(t)\right] \tag{3}$$

where $v_{n,y}(t)$ is the projection of the velocity $v_n(t)$ along the y-axis.

However the vehicle following target is not only the preceding car in this paper. According to the different vehicular behavior decisions, the vehicle following targets can be the preceding vehicle, lateral-preceding vehicle, road centerline and the gap between the preceding vehicle and lateral-preceding vehicle.

Because the subject vehicle's number is 0, the calculating formulae for the acceleration is rewritten as follows

$$\frac{dv_{0,y}(t)}{dt} = \alpha\left[V\left(\Delta y_{0,target}(t)\right) - v_{0,y}(t)\right] \tag{4}$$

$v_{0,y}(t)$ is the projection of the velocity $v_0(t)$ along the y-axis. According to the formations, $\Delta y_{0,target}(t) = y_0 - y_{target}$ is the longitudinal distance between vehicle 0 and vehicle following target. The process and formulation is shown in Fig. 3.

The influencing factors and classification of vehicular behavior are proposed above, and the target steering angle is introduced to describe the movement. Then the reasons for different behaviors should be discussed. And according to the different vehicular behaviors, the vehicle following target and corresponding target steering angle θ are decided.

The basic assumption of the analysis is that the free flow doesn't occur at the bottlenecks which means the vehicle can't drive freely at the bottlenecks. When the velocity of the vehicle is relatively high, the driver expects to pass the current road segment quickly, so he/she will give priority to lane-changing behavior and gap-following behavior. Only when he/she can't change lane or follow the gap, would he/she consider following the preceding vehicle or driving along the road centerline.

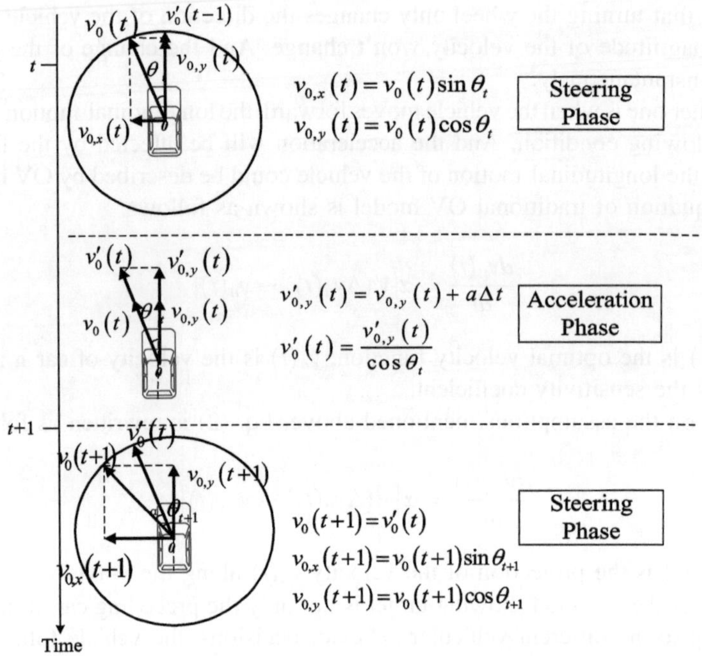

Fig. 3 Process of updating the vehicular situation based on TSA

- **Analysis of lane-changing behavior**

It's worth mentioning that this paper only considers the discretionary lane-changing behavior. And the analysis of lane-changing behavior includes two parts: the generation of the lane-changing intention and feasibility analysis. The implementation of lane-changing behavior is described by the movement equation based on TSA which is mentioned above.

The position relationship between the subject vehicle 0 and neighboring vehicle 1–8 is shown Fig. 4, when the subject vehicle 0 intents to change lane. In order to clearly describe the process, this paper only consider the leftward-lane-changing behavior, the right side is the same.

In the discretionary-lane-changing case, the reason for the generation of lane-changing intention is the driver wants to obtain more desirable driving status. And there are two major reasons.

One is the driver expects to pass the current road segment quickly, he/she wants to obtain higher speed by changing the lane. When the velocity of subject vehicle 0 is higher than the velocity of preceding vehicle 2 at time t, the driver will generate the lane-changing intention. The formula is

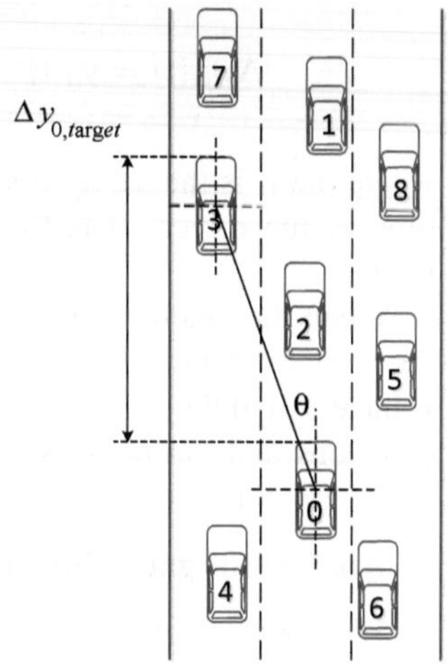

Fig. 4 Analysis of lane-changing behavior and the target steering angle θ

$$v_0(t) > v_2(t) \quad (5)$$

The other reason is longitudinal distance between the preceding vehicle 2 and the left-preceding vehicle 3 meets the lane-changing stimulus threshold. And the trigger condition is formulated as follows

$$\Delta y_{32}(t) = y_3(t) - y_2(t) > \alpha l_c \quad (6)$$

where αl_c is the lane-changing stimulus threshold which is a function relates to the length of the vehicle, and the α is the undetermined coefficient.

After considering the intention, feasibility will be analyzed. Here the gap acceptance theory is introduced. And it's generally considered that the distance for lane-changing will be longer if velocity is high or the velocity difference is large. The acceptable gap can be represented as follows

$$g_1 = \Delta y_{30}(t) - l_c \geq \max\{g_{min}, g_{min} + t_{lc} \cdot [\alpha_1 \cdot v_0(t) + \alpha_2 \cdot \Delta v_{03}(t)]\} \quad (7)$$

where g_1 is the longitudinal distance between the subject vehicle 0 and the left-preceding vehicle 3, g_{min} is the minimum acceptable gap, t_{lc} is the time for lane-changing, α_1 and α_2 are undetermined coefficients.

What's more, the security issues should be taken into account. To avoid the collision with the left-following vehicle 4, the longitudinal distance between the subject vehicle 0 and left-following vehicle 4 needs to meet the requirement of safety distance, and it can be represented as follows

$$\Delta y_{04}(t) = y_0(t) - y_4(t) > v_4(t)t_0 + l_c + l_s + \frac{v_4(t)^2 - v_0(t)^2}{2g(\varphi \pm i_0)} \quad (8)$$

where the l_s is the safety distance, l_c is the length of the vehicle, t_0 is the response time of the driver, φ is the longitudinal friction coefficient, i_0 is the road slope grade.

When the lane-changing behavior is feasible, the subject vehicle 0 will make a decision to change the lane. And now the following target is the left-preceding vehicle 3, and the $\Delta y_{0,target}$ is Δy_{30}. The acceleration can be calculated by movement equations and the target steering angle θ can be calculated according to the geometric relationship shown in Fig. 4.

- **Analysis of gap-following behavior**

The position relationship between the subject vehicle 0 and neighboring vehicle 1–8 is shown Fig. 5 when the subject vehicle 0 intents to follow the gap. In order to clearly describe the process, this paper only consider the leftward-gap-following behavior, the right side is the same.

As mentioned, if the subject vehicle 0 chooses to drive in the gap of the preceding vehicle 2 and the lateral-preceding vehicle 3 or 5, the longitudinal distance of the two vehicles needs to meet the following requirement

$$|\Delta y_{2j}(t)| = |y_2(t) - y_j(t)| \le l_c, \quad \text{where } j = 3, 5 \quad (9)$$

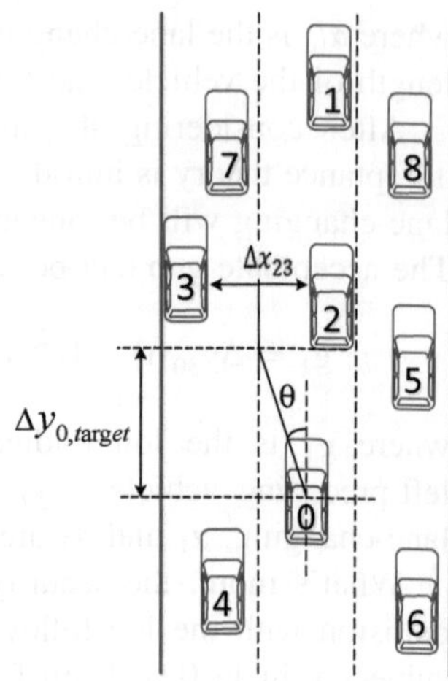

Fig. 5 Analysis of preceding-vehicle-following behavior and the target steering angle θ

Then analysis of gap-following behavior is also divided into two parts, including generation of the gap-following intention and the feasibility analysis.

There are three reasons for the generation of the gap-following intention.

The first one is the speed limit caused by preceding car, which is expressed mathematically as follows

$$v_0(t) > v_2(t) \tag{10}$$

The second reason is the lateral separation between the preceding vehicle 2 and the left-preceding vehicle 3 gives vehicle 0 the stimulation to follow the gap. As a result, the lateral distance between the preceding vehicle 2 and the left-preceding vehicle 3 meets the gap-following stimulus threshold. And the trigger condition is formulated as follows

$$|\Delta x_{23}(t)| = |x_2(t) - x_3(t)| \geq \beta w_c \tag{11}$$

where βw_c is the gap-following stimulus threshold which is a function relates to the width of the vehicle, and the β is the undetermined coefficient.

The third reason is the driving status of the next-preceding vehicle 1 and the left-next-preceding 7 have no impact on gap-following behavior. Because of the lateral separation, the next-preceding vehicle 1 and lateral-next-preceding vehicle 7 or 8 can be seen. If the velocity of vehicle 7 or vehicle 1 is slower than vehicle 3 or vehicle 2, it's believed that the following the gap can't get the desired benefits.

As a result, the generation of the gap-following intention needs to meet the following formulas:

$$v_1(t) \geq v_2(t) \tag{12}$$

$$v_7(t) \geq v_3(t) \tag{13}$$

After considering the intention, feasibility will be analyzed. There are mainly 2 requirements.

One of them is the lateral distance between the preceding vehicle 2 and the left-preceding vehicle 3 should bigger than acceptable gap—G_{accept}, and the Maximum Escape Speed (MES) of the acceptable gap needs to be faster than the speed of preceding vehicle 2 and the left-preceding vehicle 3. The formulas are as follows

$$|\Delta x_{23}(t)| = |x_2(t) - x_3(t)| \geq G_{accept} \tag{14}$$

$$\text{MES} > v_2(t), \quad \text{MES} > v_3(t) \tag{15}$$

And the Maximum Escape Speed (MES) can be calculated as follows

$$FC = \frac{G_{accept} - w_c}{2} \tag{16}$$

$$MES = -17.2(FC)^2 + 77.6(FC) - 0.7 \tag{17}$$

As the formulas shown, the acceptable gap is a function relates to the velocity of the preceding vehicle 2, left-preceding vehicle 3 and the width of the vehicle.

The other one is taking the security issues into account to avoid the collision. According to the velocity, the trajectories of vehicle can be calculated.

Based on the target steering angle, the acceleration and velocity $v_0'(t+1)$ can be calculated. $v_{0,x}'(t+1)$ and $v_{0,y}'(t+1)$ are the projection of the velocity $v_0'(t+1)$ along the x-axis and y-axis. And it's considered that the driving state of the left-following vehicle 4 will not change. As a consequence, the trajectory can be calculated.

$$\text{Trajectory of vehicle 0}\begin{cases} x_0(t+\tau) = x_0(t) + v_{0,x}'(t+1)\cdot\tau \\ y_0(t+\tau) = y_0(t) + v_{0,y}'(t+1)\cdot\tau \end{cases} \tag{18}$$

$$\text{Trajectory of vehicle 4}\begin{cases} x_4(t+\tau) = x_4(t) + v_{4,x}(t)\cdot\tau \\ y_4(t+\tau) = y_4(t) + v_{4,y}(t)\cdot\tau \end{cases} \tag{19}$$

where the τ is the time for finishing following the gap.

Then the inequations for collision are shown as follows

$$\begin{cases} \tau \le \frac{\frac{X_2+X_3}{2} - X_0(t)}{V_{0,x}'(t+1)} \\ |x_0(t+\tau) - x_4(t+\tau)| \le w_c \\ |y_0(t+\tau) - y_4(t+\tau)| \le l_c \end{cases} \tag{20}$$

It's required that τ has no solution, and the requirement for position of vehicle 4 can be worked out.

When the gap-following behavior is feasible, the subject vehicle 0 will make a decision to follow the gap. And now the following target is the gap, and the $\Delta y_{0,\text{target}}$ is $\Delta y_{20} - \frac{1}{2}w_c$. The acceleration can be calculated by OV model and the target steering angle θ can be calculated according to Fig. 5.

As before, the intention and feasibility analysis of the lane-changing behavior and the gap-following behavior are discussed. However there is special case that the intentions of lane-changing and gap-following are generated at the same time. The gap-following behavior is considered as a more-aggressive behavior, so the driver will give priority to the lane-changing behavior.

- **Analysis of preceding-vehicle-following behavior**

According to the basic assumption, when the subject vehicle can't change lane or follow the gap, preceding-vehicle-following behavior and centerline-following behavior will be considered.

As mentioned in Sect. 1, the lateral separation exists widely at bottlenecks. This paper proposes that vehicle decides to follow the preceding car behavior when the gets disturbed by the neighboring vehicles. In this case, following the preceding vehicle is the best.

In this paper, the threshold of the lateral stimulation $d_{lateral}$ is proposed. When the vehicle gets disturbed, the longitudinal distance and lateral distance should meet the requirement as follows

$$\Delta y_{0i}(t) = |x_0(t) - x_i(t)| \leq l_c \tag{21}$$

$$\Delta x_{0i}(t) = |x_0(t) - x_i(t)| \leq d_{lateral} \tag{22}$$

where $i = 3, 4, 5, 6$. It's worth mentioning that if two of these four vehicles satisfy the inequations the vehicle will decide to follow the preceding vehicle.

When the vehicle decides to follow the preceding vehicle, the following target is preceding vehicle, and the $\Delta y_{0,target}$ is Δy_{20}. The acceleration can be calculated by OV model and the target steering angle θ can be calculated according to Fig. 6.

- **Analysis of preceding-vehicle-following behavior and the target steering angle**

The centerline-following behavior at bottlenecks is the timid vehicular behavior. When the vehicle doesn't get disturbed by the neighboring vehicles, centerline-following behavior is chosen. As a result, the following target is the centerline, and the $\Delta y_{0,target}$ is Δy_{20}. The acceleration can be calculated by OV model and the target steering angle θ can be calculated according to Fig. 7.

Fig. 6 Preceding-vehicle-following behavior and the target steering angle θ

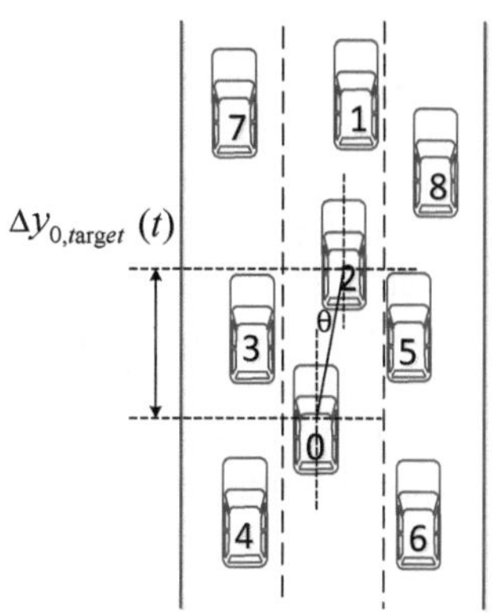

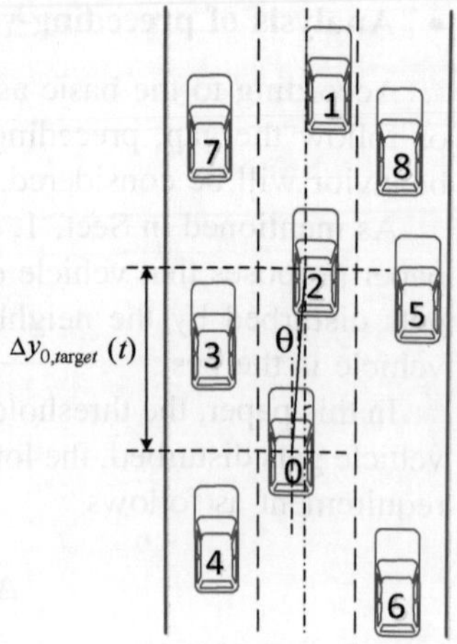

Fig. 7 Analysis of centerline-following behavior and the target steering angle θ

Based on the analysis above, the vehicular behavior decision-making process is set up (see Fig. 8).

3 Numerical Experiment

In this section, the video recognition technology is introduced into obtaining the track of the vehicle, and the data is used to verify the decision-making process. And according to the result, the decision-making process can be improved and the foundation is built for modeling properly in the future.

3.1 Data Processing

Firstly, according to the trajectories of vehicles, the neighboring vehicles can be found, and the velocity differences and position relationship can be calculated respectively. Then using the information such as velocity differences and position relationship to judge the vehicular behavior based on the decision-making process which is shown in Fig. 8. Then the decision-making sample is obtained.

Next, checking sample will be obtained by the direct judgement of the trajectories of vehicles, which is considered as the actual vehicular behavior.

Finally, making a comparison between the decision-making sample and the checking sample to verify and improve the decision-making process.

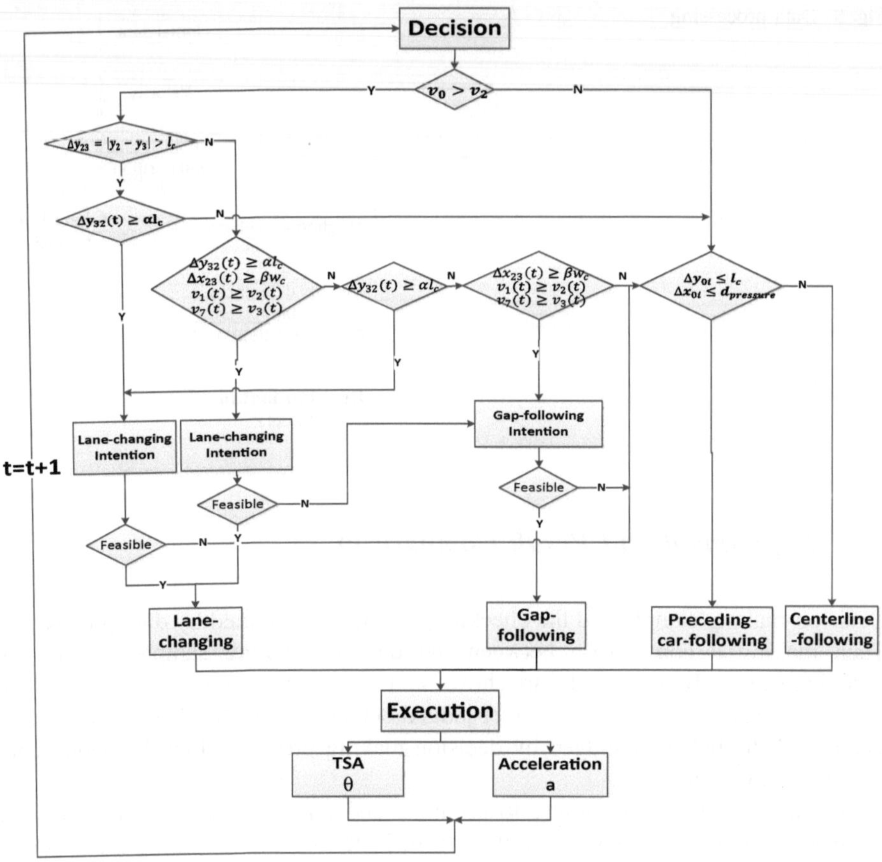

Fig. 8 Vehicular behavior decision-making process

The initial data include the car-ID and the corresponding sampling time, abscissa and ordinate values. In order to obtain the decision-making sample and the checking sample the process is shown in Fig. 9.

Note that the involved parameters mentioned above in decision-making process are shown in Table 1.

And according to the empirical value and daily experience, 10 sets of parameters are set for experiment (see Table 2). Note that according to value of parameters and decision-making process, the decision-making sample is obtained.

Because the video recognition software can't recognize all the vehicles in the video, some trajectory data are missing. And if one neighboring-vehicle's data are missing, the set of data are invalid. As a result, only 109 sets of data are considered valid after calculation.

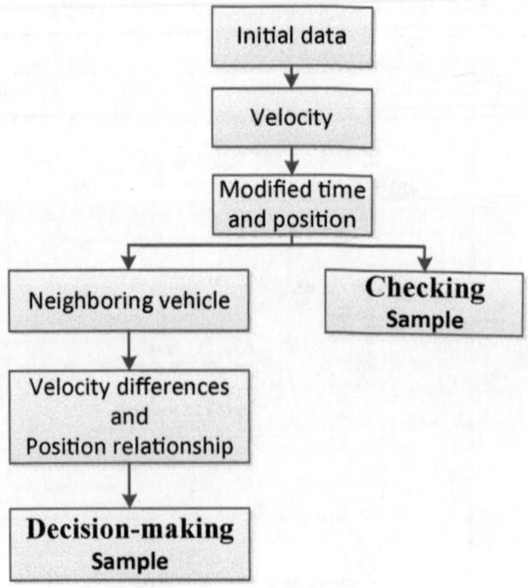

Fig. 9 Data processing

3.2 *Experiment and Result Comparison*

Decision-making sample and the checking sample are obtained by data processing. Then making a comparison between the decision-making sample A and the checking sample B. The results are shown in Tables 3, 4, 5 and 6. The 'Correct' in the table means the number of correct-judged behaviors. While 'Wrong' means the number of the behavior judged by decision-making process which is inconsistent with the actual behavior.

As Table 3 shown, according to accuracy and false-positive rate, the most of the lane-changing behaviors are well judged. Note that by comparing the 1–4 sets of results, the lane-changing stimulus threshold and the acceptable gap have an important effect on the judgement of lane-changing behavior.

As for the gap-following behavior, according to accuracy, with the decrease of gap-following stimulus threshold the accuracy is higher. Thus it could be seen that,

Table 1 The meaning of the parameters

Decision	Parameter	Meaning
Lane-changing intention	α	Coefficient for lane-changing stimulus threshold
Lane-changing feasibility	g_{min}	Minimum acceptable gap (pixel)
	t_{lc}	Time for lane changing (s)
	α_1	Coefficient for velocity
	α_2	Coefficient for velocity differences
	φ	friction coefficient
	t_0	Reaction time (s)
Gap-following	β	Coefficient for gap-following stimulus threshold
Preceding—car—following	$d_{lateral}$	Threshold of the lateral stimulation

Table 2 The value of parameters

Parameters	α	g_{min}	t_{lc}	α_1	α_2	φ	t_0	m	n	β	$d_{lateral}$
NO.1	1	88	3	0	0	0.5	1	1	1	1.3	20
NO.2	0.5	88	3	0	0	0.5	1	1	1	1.3	20
NO.3	0.8	88	3	0	0	0.5	1	1	1	1.3	20
NO.4	0.8	66	3	0	0	0.5	1	1	1	1.3	20
NO.5	0.8	66	3	0	0	0.5	1	1	1	1.2	20
NO.6	0.8	66	3	0	0	0.5	1	1	1	1.1	20
NO.7	0.8	66	3	0	0	0.5	1	1	1	1.15	20
NO.8	0.8	66	3	0	0	0.5	1	1	1	1.15	15
NO.9	0.8	66	3	0	0	0.5	1	1	1	1.15	25
NO.10	0.8	66	3	0	0	0.5	1	1	1	1.15	10

Table 3 Judgement of lane-changing behavior

	Parameters				Judgement of lane-changing behavior			
	α	g_{min}	β	$d_{lateral}$	Correct	Wrong	Accuracy (%)	False-positive rate (%)
NO.1	1	88	1.3	20	20	0	86.96	0.00
NO.2	0.5	88	1.3	20	20	3	86.96	3.49
NO.3	0.8	88	1.3	20	21	0	91.30	0.00
NO.4	0.8	66	1.3	20	23	0	100.00	0.00
NO.5	0.8	66	1.2	20	23	0	100.00	0.00
NO.6	0.8	66	1.1	20	23	0	100.00	0.00
NO.7	0.8	66	1.15	20	23	0	100.00	0.00
NO.8	0.8	66	1.15	15	23	0	100.00	0.00
NO.9	0.8	66	1.15	25	23	0	100.00	0.00
NO.10	0.8	66	1.15	10	23	0	100.00	0.00

Table 4 Judgement of gap-following behavior

	Parameters				Judgement of gap-following behavior			
	α	g_{min}	β	$d_{lateral}$	Correct	Wrong	Accuracy (%)	False-positive rate (%)
NO.1	1	88	1.3	20	2	4	28.57	3.92
NO.2	0.5	88	1.3	20	2	3	28.57	2.94
NO.3	0.8	88	1.3	20	2	4	28.57	3.92
NO.4	0.8	66	1.3	20	2	3	28.57	2.94
NO.5	0.8	66	1.2	20	5	3	71.43	2.94
NO.6	0.8	66	1.1	20	7	5	100.00	4.90
NO.7	0.8	66	1.15	20	7	3	100.00	2.94
NO.8	0.8	66	1.15	15	7	3	100.00	2.94
NO.9	0.8	66	1.15	25	7	3	100.00	2.94
NO.10	0.8	66	1.15	10	7	3	100.00	2.94

Table 5 Judgement of preceding-vehicle—following behavior

	Parameters				Judgement of preceding-vehicle—following behavior			
	α	g_{min}	β	$d_{lateral}$	Correct	Wrong	Accuracy (%)	False-positive rate (%)
NO.1	1	88	1.3	20	23	26	74.19	33.33
NO.2	0.5	88	1.3	20	23	26	74.19	33.33
NO.3	0.8	88	1.3	20	23	26	74.19	33.33
NO.4	0.8	66	1.3	20	24	26	77.42	33.33
NO.5	0.8	66	1.2	20	24	27	77.42	34.62
NO.6	0.8	66	1.1	20	24	28	77.42	35.90
NO.7	0.8	66	1.15	20	24	27	77.42	34.62
NO.8	0.8	66	1.15	15	21	22	67.74	28.21
NO.9	0.8	66	1.15	25	26	31	83.87	39.74
NO.10	0.8	66	1.15	10	19	12	61.29	15.38

the decision-making process is quite reasonable for gap-following behavior. Note that there has false positive. Because gap-following behavior is an aggressive behavior, even if the driver can follow the gap, the gap-following behavior may don't be executed. In further study, the psychological factors should also be considered (see Table 4).

As shown in Tables 5 and 6 the preceding-vehicle—following behavior and centerline-following behavior are not well judged. In further study, the other factors should also be considered carefully.

Table 6 Judgement of centerline-following behavior

	Parameters				Judgement of centerline-following behavior			
	α	g_{min}	β	$d_{lateral}$	Correct	Wrong	Accuracy (%)	False-positive rate (%)
NO.1	1	88	1.3	20	10	24	20.83	39.34
NO.2	0.5	88	1.3	20	10	22	20.83	36.07
NO.3	0.8	88	1.3	20	10	23	20.83	37.70
NO.4	0.8	66	1.3	20	11	20	22.92	32.79
NO.5	0.8	66	1.2	20	11	16	22.92	26.23
NO.6	0.8	66	1.1	20	11	11	22.92	18.03
NO.7	0.8	66	1.15	20	11	14	22.92	22.95
NO.8	0.8	66	1.15	15	14	19	29.17	31.15
NO.9	0.8	66	1.15	25	9	10	18.75	16.39
NO.10	0.8	66	1.15	10	16	29	33.33	47.54

4 Conclusion

This paper analyzes the vehicular behaviors at the traffic bottleneck based on the concept of target steering angle. And a three-lane experimental scene is set up to describe the vehicular behaviors.

According to the data collected by the video recognition technology, it's found that the lateral separation widely exists. Then the influencing factors and classification of vehicular behavior are proposed, and the target steering angle is introduced to describe the movement. Then the reasons for different behaviors are discussed. A series of rules are established for determining TSA by considering the lateral separation characteristics. And based on the analysis, a vehicular behavior decision-making process is set up. As for the experiment, video recognition technology is introduced to obtain the data to evaluate the decision-making process.

To our knowledge, the vehicular behaviors at bottlenecks are complex, while this paper proposed a relatively easy decision-making process. It is found out that affecting factors need to be studied further. And the data processing method should be improved.

Acknowledgments This work was supported by the Science and Technology Planning Project of Guangdong Province, China (NO.2014B010118002).

References

1. Gunay, B.: Car following theory with lateral discomfort. Transp. Res. Part B Methodol. **41**(7), 722–735 (2007)
2. Reuschel, A.: Fahrzeugbewegungen in der Kolonne. Oesterrich. Ingr. Arch. **4**, 193–215 (1950)
3. Pipes, L.A.: An operational analysis of traffic dynamics. J. Appl. Phys. **24**(3), 274–281 (1953)
4. Newell, G.F.: Nonlinear effects in the dynamics of car following. Oper. Res. **9**(2), 209–229 (1961)
5. Bando, M., et al.: Dynamical model of traffic congestion and numerical simulation. Phys. Rev. E Stat. Phys. Plasmas Fluids Relat. Interdiscip. Top. **51**(2), 1035–1042 (1995)
6. Helbing, D., Tilch, B.: Generalized force model of traffic dynamics. Phys. Rev. E **58**(1), 133–138 (1998)
7. Jiang, R., Wu, Q.S., Zhu, Z.J.: Full velocity difference model for a car-following theory. Phys. Rev. E **64**(1), 017101 (2001)
8. Wang, T., Gao, Z., Zhao, X.: Multiple velocity difference model and its stability analysis. Acta Physica Sinica **55**(2), 634–640 (2006). doi:10.3321/j.issn:1000-3290.2006.02.028
9. Jin, S., Wang, D., Tao, P., et al.: Non-lane-based full velocity difference car following model. Physica A **389**(21), 4654–4662 (2010)
10. Jin, S., Wang, D., Xu, C., et al.: Staggered car-following induced by lateral separation effects in traffic flow. Phys. Lett. A **376**(3), 153–157 (2012)
11. He, Z., Sun, W., Zhang, L., Xu, F., Zhuang, L.: Modeling and analysis of vehicular behavior at bottlenecks. Acta Physica Sinica **62**(16), 168901 (2013). doi:10.7498/aps.62.168901
12. Jin, S., Wang, D., Yang, X.: Non-lane-based car-following model with visual angle information. Transp. Res. Rec. J. Transp. Res. Board **47**(2249), 7–14 (2011)

Research on the Method of Driver's Eye Location Based on MATLAB

Derong Tan, Changmin Lv, Hongjia Zhang and Yang Yu

Abstract Accurate location of driver eyes has a positive meaning for eye state detection. This paper presents a new algorithm which is fusion of a mixed color model, integral projection and Prewitt algorithm. First, establishes a mixed color model to locate faces in color space, and finds face area through integral projection and formwork design based on MATLAB. Then eye area is detected through integral projection and a curve optimal. In the end, localizes eyes accurately using Prewitt algorithm and morphologic processing. The effectiveness of the algorithm is verified by the simulation analysis.

Keywords Eye location · Mixed color model · Integral projection · Prewitt edge detection

1 Introduction

With the development of the transportation industry, the phenomenon of fatigue driving is getting more serious. According to data released by the French National Police, Personal injury each year in traffic accidents due to fatigue driving account for 35.5 %. In China, each year the traffic accidents caused by fatigue driving account for 20 %, more than 40 % of serious traffic accident and accounting for 83 % in traffic accidents mortality [1]. Driver fatigue can be judged from the eye. Therefore, fast and accurate position of the driver's eyes has a great significance for analyzing the characteristics of the driver's eye and identification of the driver's condition.

Currently, national studies have proposed many methods for people's eye location, including Hough transformation, edge detection feature extraction method, the method of template matching and so on, but all the following methods

D. Tan (✉) · C. Lv · H. Zhang · Y. Yu
School of Transportation and Vehicle Engineering,
Shandong University of Technology, Zibo 255049, China
e-mail: tdrong163@163.com

© Springer Science+Business Media Singapore 2017
H. Lu (ed.), *Proceedings of the Second International Conference on Intelligent Transportation*, Smart Innovation, Systems and Technologies 53,
DOI 10.1007/978-981-10-2398-9_17

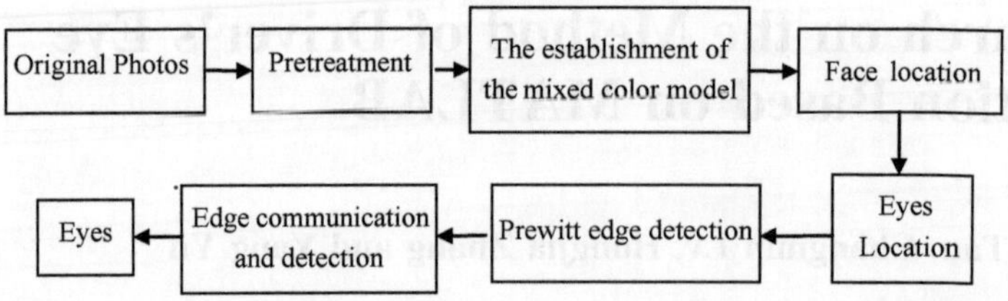

Fig. 1 The flow chart of eyes location

have their shortcoming [2]. This paper presents an idea which combines the model of mixed skin color, integral projection and Prewitt Edge detection together to achieve driver's eye quickly location. First, the collected color image pre-processing can support us to achieve the effect of noise reduction and image enhancement. Then determine color range based on the skin color model established in KL-YCBCR color space, and determine the face area by vertical integral projection and face template design. After the face region noise reduction, determine the human eye area by the use of horizontal integral projection and curve optimization. Finally, we use Prewitt operation to detect eye area, which is followed by the use of the edge region communication procession to achieve the eye precise location (Fig. 1).

2 Image Preprocessing

Image pretreatment was applied to improve the image definition and contrast Ratio. It can ensure accuracy of the eyes detection. The paper is based on continuous wavelet transform theory to handle pictures for the image pretreatment. The image is developed on separate hierarchy by the wavelet basis, and stopped according to the nature of the image and ideal standards. It can also distinguish component of details and approximate component. So the wavelet analysis is applied in image de-noising [3]. The formula (2) is definition formula of wavelet transform. Meanwhile, the wavelet analysis weighs time-domain processing and frequency-domain processing. So the wavelet analysis is also widely applied in image enhancement.

Wavelet basis function is based on the translation and expansion of basic wavelet:

$$\psi_{a,b}(x) = \frac{1}{\sqrt{a}}\psi\left(\frac{x-b}{a}\right) \tag{1}$$

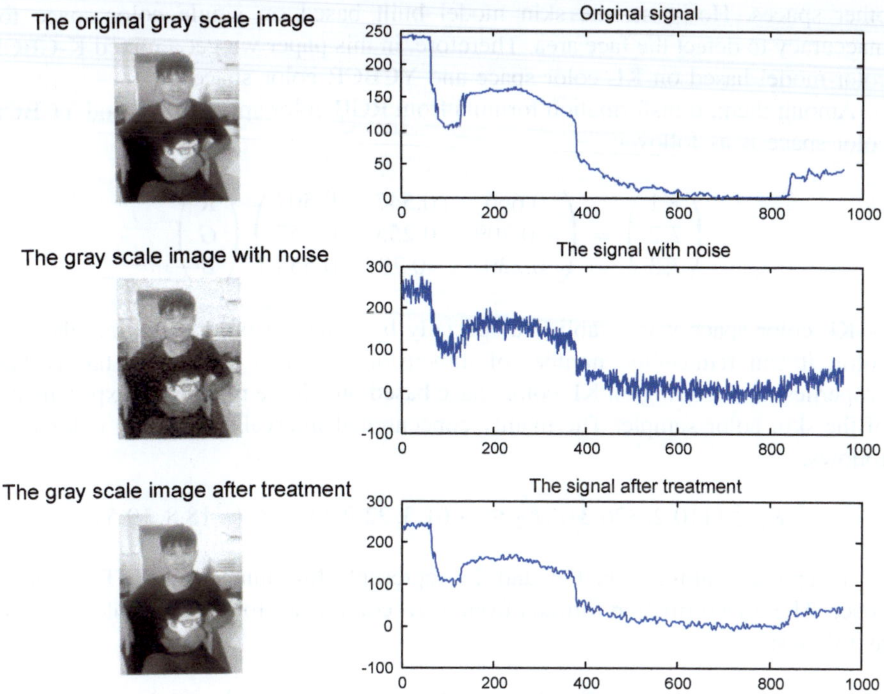

Fig. 2 The treatment process of wavelet de-noising

Continuous wavelet transform is integral wavelet transform, definition formula:

$$W_f(a,b) = \{f, \psi_{a,b}(x)\} = \int_{-\infty}^{+\infty} f(x)\psi_{a,b}(x)dt = \frac{1}{\sqrt{a}} \int_{-\infty}^{+\infty} f(x)\psi_{a,b}\left(\frac{x-b}{a}\right)dx \quad (2)$$

where, a is dilation factor, b is displacement factor.

The wavelet transform is good to improve image definition. Figure 2 is comparison of the gray scale image in noise reduction processing. The gray scale image with noise treated by wavelet transform is same as the gray scale image.

3 Established the Mixed Color Model

Its surroundings have little effect on the skin color. To establish the correct color model is the key for face detection. Skin color pixels have certain characteristics in different color space. Using this feature, a reasonable color model can be established and the face area is successfully segmented. The color spaces used in building skin color model include RGB, YCBCR, HIS, KL, YUV, YIQ, TSL and other spaces. However, the skin model built based on single color space too inaccuracy to detect the face area. Therefore, in this paper we get a mixed K-CBCR color model based on KL color space and YCBCR color space.

Among them, transformation formula from RGB color space to KL and YCBCR color space is as follows:

$$\begin{pmatrix} K1 \\ K2 \\ K3 \end{pmatrix} = \begin{pmatrix} 0.666 & 0.547 & 0.507 \\ -0.709 & 0.255 & 0.657 \\ 0.230 & -0.797 & 0.558 \end{pmatrix} \begin{pmatrix} R \\ G \\ B \end{pmatrix} \tag{3}$$

KL color space was established specially by others in order to extract the skin color, it can reduce the number of dimensions. Human skin color has certain properties of clustering on KL color space based on a large number of experimental of the skin color sample. The mainly concentrated interval of the skin color is as follows:

$$K_1 \in (110.2, 376.3) \cap K_2 \in (-61.3, 32.9) \cap K_3 \in (-18.8, 19.5)$$

Y represent luminance, CB and Cr represent chrominance in YCBCR color space. The transformation formula from RGB color space to YCBCR color space is as follows:

$$\begin{pmatrix} Y \\ Cb \\ Cr \end{pmatrix} = \begin{pmatrix} 0.29900 & 0.58700 & 0.11400 \\ 0.16874 & -0.33126 & -0.50000 \\ 0.50000 & -0.41869 & -0.08131 \end{pmatrix} \begin{pmatrix} R \\ G \\ B \end{pmatrix} \tag{4}$$

The statistics made by professor Anil K. Jain about skin color pixels over 85357 show that the skin color has also certain properties of clustering on YCRCB color space [4]. Figure 3 showed the distribution of skin color pixels in CBCR Plane.

Fig. 3 The statistics of skin color pixels

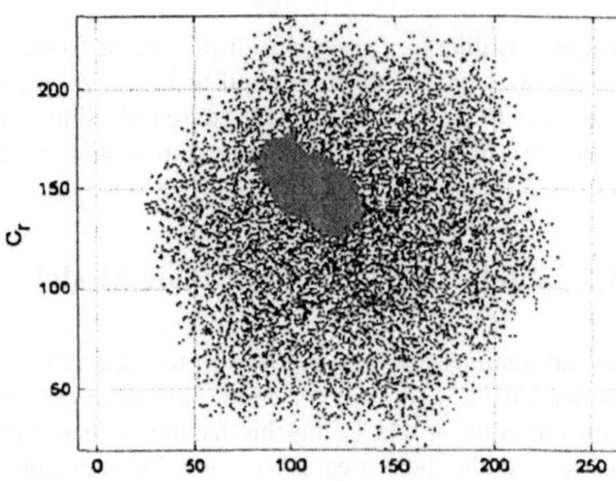

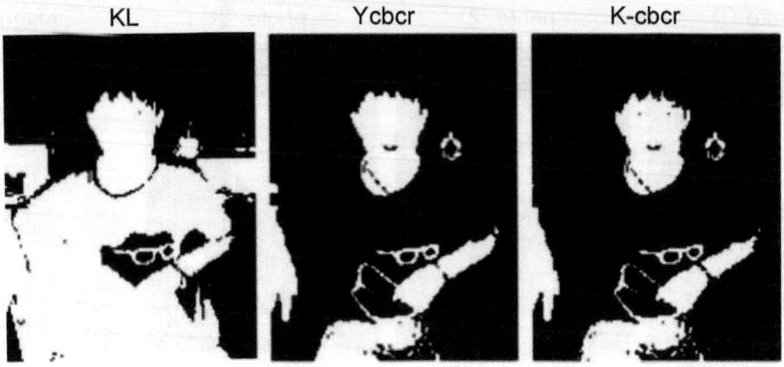

Fig. 4 The results comparison of three models

It is hard to detect out skin color range based on the skin color model established in single color space, such as if Y component and CBCR component in YCBCR color space cannot be separated completely, detection results will be influenced by image brightness. The mixed color model was established in this paper based on above two types of the color, and the detection effect is good, Fig. 4 showed the detection results of three models.

We can see from Fig. 4 that the detection result based on KCBCR model is better and the detection result based on KL model is poorest [5].

4 Eye Location

In last section, the mixed color model was established to detect skin color region of picture. Then the opening and closing operation and area extraction treatment have been applied to improve skin color region's smoothness and extract the larger the larger skin color region. And the face region was determined by vertical integration projection and template matching operation. To determine rough eyes location by horizontal integral projection and curve optimal operation as a basis for the precise eyes location [6, 7]. Finally, Prewitt algorithm was applied to locate eyes precisely.

4.1 Face Location

The center position of face was determined by vertical integration projection to reduce eyes' searching range. The calculation formula for the vertical integral projection is as follows:

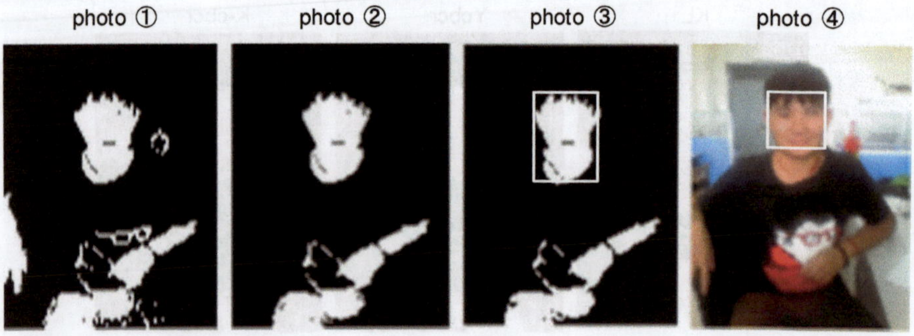

Fig. 5 The results of face location

$$H(x) = \frac{1}{x_2 - x_1} \sum_{x_1}^{x_2} G(x, y) \tag{5}$$

where, $G(x, y)$ is the gray value in coordinates (x, y).

According to a lot of integral projection curves of the driver images, the approximate location of human face could be found by Integral Projection. In order to further accurately determine the location of human face, the method combining integral projection and face matching was applied to detect human face. The method only needs the face position maximum value of driver image's vertical integral projecting. This location was proved close to face center line by many experiments.

By design temple, further screening image after de-noising, and determining the location of human face, the processes of screening are as follows: (1) the number of connected region is calculated, selecting a rectangle of the two larger connected region. (2) height-width ratio (h/a) of rectangle is calculated, selecting the location of the value less than 0.8 or more than 2. (3) the relative position between maximum abscissa of vertical integral projecting x_1 and the center of rectangular width x_2 is determined, selecting the rectangle of $|x_2 - x_1|/a < 0.3$.

Figure 5 shows the result of face location based on integral projection and face matching. Among them, photo ① is the result of the mixed color model, the effect of photo ② is better after opening and closing operation and area extraction treatment. Photo ③ and photo ④ are the final results after the template matching operation.

4.2 Rough Eyes Location

Analyzing the face edge image gets the facial region. The images noise reduction processing have been applied to improve the quality of the facial region and reduce the interference factors. The horizontal integral projection curve is determined according to gray level projection [8]. The calculation formula for the horizontal integral projection is as follows:

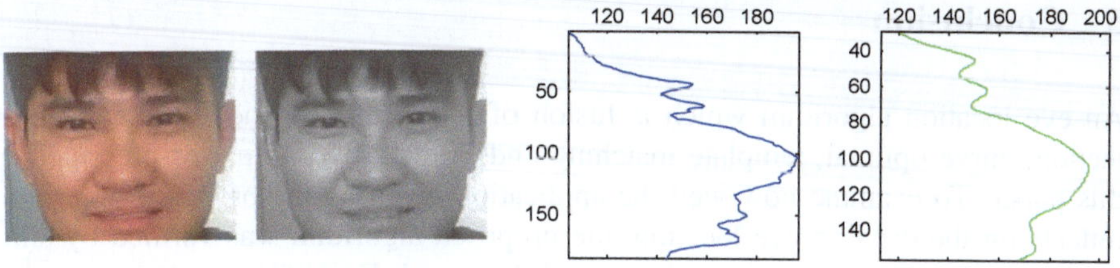

Fig. 6 Integral projection and curve optimal

$$V(y) = \frac{1}{y_2 - y_1} \sum_{y_1}^{y_2} G(x, y) \tag{6}$$

From the horizontal integral projection curve we can see that the eyes' location on face correspond with the valleys in the cycle curves, but the curve is unsmoothed, so computer cannot find the precise range of the valleys. Next we carried out an optimal treatment on the horizontal integral projection curve in order to ensure the Smoothing of curve. As the Fig. 6 show that it is easy to determine the extreme point of curve after an optimal treatment. Figure 8 showed the treatment process of the rough eyes location, it provide a basis for the precise eyes location.

4.3 *Precise Eyes Location*

Rough eyes area was determined in Sect. 4.2. Then the area was ashy in the imaging processing. And we process the gray Image by the Prewitt edge detection algorithm. In addition, the opening operation was used to determine the eye edge accurately, the method could improve detection effect of Prewitt algorithm. Next, we connect eye edge line to form the connected domain. And the two the maximum connected domain is eyes area. The algorithm was verified by 100 pictures from network based on the MATLAB platform, it was confirmed that the proposed method was capable of accurate detection, the precision achieved 93 %. Figure 7 showed the treatment process of the precise eyes location and the result measured by the proposed algorithm.

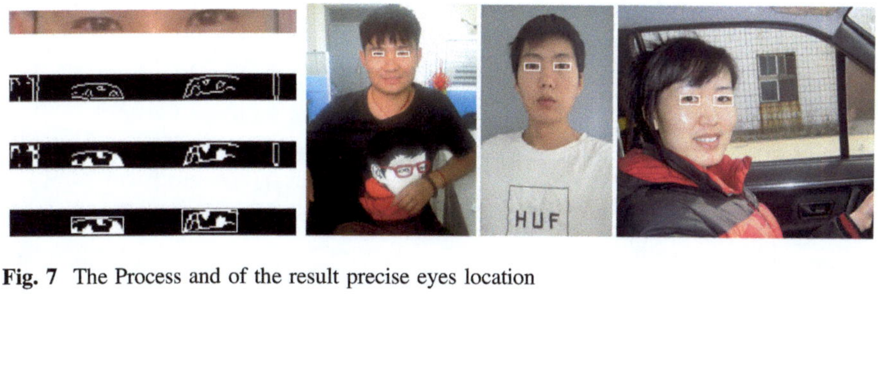

Fig. 7 The Process and of the result precise eyes location

5 Conclusion

An eye location algorithm which is fusion of a mixed color model, integral projection, curve optimal, template matching and Prewitt algorithm was presented in this paper. To examine how well the application given by the proposed algorithm reflects on the driver's eye location, the proposed algorithm was verified by simulation, and the results were analyzed and discussed. From the simulation results, our findings indicated that the algorithm can locate driver's eyes accurately and quickly, it have positive significance for the eyes' state detection and driver's fatigue discrimination.

References

1. Niu, Q.: Research on driver fatigue detection based on hybrid measures. Jilin University (2014)
2. Huang, X., Yang, W., Yang, H.: Eye location algorithm based on skin color model and gray projection. J. Chongqing Univ. Technol. (Natural Sci.) **12** (2014)
3. Dong, C., Lai, Z., Yu, X.: Matlab image processing and application. National Defence Industrial Press (2004)
4. Cui, J.: Research of fatigue driving detection algorithm based on eye detection. Dalian Maritime University (2013)
5. Zhang, Y.: Matlab image processing. Posts & Telecom Press (2014)
6. Ye, Q.: Research of fatigue detection system based on human eyes detection Guangdong. University of Technology University (2014)
7. Feng, Jianqiang, Liu, Wenbo, Shenglin, Yu.: Eyes location based on gray-level integration projection. Comput. Simul. **22**(4), 75–76 (2005)
8. Hu, Z.: Research of technique for driver' s fatigue supervising based on eye detecting. Beijing Jiaotong University (2009)

Catenary Poles Detection and Occlusion Reasoning for Electric Railway Infrastructure Visual Inspection

Peng Tang and Weidong Jin

Abstract Machine vision based automated inspection and monitoring for railway Infrastructures, such as pantograph overhead lines, is a promising technical trend to increase the efficiency and ease the manpower burdens. Vision data are naturally informative and comply with human sensing and cognition. However, automatically detecting and locate the infrastructure facilities from variance cluttered background in rail-lines inspection videos is still a challenging task due to the ill-pose essential. We propose a catenary poles and gantries segmentation framework by combining the appearance and motion patterns via a sequential Bayesian approach. The poles and supporting arms of power supply lines are detected firstly to yield the region of interest for detailed processing. Then the motion hypothesis of foreground and background are estimated from the edge flow extracted from local curve and line segments, so as to occlusion reasoning. After that, the pole model and background model are implemented to classify the candidates in hypothesis. Finally, promising experimental results demonstrate the potentials of the proposed poles segmentation method with respect to various insignificant patterns and cluttered backgrounds.

Keywords Railway catenary poles · Visual inspection · Occlusion reasoning

1 Introduction

In recent decades, railway transportation in China undergoes a great change in both running speed and operating intensity. Consequently, a challenge of railway network maintenance shall be faced, as the high-speed operation accelerates endurance

P. Tang (✉) · W. Jin
School of Electrical Engineering, Southwest Jiaotong University, Chengdu 610031, People's Republic of China
e-mail: tang.peng@home.swjtu.edu.cn

W. Jin
e-mail: wdjin@home.swjtu.edu.cn

© Springer Science+Business Media Singapore 2017
H. Lu (ed.), *Proceedings of the Second International Conference on Intelligent Transportation*, Smart Innovation, Systems and Technologies 53,
DOI 10.1007/978-981-10-2398-9_18

failures and raises defect risks. Therefore, the conditions of railway infrastructure are highly concerned and must be inspected frequently to guarantee operation safety and reduce maintenance costs. Rail defect detection related non-destructive testing (NDT) techniques have attracted much attentions from various researchers [1, 2] in recent years. To partially replace the traditional manual field-inspection, long distance visual surveillance has been introduced into practice, which highly facilitates the intelligent level of railway management. However, the extensive construction and expansion of high-speed railway results in so much surveillance video to be processed, which gradually exceed the capability of human inspection. In fact, humans are good at searching for the unusual and locating faults in variant environments, and also rapidly tire when large number of items have to be scrutinized. Therefore manual inspection is generally slow, subjective and inaccurate. The limitation of bare-eye monitoring has urged deeper study on automatic visual inspection for electric railway infrastructure. The goal of such system is utilizing computers to analyse the visual raw data, and to report rail-road facility defects and potential anomalies if detected, so as to maintain railway healthy effectively.

As visual data are compliant with human perception, video based inspection is considered as one of the most attractive techniques for complicated tasks. The mature of video hardware for capturing, transmission and storing has paced the way to widely application of video surveillance. In a visual inspection system, a high-speed digital camera that is installed in a operational train is used to capture sight views of a locomotive as the train moves along the track. Then, the obtained images are analysed automatically on-line or off-line by a specialized image processing software. Visual inspection has been developed with the great progress of computer vision techniques in recent years. However, visual data are implicit expression of real world scenes, so the inferencing is prone to fail in complex environments, such as cluttered background or occlusions.

This paper focused on the problem of distinguishing electric railway catenary supporting equipment, such as vertical catenary poles and portals, from various background images obtained from locomotive-mounted cameras. Given an sequence of surveillance images, algorithm initialization is achieved by estimating sparse motion fields on image edges. then interpolate the sparse edge flows into dense motion fields, and consequently refine the motions and background estimation. Our main contribution is in a railway-specific robust algorithm for both motion and foreground estimation in the presence of background disturbance, which is fully automatic and adaptive to variance environments.

2 Related Researches

Automated visual inspection systems are highly concerned by railway engineers and academic researchers [3]. As an informative non-contact sensing technique, the video cameras are used in various railway applications. A major branch of researches focus on the problem of measuring profiles or infrastructure of the rail tracks, such as discrete surface defects [2] of rail heads, rail damage in high-speed railway turnout [1], and ballast condition [4]. The pantograph-catenary system, as an critical component of train electrification, is another arising research area for vision assist automatic inspection [5], including the loosen of dropper, the abnormal of suspension and supporting device, the distance deviation of variant cables, the invasion of limit, abrasion on pantograph slide [6], etc. Besides the applications mentioned above, image camera are also implemented in other tasks, such as wheel inspection, locked wheel detection, brake pad thickness measurement, overheating of the breaking system and fire detection.

The image layer decomposition by motion parallax is another researching approach. Several papers proposed to automatically detect and remove foreground objects as obstructions from a batch of images. Mu et al. proposed to remove fence patterns based on visual parallax [7]. Xue extended this idea to a unified computational approach for taking photos through reflecting or occluding element [8].

3 Catenary Poles Detection by Motion Layer Decomposition

Since inspection video frames $I_1, \ldots, I_t, \ldots$ arriving consecutively, where $I_t \in \mathbb{R}^m$ is a composition of foreground and background layers due to the presence of catenary poles and gantries, to extract foreground objects is actually to update inference on an unknown parameter L online, where L is an alpha blending mask, which assigns a blending factor to each pixel, so $L(x) \in \{\mathcal{F}, \mathcal{B}\}$ represents the label of pixel $x = (x, y)$. The pixels with label $\mathcal{F}$ shall be further examined to analysis the equipment operating condition. The estimation of a confidence value of each pixel x that belongs to the foreground or the background can be calculated using sequential Bayesian approach, as

$$\begin{aligned} \Psi_t(x) &= P(L_t(x)|I_{1:t}(x)) \\ &= P(I_t(\mathrm{x})|L_t(\mathrm{x})) \sum_{L_{t-1}(x)} P(L_t(\hat{x})|L_{t-1}(x))\Psi_{t-1}(\hat{x}) \end{aligned} \tag{1}$$

where $\Psi_{t-1}(\hat{x})$ is the previous estimation. The pixel position $\hat{x} = W(x;\hat{p})$ comply with the field of expansion rule, where $W(\cdot, p)$ denote the parameterized set of allowed warps and $p = (p1, \ldots, p_n)^T$ is a vector of parameters. For perspective warps, there are eight parameters $p = (p_1, p_2, p_3, p_4, p_5, p_6, p_7, p_8)$. The warp defined by $W(x;p)$ maps the pixel $\mathbf{x}$ in the template coordinate to the sub-pixel location. So the goal of the frame warping is to minimize the sum of squared error between the current frame I_t and the previous frame I_{t-1} warped back onto the former coordinate frame:

$$\hat{p} = \arg\min_{p} \sum_{x} \sigma(L(x)) \| I_t(x) - I_{t-1}(W(x;p)) \|_2^2 \quad (2)$$

Decomposing a single frame I into the foreground layer I_f and the background layer I_b is ill-posed. However, since the foreground object is relatively closer to the camera than the background, the projected motion of foreground objects on the image plane will be larger than those of background objects due to the visual parallax. The additional constrains based on prior of decomposed images $\|\nabla L\|_1$ is included to reduce the ambiguity, and $\|\nabla \cdot\|$ are the gradients of the image. Finally, we use the maximum a posterior (MAP) strategy to search for the best estimation.

$$L = \arg\max_{L,p} (\log \Psi_t + \lambda \|\nabla L\|_1) \quad (3)$$

4 Problem Simplification and Optimization

The core problem is to optimization problem for recovering the mask components, L_t from an input image sequence, $\{I_t\}_{t=1}^{n}$. Some assumption will firstly be made to simplify the problem complexity.

4.1 *Railroad Manhattan Scene Model*

As the in-vehicle vision is abundant for inspection, a sketchy scene structure should be retrieved before detecting and identify objects, including the rail tracks and electric power transmission facilities, so as to filter out irrelevant observations. A typical simplification for man-made environments is considering the scenes as a Manhattan world [9], which assumes the foreground objects are embedded in planes with mutually orthogonal orientations. Furthermore, the inspection camera is pointing approximately to the vanish-point of straight rail tracks, and moving together with the locomotive alone rail line. So in the view field, all stationary objects shall appear to diverge from a vanish-point, which is called the *focus of expansion* (FOE for short). Thus, the image contains a FOE corresponding to locomotive moving direction.

The scene observed by locomotive camera is composed of the rail track on the ground plane, the infrastructure of electric power transmission that is perpendicular to the ground and various background, as in Fig. 1. Obviously, extracting railway region is fundamental. We use multi-scale Hough transform [10] to detect the tangent lines of rail tracks, and calculate the vanish point as the FOE without calibration or fundamental matrix. Given a image point x, its flow velocity and FOE comply with the time-to-adjacency rule [5]:

Fig. 1 Locomotive view content composition

$$\frac{\Delta r}{r} = c, \text{w.r.t} x = x_{FOE} + r \tag{4}$$

where Δr is defined as the distance from the FOE and c is a constant for points from the object with same depth.

4.2 *Motion Estimation by Edge Flow*

As the inspection videos are recorded continuously, each frame can be reasonably mapped to the coordinate system of consecutive frames according to reasonable small motion flows. However, due to the parallax phenomena, the motion magnitude of catenary poles pixels are notably larger, which provides the possibility to judge the catenary poles by distinguishing the saliently-moving pixels. Based on this phenomenon, the edge flow algorithm is used to estimate the sparse motion fields at each edge pixel identified in the image by Canny edge detector, as show in Fig. 2. The motion directions are determined by depth vanish point mentioned in Manhattan scene model, so the amplitude is the only parameter that should searched. As an observed image edge will generally belong to neither the foreground or the background layer, motion vectors estimated from pixels with salient gradients are generally more robust.

The sparse motion vectors obtained by edge flow are then separated into two sparse motion fields, one for each layer. Assuming the background pixels are more dominant, a perspective transformation to the sparse motion field is fitted firstly using RANSAC, and assign all the edge pixels that best fit this transformation to the background layer. Then, another perspective transformation to the rest of the edge pixels are fitted again using RANSAC, and assign the pixels best fitting the second transformation to the reflection layer. Finally, dense motion fields for both layers are calculated using visual surface interpolation.

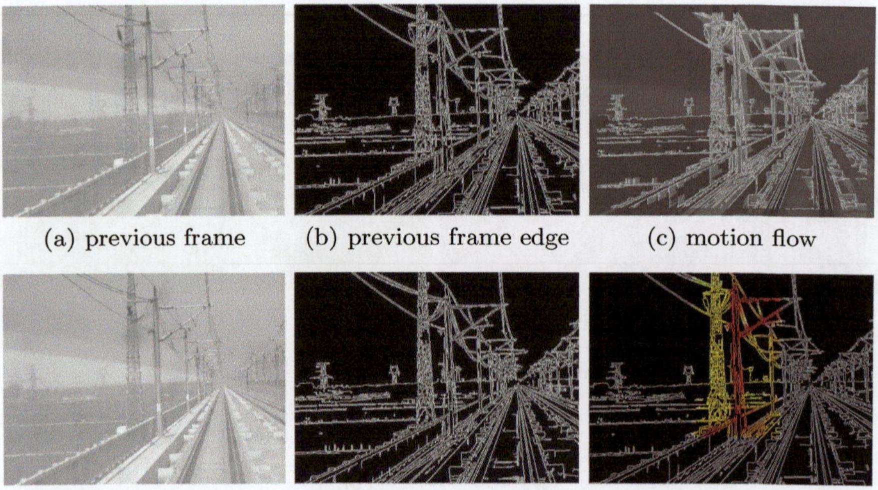

(a) previous frame (b) previous frame edge (c) motion flow

(d) current frame (e) current frame edge (f) pseudo color visualized in ROI

Fig. 2 Edge flow decomposition

4.3 Prediction and Updating

Since the model encompasses different properties of an object in a probabilistic way, which includes both geometrical constrain and objects with distinctive appearance. For each new observation, we first predict the warping by using previous state parameters, then update the estimation from the observations. Finally, we calculate the motion of detected edge pixels by solving a discrete Markov random field (MRF) with Graph Cuts algorithm [7].

5 Experiments

5.1 Datasets Preparation

To verify the effectiveness of our proposed framework, a set of experiments were carried out based on the image datasets collected from visual inspection project [11]. Image data were generated from a industrial camera with wide-angle lens mounted in the front of the electrical locomotive, which practically runs on the rail-lines in China. The bitmap colour images used in our experiments are of frame size 1620 $\times$ 1220 pixel, 24 bit pixel, frame rate of 5 frames per second, which includes challenging scenes such as disturbing background objects, waving branches, shadows, and weak boundaries etc (Table 1).

Table 1 Collected experimental datasets

	#Frames	Sequential	Marked	Light
Dataset1	28	Yes	No	Dark
Dataset2	33	Yes	No	Bright
Dataset3	42	No	Yes	Dark
Dataset4	43	No	Yes	Dark

5.2 *Experimental Results*

We demonstrate the sufficiency of proposed framework to segment pole structure from cluttered background. Our method is implemented in C++ codes under OpenCV 2.4.9 environment, and runs in a dual 2.6 GB Intel Xeon E5-2630v2 CPU processor machine using 32 GB of RAM. Optimizations may be performed to reduce hardware requirements in our future work so as to comply our algorithm with the need of real-time video surveillance.

By considering the motion prior that is determined by previously extracted FOE, the detection for pole structure is implemented as an basic processing step as shown in Fig. 3. The balance parameters are given as $\lambda = 0.2$. The search for image patch size for edge flow is up to 64 pixel in each pyramid. We validated the accuracy of

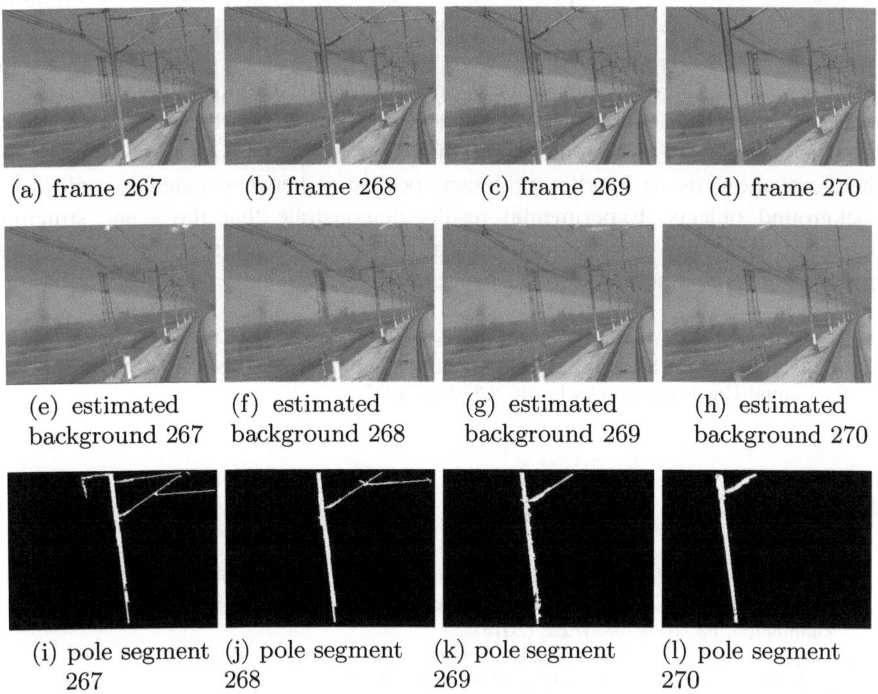

(a) frame 267 (b) frame 268 (c) frame 269 (d) frame 270

(e) estimated background 267 (f) estimated background 268 (g) estimated background 269 (h) estimated background 270

(i) pole segment 267 (j) pole segment 268 (k) pole segment 269 (l) pole segment 270

Fig. 3 Catenary pole segmentation and background estimation

Table 2 Nest twig detection recall and precision

		Dataset1	Dataset2	Dataset3	Dataset4
LK OptFlow	Precision	0.67	0.70	0.64	0.70
	Recall	0.75	0.72	0.76	0.66
Proposed	Precision	0.86	0.89	0.83	0.81
	Recall	0.87	0.91	0.88	0.75

the pole structure models by extracting the region of interest for a set of images in Dataset3. The pole-arm structure detection precision and recall on our collected experiment dataset are given in Table 2. It can be seen that due to the multiple miss detection, the performance of the motion detect by LK optical flow [12] can be considered unsatisfactory. Our algorithm outperformed compared with the object subtraction algorithm. Because of the use of the motion prior, the analysis was performed in a hyper feature space which results in a much fine detection outcome even in a high non-significant condition.

6 Conclusion

In this work, an implicit representation of railway catenary poles is formulated using sequential Bayesian model to capture the motion-appearance correlated essential. As a latent parameters, the pole model is introduced as the prior guidance to narrow down the detection scope. Consequently the feature hypothesis are collected by enhanced locally distinct patches. Finally the hypothesis are elaborately classified based on appearance and motion jointly. The resulting framework therefore allow us to handle significant occlusion between catenary poles and background objects. Experimental results demonstrate that the scene structure enhanced Bayesian detector outperform classical motion segmentation in the presence of large amounts of disturbances and variance.

Acknowledgments The work described in this paper was supported by the National Natural Science Foundation of China Project under Grant 61134002 and the Fundamental Research Funds for the Central Universities under Grant 2682014CX027.

References

1. Chen, R., Wang, P., Xu, H. Integrated monitoring system for rail damage in high speed railway turnout. In: 2013 Fourth International Conference on Digital Manufacturing and Automation, pp. 704–708. IEEE (2013)
2. Li, Q., Ren, S.: A real-time visual inspection system for discrete surface defects of rail heads. IEEE Trans. Instrum. Meas. **61**(8), 2189–2199 (2012)

3. Li, B., Tian, B., Li, Y., Xiong, G., Zhu, F.: A vision-based serial number recognition algorithm for HSR trains by nearest neighbor chains of connected components. In: IEEE International Conference on Vehicular Electronics and Safety, pp. 36–41. IEEE (2013)
4. Labarile, A., Stella, E., Ancona, N., Distante, A.: Ballast 3D reconstruction by a matching pursuit based stereo matcher. In: IEEE Intelligent Vehicles Symposium, pp. 653–657. IEEE (2004)
5. Aydin, I., Karakose, M., Akin, E.: A robust anomaly detection in pantograph-catenary system based on mean-shift tracking and foreground detection. In: 2013 IEEE International Conference on Systems, Man, and Cybernetics, pp. 4444–4449. IEEE (2013)
6. Ma, L., Wang, Z.-Y., Gao, X.-R., Wang, L., Yang, K.: Edge detection on pantograph slide image. In: 2nd International Congress on Image and Signal Processing, pp. 1–3. IEEE (2009)
7. Mu, Y., Liu, W., Yan, S.: Video defencing. Circ. Syst. Video Technol. IEEE Trans. **24**(7), 1111–1121 (2014)
8. Xue, T., Rubinstein, M., Liu, C., Freeman, W.T.: A computational approach for obstruction-free photography. ACM Trans. Graph. (TOG) **34**(4), 79 (2015)
9. Delage, E., Lee, H., Ng, A.Y.: Automatic single-image 3D reconstructions of indoor manhattan world scenes. In: Robotics Research, pp. 305–321. Springer (2007)
10. Block, M., Rojas, R.: Local Contrast Segmentation to Binarize Images, pp. 294–299. IEEE (2009)
11. Transport power supply department of MOR, China Academy of Railway Sciences, Southwest Jiaotong University. In: The High Speed Railway Power Supply Safety Inspection and Monitoring System (6C) General Technical Specification, 1st edn. July 2012
12. Baker, S., Matthews, I.: Lucas-kanade 20 years on: a unifying framework. Int. J. Comput. Vision **56**(3), 221–255 (2004)

Detection and Tracking of Vehicles Based on Video and 2D Radar Information

Min Wang, Liangwei Jiang, Wenjie Lu and Qing Ma

Abstract For practical application, over-speed behavior of a vehicle is generally detected by the combination of the radar speedometer and the camera capture. Little information, just a single image of the vehicle and the velocity is recorded as the evidence. If other vehicles are in the sight of the camera, this method cannot identify which one the measured velocity belongs to. The velocity is even inaccurate due to the influence. Focusing on this problem, we introduce a detection and tracking system of vehicles based on the combination of the video and 2D radar information. Multiple vehicles in the sight of the camera are detected and tracked, whose radar information is calibrated and embedded into the image coordinate system. The trajectory and velocity of each vehicle is recorded, and the over-speed judgment of multi-target can be solved. The system is evaluated in the experiments and the practical application. The results show that the proposed fusion algorithm of the video and radar information has a good performance on the detection, tracking and velocity measurement of vehicles.

Keywords Traffic video · Video monitoring · Vehicle detection · Vehicle tracking · Multi-target radar

M. Wang (✉) · L. Jiang · W. Lu
Key Laboratory of Ministry of Public Security for Road Traffic Safety, Wuxi, China
e-mail: wangmin_tmri@126.com

L. Jiang
e-mail: jlw18@126.com

W. Lu
e-mail: luwenjie0122@msn.cn

Q. Ma
Traffic Management Research Institute of Ministry of Public Security, Wuxi, China
e-mail: 13338110260@126.com

© Springer Science+Business Media Singapore 2017

H. Lu (ed.), *Proceedings of the Second International Conference on Intelligent Transportation*, Smart Innovation, Systems and Technologies 53,
DOI 10.1007/978-981-10-2398-9_19

1 Introduction

Over-speed vehicle capture system is applied widely in China, which is an important evidence collecting method of the over-speed behavior of vehicles. Generally, the over-speed vehicle capture system is applied based on the mechanism of loop-camera [1], radar-camera [2], virtual loop-camera [3], or interval velocity measurement [4]. The method of loop-camera calculates the vehicle velocity based on the travel time between the two neighbored loops. The measuring accuracy is high, but the loop is always fragile and not easily installed. Radar-camera combination is worked duo to the Doppler effect, which can be used for the calculation of the vehicle. Radar sensor is worked reliably and can be fixed at the road side conveniently. The method of virtual loop-camera builds up virtual loops in the image coordinate system, and similar mechanism of velocity measurement to the loop-camera is applied. The advantage of this method is that just a camera can finish the velocity measurement and capture the frame reliably, and the disadvantage is the complicated calibration of the camera. Also, the accuracy of the velocity measurement depends on the performance of the applied image processing algorithm. The system of the interval velocity measurement calculates the average velocity of the vehicle through a certain road section. The accuracy of the system is high, but it has to be installed on two checkpoints for a high cost.

Considering the accuracy, performance-price ratio, reliability and maintainability of the velocity measurement, the method of radar-camera have certain advantages. Also, it is a widely used over-speed vehicle capture method on the highway. However, the general narrow-beam radar applied on single-lane is always influenced by the vehicle on the lane beside, and the measured velocity is inaccurate. Hence, it is hard to solve the problem of the velocity measurement for multi-target in the high traffic flow. Currently, a new kind of 2D radar is developed in the transportation industry. This kind of radar, with a beam covering several traffic lanes, can obtain the velocity and location information of multiple vehicles on the road based on the feature analysis of time delay and frequency shift of the echo signal. However, the result of radar sensor is not visualized for the manager and is also difficult to understand. It should be integrated with the captured image, and then can be treated as the evidence of the over-speed behavior.

Therefore, we try to build up a velocity measurement system of vehicle based on the video and 2D radar information. The localization of the vehicle target is integrated in the coordinate systems of the camera and radar based on the coordinate calibration, and the accurate location and velocity is obtained, which can be used to handle the multi-vehicle velocity measurement problem. On the other side, a vehicle tracking mechanism is built up based on the analysis of the video and radar information. The trajectory of each vehicle target is tracked and recorded in real-time, which can be used for further analysis of the driving behavior.

2 System Design

Due to the development of the technology of pattern recognition, image processing and sensor, the automation and intelligence of the traffic monitoring device is growing, which relays on the high requirement of computational capability and performance of the hardware. In order to relieve the pressure on the computation, network and storage of the system from the signal processing and data transmission, the embedded processing technology begins to be applied in the traffic monitoring device. In this paper, we also proposed the velocity measurement method with the fusion of video and radar information based on the embedded processing technology. Most computation is completed in the device end, and the host computer just need to receive the video flow and the recognition result for illustration. The operating load of the host computer is reduced and the system is also easy to be installed.

The proposed system consists of the intelligent camera, multi-target radar and hosted computer. The intelligent camera is integrated with an embedded processing platform, including CCD sensor, Field Programmable Gate Array (FPGA), H. 264 encoder, Digital Signal Processing (DSP), RAM, FLASH and other peripheral device such as digital IO, COM and Ethernet. The multi-target radar is connected to the camera by RS485, and the hosted computer uses TCP/IP to communicate with the camera.

The functional structure of the system is shown in Fig. 1. The FPGA drives the image sensor, captures and pre-process the image data, which will be transferred to the H. 264 encoder for video compress, or to the DSP for further processing. Radar sensor transfers the radar information to the Universal Asynchronous Receiver/Transmitter (UART) for decoding, and the vehicle velocity and position is obtained. The image data acquisition in the DSP caches the image data and provides the image format conversion. This image data can be further compressed by the

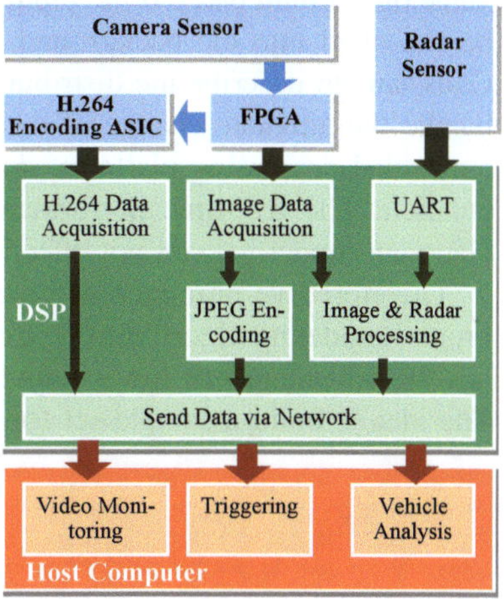

Fig. 1 The functional structure of the proposed vehicle monitoring system based on the video and radar information

JPEG encoder. The image and radar processing is the key part of the system for realizing the algorithm function. It is used to integrate the video and radar information, and obtain the current position, trajectory and velocity of the vehicle in the image. The compressed video/image data, analysis result and other control command is send and received by the network. The hosted computer receives the video flow, image date and analysis result for illustration and recording.

The image and radar processing detects the vehicle target in the image, and integrates the radar information for accurate vehicle localization, velocity measurement and trajectory tracking based on the space calibration.

3 Video Detection of Vehicles

Due to the development of the technology of pattern recognition, image processing and sensor, the automation and intelligence of the traffic monitoring device is growing, which relays on the high requirement of computational capability and performance of the hardware. In order to relieve the pressure on the computation, network and storage of the system from the signal processing and data transmission, the embedded processing technology begins to be applied in the traffic monitoring device. In this paper, we also proposed the velocity measurement method with the fusion of video and radar information based on the embedded processing technology. Most computation is completed in the device end, and the host computer just need to receive the video flow and the recognition result for illustration. The operating load of the host computer is reduced and the system is also easy to be installed.

Extracting vehicles from the road image is a challenge of the system. The most widely used object detection method for monitoring cameras is the background modeling. Single Gaussian [5] and mixtures of Gaussians [6] models are always applied. Sing Gaussian model uses a Gaussian distribution to simulate the possible value of a certain pixel. If the input satisfies this model, the corresponding pixel will be classified into the background. Model of mixtures of Gaussians uses multiple Gaussians to describe the distribution of the pixel value. This model is suitable to handle the alternate change of the pixel. However, these methods do not suit the embedded processing platform, duo to the high requirement of floating-point calculation. The real-time performance will be low in the system, and the single Gaussian model is even not effective for the road environment.

Therefore, we introduce the vehicle detection method based on the Vibe background modeling algorithm [7]. Compared with the Gaussian models, this method has the advantages of low computation complexity and accurate detection result. The idea is build a sample set for each pixel, where the samples are the previous values of the pixel or those of its neighbors. The input new value of a certain pixel is compared in the sample set to determine whether it belongs to the background.

In the initialization of the background model, general methods need a period of time to count the values of the pixel for the sampling. However, the proposed method just needs a single frame for this procedure. The sample set $\{v_1^i, v_2^i, \ldots v_m^i\}$ of pixel x^i is filled by the values of the pixels in the neighborhood Ω. It is because that the values of the pixels adjacent have the similar spatial-temporal distributions. It is suitable to use the spatial distribution to approximate the temporal distribution of the pixel value.

In the updating of the background model, the method updates the samples based on counting and randomness. If a pixel x^i is continuously classified as foreground for more than N_0 times, it will be a background pixel, and will replace a sample in the sample set of its own or its neighbors with the probability of $1/\phi$. Hence, the probability of an arbitrary sample in this model can be survived during the time interval dt is,

$$P(t, t+dt) = \left(\frac{N_0 - 1}{N_0}\right)^{(t+dt)-1} = \exp\left\{-\ln\left(\frac{N_0}{N_0 - 1}\right) \cdot dt\right\},$$

which is independent to the system time t. Therefore, this updating is stable and desirable.

The procedures of the algorithm are detailed below.

Initialization	Background modeling. Sample set $\{v_1^i, v_2^i, \ldots v_m^i\}$ of pixel x^i is randomly sampled in the neighborhood Ω
Step 1	Foreground extraction. Calculate the distances between the value v_{new}^i of pixel x^i with $\{v_1^i, v_2^i, \ldots v_m^i\}$, and count the number n^i of the distances smaller than the threshold. If $n^i \geq n_0$, x^i will belong to the background; otherwise, the foreground
Step 2	Model updating. Count the number N of times of x^i continuously classified as foreground. If $N \geq N_0$, v_{new}^i will randomly replace a sample in $\{v_1^i, v_2^i, \ldots v_m^i\}$ with a probability of $1/\phi$, it also will replace a sample of its neighbor similarly
Step 3	Morphological operation. Make several times of expansion-corrosion to the foreground image, and remove the discrete foreground pixels
Step 4	Output the results. Calculate the size of each foreground area. If the size is larger than the threshold, the corresponding area will be output as the vehicle position

The proposed vehicle detection method based on Vibe can extract the foreground area from the monitoring video accurately in Fig. 2b. And the whole vehicle area can be found based on the morphological operation in Fig. 2c.

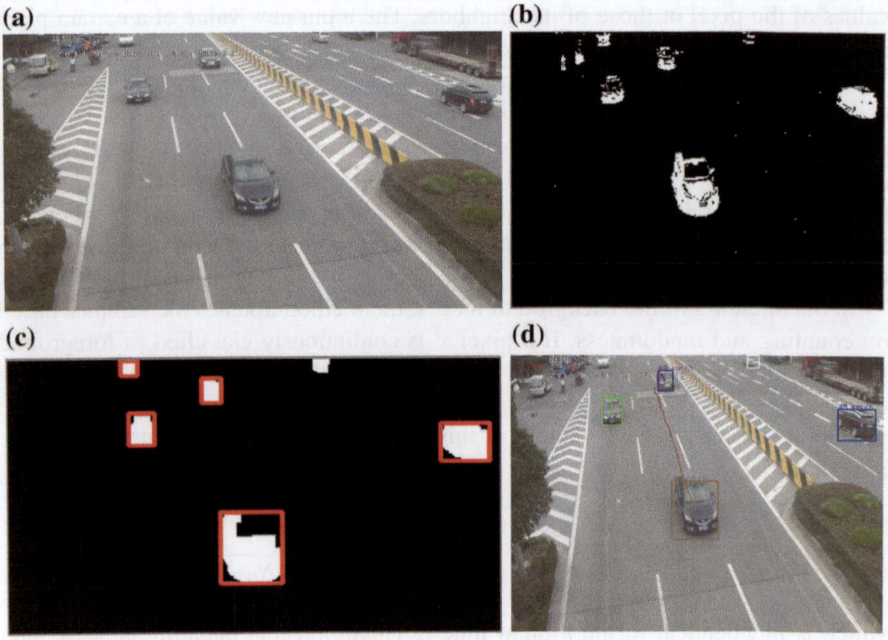

Fig. 2 The vehicle detection in the image. **a** is the original frame in the video. **b** illustrates the extracted foreground area. **c** illustrates the extracted vehicle areas marked *red rectangles*. **d** is the detection result

4 Radar Information Calibration

The vehicle position obtained by multi-target radar needs to be mapped to the vehicle area in the video. Then the radar information will help to handle the occlusion of the vehicle appearance in the video, and the video detection result will help to find the false-positive target caused by the radar clutter. This processing is realized based on the calibration of the radar information in the coordinate system of the camera.

The multi-target radar we used has transformed the vehicle coordinate from the coordinate system $O_1X_0Y_0Z_0$ of radar to that of the road (O_1XYZ). These two coordinate systems may have angles on the directions perpendicular to the *x*- and *y*-axis due to the installation of radar. The target position is denoted by the distances (D_h and D_v) between the vehicle and the radar on the directions perpendicular to or along the road respectively. So the target position P_{O_1XYZ} in O_1XYZ is (D_h, H, D_v), where H is the installation height of radar.

As shown in Fig. 3, we should further transform the target position from O_1XYZ to the coordinate system of the camera. This procedure can be divided into several steps.

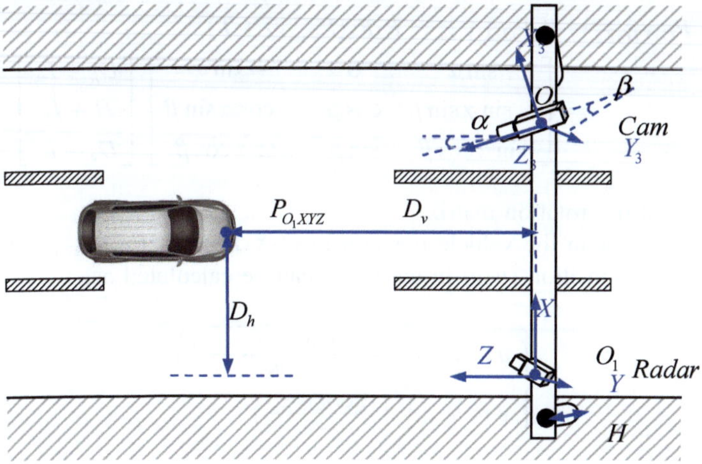

Fig. 3 The coordinate of the vehicle in the coordinate system of the radar and that of the camera

First, shift the coordinate system from O_1XYZ to $O_2X_1Y_1Z_1$, the vehicle position is changed to,

$$P_{O_2X_1Y_1Z_1} = P_{O_1XYZ} + T = \begin{bmatrix} D_h \\ H \\ D_v \end{bmatrix} + \begin{bmatrix} L_x \\ L_y \\ L_z \end{bmatrix},$$

where T is the shift matrix, and L_x, L_y and L_z are the distances between the camera and radar in the three directions of the world coordinate system.

Then, considering the installation angle α of the camera on the horizontal direction, we rotate the coordinate system $O_2X_1Y_1Z_1$ on the $X_1O_2Z_1$ plane, and obtain,

$$P_{O_2X_2Y_2Z_2} = R_1 P_{O_2X_1Y_1Z_1} = \begin{bmatrix} \sin\alpha & 0 & -\sin\alpha \\ 0 & 1 & 0 \\ \sin\alpha & 0 & \cos\alpha \end{bmatrix} \begin{bmatrix} D_h + L_x \\ H + L_y \\ D_v + L_z \end{bmatrix},$$

where R_1 is the rotation matrix.

Further, considering the installation angle β of the camera on the vertical direction, we rotate the coordinate system $O_2X_2Y_2Z_2$ on the $Y_2O_2Z_2$ plane and obtain,

$$P_{O_2X_3Y_3Z_3} = R_2 P_{O_2X_2Y_2Z_2} = R_2 R_1 (P_{O_1XYZ} + T)$$

$$= \begin{bmatrix} \sin\alpha & 0 & -\sin\alpha \\ -\sin\alpha\sin\beta & \cos\beta & -\cos\alpha\sin\beta \\ \sin\alpha\cos\beta & \sin\beta & \cos\alpha\cos\beta \end{bmatrix} \begin{bmatrix} D_h + L_x \\ H + L_y \\ D_v + L_z \end{bmatrix},$$

where R_2 is also a rotation matrix.

At last, we obtain the vehicle position in $O_2X_3Y_3Z_3$ as $P_{O_2X_3Y_3Z_3} = (x_3, y_3, z_3)$, then the vehicle position (u, v) in the video can be calculated as,

$$\begin{cases} u = w \cdot \left(\frac{1}{2} + \frac{\arctan\left(x_3/\sqrt{y_3^2 + z_3^2}\right)}{a_w} \right) \\ v = h \cdot \left(\frac{1}{2} + \frac{\arctan(y_3/z_3)}{a_h} \right) \end{cases},$$

where a_w and a_h are the horizontal and vertical view angles of the camera, and w, h are the horizontal and vertical resolutions of the video frame, respectively. Therefore, we find the correspondence between the video detection results and radar information. If two radar targets are mapped to the same vehicle image area, we will divide the image area as two occluded vehicles. If none vehicle image area is existed in the mapped coordinate of the radar target, we will ignore it.

5 Experimental Result

We applied the proposed vehicle detection and tracking system based on the video and radar information on the road checkpoint for evaluation. 3 lanes are covered by the system for vehicle localization, velocity measurement and tracking. A complete software interface is realized on the hosted computer, including lane calibration, trajectory indication and some behavior recognition, such as over-speed, illegal parking, go in the wrong lane and so on.

The system can stable running for a long time without any shutdown or reboot. The speed of the video processing is more than 21 fps, which satisfies the real-time requirement. The localization, velocity measurement and tracking results are shown in Fig. 4. The marked speeds in the images are from the radar information. The results show that the proposed system can solve the problem of the multi-vehicle monitoring on the road.

Fig. 4 The localization, velocity-measurement and tracking results of the vehicle on road. The *yellow quadrangles* are used to mark the traffic lanes

6 Conclusion

This paper introduces a vehicle monitoring system based on the integration of the video and radar information, which can detect and track multiple vehicles on the road. The velocity measurement problem of the multi-vehicle can be solved, and the completeness and continuity of the vehicle monitoring is improved. This paper focuses on the video detection of the vehicle and the radar information calibration methods based on an embedded processing platform. The video detection result and the radar information are integrated with verification. The accuracy of the localization and velocity measurement is improved, which is significant for practical application.

Acknowledgments The work of this paper was supported in part by the National Natural Science Foundation of China under Grant 61303174, the National Key Technology Program of China under Grant 2014BAG01B03, and the China Postdoctoral Science Foundation funded project under Grant 2015M571648.

References

1. Ki, Y.K., Baik, D.K.: Model for accurate speed measurement using double-loop detectors. IEEE Trans. Veh. Technol. **55**(4), 1094–1101 (2006)
2. Gupte, S., Masoud, O., Martin, R.F.K., Papanikolopoulos, N.P.: Detection and classification of vehicles. IEEE Trans. Intell. Transp. Syst. **3**(1), 37–47 (2002)
3. Yan, Y., Shi, Y.C., Ma, Z.Q.: Research on vehicle speed measurement by video image based on Tsai's two stage method. In: Proceedings of 5th International Conference on Computer Science and Education (2010)
4. Wang, M., Ma, Q.: Dynamic prediction method of route travel time based on interval velocity measurement system. In: Proceedings of IEEE International Conference on Service Operations and Logistics, and Informatics, Qingdao (2014)
5. Wren, C.R., Azarbayejani, A., Darrell, T., Pentland, A.P.: Pfinder: real-time tracking of the human body. IEEE Trans. Pattern Anal. Mach. Intell. **19**(7), 780–785 (1997)
6. Stauffer, C., Grimson, W.E.L.: Adaptive background mixture models for real-time tracking. In: Proceedings of IEEE Computer Society Conference on Computer Vision Pattern Recognition (1999)
7. Barnich, O., Droogenbroeck, M.V.: ViBe: a universal background subtraction algorithm for video sequences. IEEE Trans. Image Process. **20**(6), 1709–1724 (2011)

Research on Route-Choice Behavior of Unexpected-Destination Trip Under Random Dynamic Conditions

Xiao-yuan Wang, Jing-lei Zhang, Hai-bo Wang, Chao Yin and Cui-cui Yu

Abstract The study of route-choice behavior has been an important part of the study of urban-road transport system. In previous studies, the emergencies of the road network have been studied mostly from different angles under random dynamic conditions. But the condition of route-choice behavior of unexpected-destination trip was rarely mentioned. So in this paper, "unexpected radiation field" is put forward, at the same time, the characteristic of it is also studied. On the basis of this, route-choice behavior of unexpected-destination trip under random dynamic conditions is researched with a combination of "radiation field method" and discrete dynamic Bayesian network. Studies display that the location and important of the unexpected destination affect the route choice behavior closely. The simulation results show that, the model had a positive meaning, and enabled to simulate and reproduce the decision-making process more appropriately.

Keywords Driver behavior · Route choice · Radiation field · Discrete dynamic bayesian network · Uncertain inference · Unexpected-destination

1 Introduction

In real life, Drivers often face the condition of multi route. The drivers need to choice a route from multi route to complete the trip, this is the problem of route-choice. According to the quantities of destination, the drivers' trip has 2 ways, single destination trip and multi destination trip. According to the different destination the drivers decided and arrive, the multi destination trip includes expected trip and unexpected trip. The expected trip is the destination that drivers want to arrive, it includes original destination and derived destination. The unexpected

X. Wang (✉) · J. Zhang · H. Wang · C. Yin · C. Yu
School of Transportation and Vehicle Engineering, Shandong University of Technology, Zibo 255049, China
e-mail: wangxiaoyuan@sdut.edu.cn

© Springer Science+Business Media Singapore 2017

H. Lu (ed.), *Proceedings of the Second International Conference on Intelligent Transportation*, Smart Innovation, Systems and Technologies 53,
DOI 10.1007/978-981-10-2398-9_20

destination means it has a new destination for some reason, such as the husband is asked to buy some vegetables by the wife on the way home etc. The unexpected destination is different from the derived destination, the derived destination means during the trip to expected destination, it has derived another destination. The study of route-choice behavior has been an important part of the study of urban-road transport system, it is also very important for the problems such as the traffic assignment and relieving traffic congestion etc.

The urban traffic environment is random dynamic, such as the time-varying traffic flow, accidental traffic accident, unexpected traffic jam, all of these can cause the random trip time [1]. In study of the random and dynamic part, some scholars such as Wang zheng wu studied the topic "the multi destination and route-choice from the uncertain condition", they analyzed the uncertain attributes of the road and the route-choice features of the drivers, and they established the route-choice model on the basis of the prospect theory and the improved CSA [2]. Some scholars such as Tang Liansheng studied the range and background of the traffic route issues from the unexpected condition [3]. Some scholars such as Guo Hanying studied the reliability of the road network from the unexpected condition, and discussed the relationship between the traffic flow and runtime in terms of the relationship between the traffic capacity and runtime [4]. Some scholars such as Feng Peiyu put forward the traffic-evacuation strategy for the unexpected situation on different influencing factors and range, on the basis of the features of the urban traffic [5].

In previous studies, the emergencies of the road network have been studied mostly from different angles under random dynamic conditions. But the condition of route-choice behavior of unexpected-destination trip was rarely mentioned. So in this paper, "unexpected radiation field" is put forward to analyzed the influence of the unexpected destination over the expected destination in terms of the drivers' trip destination, and route-choice behavior of unexpected-destination trip under random dynamic conditions is researched with a combination of "radiation field method" and discrete dynamic Bayesian network.

2 Dynamic Bayesian Network and Inference

2.1 Dynamic Bayesian Network Model

Dynamic Bayesian network is a directed acyclic graph version, each gridding node has directional and loop-free node. It is a relatively static graphic mode, the connection between nodes, conditional probability, is the basis of the inference. The probability estimate for any nodes of the Bayesian network has to consider the nodes aggregation before X(father node F) and after X(child node C) [6–8]. But the acyclic Bayesian network did not consider the influence of time on variables, the Bayesian network varying with time is called dynamic Bayesian network(Dynamic Bayesian Networks. DBN) [9–11]. Dynamic Bayesian network is a novel and flexible diagram

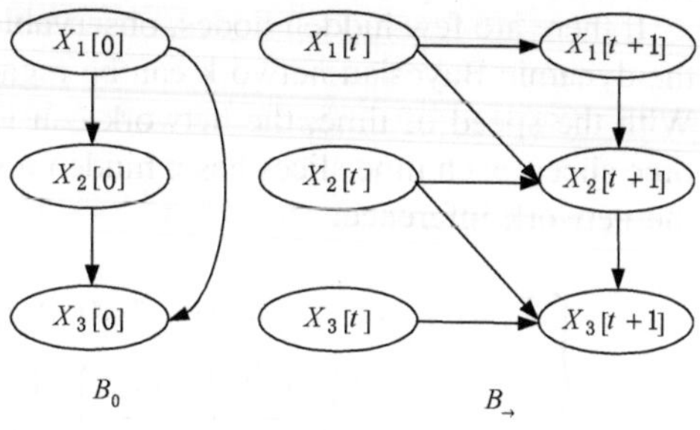

Fig. 1 Dynamic bayesian network structure diagram

expression for the state-space model with lots of random process. It has 2 Bayesian network: initialized network B_0 and transfer network $B_{\rightarrow}$. B_0 is the initializing probability distribution in this process $p(x[0])$. $B_{\rightarrow}$ is the state transition probability $p(x[t]|x[t-1])$ for the time points 0, 1, …, t, from Time t −1 to Time t. The graphic mode is as the Fig. 1.

2.2 *Model Dynamic Inference*

The basis of the Bayesian network and dynamic Bayesian network inference is Bayesian formula:

$$p(x|y) = \frac{p(yx)}{p(y)} = \frac{p(yx)}{\sum_x p(yx)} \tag{1}$$

Dynamic Bayesian network is the extension of Bayesian network in the space of time sequence, the essence of the inference is the same as the acyclic Bayesian network. The discrete acyclic Bayesian network with n hidden terminal and m observed terminal applies the independent condition of the Bayesian network to get the computing formula of the inference:

$$P(x_1, x_2, \ldots, x_n | y_1, y_2, \ldots, y_m) = \frac{\prod_j P(y_j|pa(Y_j)) \prod_i P(x_i|pa(X_i))}{\sum_{x_1 x_2 \cdots x_n} \prod_j P(y_j|pa(Y_j)) \prod_i P(x_i|pa(X_i))}, \tag{2}$$
$$i = 1, \ldots, n; j = 1, \ldots, m$$

In this computing formula, x_i is a value of X_i. $pa(Y_j)$ is the father node group of Y_j, in the $\sum$ of the denominator, $x_1, x_2, \ldots, x_n$ is a hidden node group, which means the joint distribution sum for the observation variables group and the hidden variables group, actually it is used to calculate the distribution of the certain observation variables group.

If there are few hidden nodes, observable nodes, level and time, the time slice in the dynamic Bayesian network can be regarded as a big acyclic Bayesian network. With the speed of time, the network can get a dynamic Bayesian network with T time slices, each time slices has n hidden nodes and m observable nodes, and here is the network inference:

$$\left\{\begin{array}{l} P(x_{11},\ldots,x_{1n},\ldots,x_{T1},\ldots,x_{Tn}|y_{11_0},\ldots,y_{1m_0},\ldots,y_{T1_0},\ldots,y_{Tm_0}) = \\ \sum_{y_{11},y_{12},\ldots,y_{Tm}} \frac{\prod_{i,j} P(y_{ij}|pa(Y_{ij}))\prod_{i,k} P(x_{ik}|pa(X_{ik}))\prod_{i,j} P(Y_{ij_0}=y_{ij})}{\sum_{x_{11},x_{12},\ldots x_{T1},\ldots,x_{Tn}}\prod_{i,j} P(y_{ij}|pa(Y_{ij}))\prod_{i,k} P(x_{ik}|pa(X_{ik}))}, \end{array}\right. \quad (3)$$

$$i=1,\ldots,T, j=1,\ldots,m, k=1,\ldots,n$$

In this computing formula: x_{ij} is a value of X_{ij}, the first subscript stands for the i,time slice, the second subscript stands for the j hidden node in the time slice; y_{ij} is the value of the observable node Y_{ij}; $pa(Y_{ij})$ is the father nodes group of Y_{ij}, Y_{ij_0} is the Observational State of Y_{ij}, the j in the i time slice; in the $\sum$ of the denominator, $x_1, x_2, \ldots, x_n$ is a hidden node group; $p(Y_{ij_0} = y_{ij})$, the continuous measurements of Y_{ij}, belongs to the membership grade of State. y_{ij} [12].

3 Radiation Field Method

"Radiation field" is a circle area, the center is the drivers' trip destination, radius is the domain of attraction of the trio destination for drivers. As the tessellating network is showed in Fig. 2, one of the pentagram symbols is the original place, and the other is the destination. R is the radius of the radiation field, the circle area is the influential area of the radiation field.

According to the research, here are some characteristics of the radiation filed:

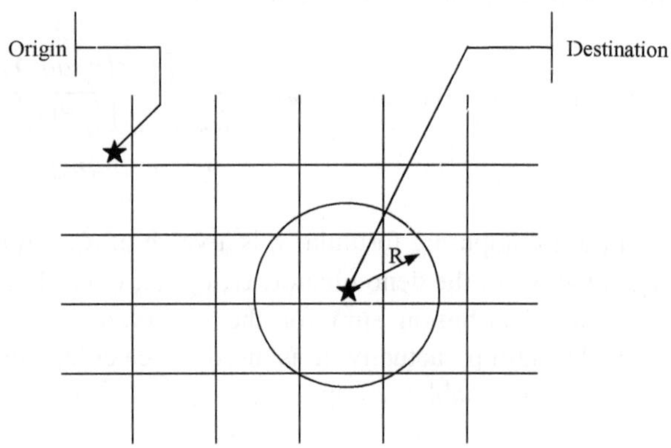

Fig. 2 Radiation field diagram

1. The extent of the radiation field depends on the importance of the destination. The more important the destination is, the more attractive it is for the drivers and the bigger the radius is, the bigger the covering area of the radiation field it is.
2. In the same radiation field, the closer drivers are to the destination, the more attractive it is.
3. Additivity. In the same network, there are always multi-destination trip, so it is reasonable that some of the 2 radiation field are coincident if the 2 destinations are close enough. At this time, the radiation field are superimposed, in the coincident places, the drivers are attracted by the 2 destinations at the same time.
4. The 2 destinations are competitive. The 2 destinations are both attractive to drivers, so which one should be arrived fist is an issue. At the same time, the importance of the integrated destination and the location of the 2 destinations should be considered.

In the radiation field, in order to get how much the influence of the destination for drivers, the influence of the destination for drivers should be analyzed, this is called radiation field method. The attraction can be showed by the data of importance of destination in each road, so as to provide more effective evidence for the route choices.

The importance of the destination can be marked with the integers from 1 to 5. According to the result of the questionnaire and the importance of different destinations, the statistical value of the radiation field can be given.

Assuming that there are n destinations in the road network, the importance of Destination i is S_i. In its influential area, the attraction of the destination for drivers is S_{ij} on the j road, the influential radius of the radiation field is called R_i, d_{ij} is the distance from the middle of the road to the destination. So as follows:

$$S_{ij} = \left(1 - \frac{d_{ij}}{R_i}\right) S_i \tag{4}$$

Then, here is the total attraction of the j is S_j

$$S_j = \sum_{i=1}^{n} \left(1 - \frac{d_{ij}}{R_i}\right) S_i \tag{5}$$

4 The Engendering of the Unexpected Radiation Field and Characteristics

4.1 Unexpected Radiation Field

As the route-choice makers, drivers should process messages by noticing outside information and own Characteristics. Directivity and concentricity are 2 basic

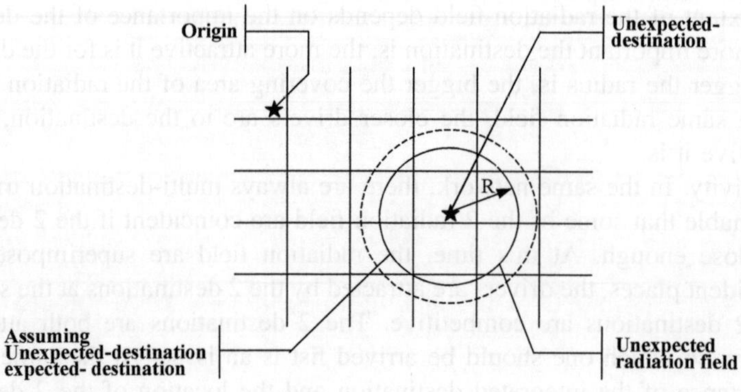

Fig. 3 Unexpected radiation field diagram

features. The directivity of heed means psychological activity reflect some objects and leave the rest of objects by its own choice. When some of the cerebral cortex can produce a positive exciting Centre, the adjacent domain of the cerebral cortex can inhibit these excitements because of the negative induction, so drivers cannot get clear reflection. The stronger negative induction is, the more concentrative heed is. So, when people concentrate on one thing, other things will be ignored. The positive exciting Centre can be transfer from one part to the other part. The transfer of the positive exciting Centre is physiological mechanism of heed transfer. New and special things are very easy to attract people's heed [13].

The unexpected radiation field is a radiation field created by the unexpected destination. Because of the unexpected destination, the heed choice is changed, so the new destination has got more attention. So, we can get a conclusion that the unexpected radiation field is much larger than the expected radiation field (Fig. 3).

Inlimine, the drivers make decision by paying attention on original destination, traffic environment and own information. When there is an unexpected destination, pay more attention on new destination, the original destination attention is attenuation, but it is still advanced processed and reflected into consciousness. During the trip, the traffic environment is complex and changeable, drivers are guided by trip destination, and pay attention on processing surrounding circumstance and own information, and ultimately reflected on the vehicle driving. New destinations can cause drivers' nervousness, and also lead drivers pay more attention, which causes that unexpected radiation field radius is a bit large.

4.2 *The Characteristics of Unexpected Radiation Field*

By the research, "unexpected radiation field" has some characteristics below except the general characteristics:

1. Unexpected radiation field radius is a bit large. The unexpected radiation field radius is much larger than the original radiation field, and the more important unexpected target is, the bigger radiation field radius is.
2. The unexpected radiation field radius size is based on the importance of unexpected destination and also related to driving tendency (risk-taking type, cautious type and conservative type). In order to finish the trip, conservative driver pay more attention on new target than risk-taking driver, which leads the radiation field radius of the former is larger than the other.
3. The unexpected radiation field have some impact on the transfer matrix of driving tendency. Unexpected destination trip can raise drivers' nervousness and can transfer drivers into risk-taking type. Besides, the influence is larger and larger if the unexpected time is closer to unexpected target.

5 Drivers' Route-Choice Modeling in Unexpected Destination Trip

The way to get the fake radiation field radius is to get the statistical value by questionnaire. Showed by Table 1.

Drivers' route-choice is influenced by 3 sides: human-vehicle-environment. Drivers' characteristics has been divided into some aspects: physiological characteristics (gender, age and constitution), psychological characteristics (character, emotion, motivation and temperament) and some characteristics acting on physiological characteristics (urgent task or not, skill, habit preference, road network familiarity). The drivers characteristics influence for driving behavior is called driving tendency. Zhang Yuan yuan studied driving tendency recognition in different traffic flow, driving tendency can be categorized into three types: risk-taking type, cautious type and conservative type, and he thinks drivers' recognition and processing for environment (it is drivers' behavior characteristics) is the result of physiology and emotion etc. (it is driving tendency) [14]. The vehicle characteristics include vehicle classes and functions. The environment characteristics can be categorized into in-bus environment (includes temperature and noises) and out-bus

Table 1 Expertise of false radiation field radius

Importance of traffic sections S_j	Normal $(S_j = \{1,2\})$	Important $(S_j = \{3,4\})$	Very important $(S_j = 5)$
Assuming expected-destination field radius R_i	2 intersections distance	3 intersections distance	4 intersections distance
Unexpected radiation field radius R_i'	3 intersections distance	4 intersections distance	6 intersections distance

Notes The unite of the radiation field radius is the number of intersections that needs drivers to make a decision

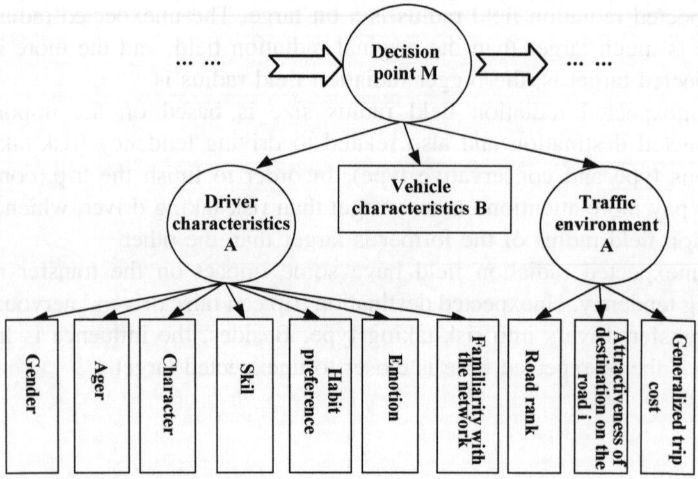

Fig. 4 Discrete dynamic bayesian network of the route choice behavior

environment (includes Road conditions, traffic disturbance, traffic guidance and weather).

Based on analysis above, we choose some factors which contribute to route-choice greatly to build discrete dynamic Bayesian network, showed by Fig. 4. In order to show drivers characteristics, we choose gender, age, characteristics, skill, habit preference, road network familiarity and emotion to study drivers tendency. Out-car environment is the only factor we should consider in environment aspect, road class decides the trip condition and comfort etc. Traffic disturbance, traffic management, traffic guidance and intersection numbers influence the trip time and cost etc., it is called generalized travel cost. So attraction of road class, generalized travel cost and destination in each road on drivers is used to show environment characteristics.

In this model, the task is divided into important, normal and unimportant; Drivers tendency is divided into 3 parts: risk, cautious and conservative type in Paper [14]. Vehicle characteristics is divided into good, normal, bad, and the traffic environment is divided into advantageous, normal and disadvantageous parts (Tables 2 and 3).

Discrete dynamic Bayesian network and radiation field method are used to build model and prove the phenomenon that the unexpected radiation field radium is larger. In dynamic Bayesian network, conditional probability and transfer matrix of all factors is given by Expertise value. This space is too small to describe more about it.

Because the unexpected radiation field influences the transfer matrix of driving tendency, drivers are forced to transfer into risk-taking type to some extent. Weight k is given here, and driving tendency is given by fuzzy reasoning, so it is just a probability to become a tendency. As a consequence, the Value distribution of

Table 2 Driver characteristics variable-state collection

Gender	Male, Female
Age	Youth(age 18–25), middle age(age 25–65), old age(age 65–70)
Personality	Extrovert(stable extrovert). Normal(unstable extrovert, unstable extrovert).Introvert(stable introvert)
Skill	Poor($\leq$ 0.5 year), normal(0.5–3 year), well($\geq$ 3 year)
Habit preference	Yes, no
Familiarity with the network	Familiar, normal, unfamiliar
Emotion	Good, normal, bad

Table 3 Road characteristics

Traffic guidance	Yes, no
Road characteristics	High, middle, low

driving tendency probability is in $(0, 1)$, then we can get $k \in [1, 10)$ to amend driving tendency.

6 Experimental Verification

In order to make a better contrast of the drivers' route-choice decisions under the unexpected goal, eight drivers should be collected (Including six male drivers and two female drivers whose age distribution is about 25–45, as well as their driving experience distribution is about 3–20). Before the experiment, the drivers should make a questionnaire on their psychology. Take District Zhang dian, Zibo City for example, only one original destination is regulated. However, there appeared the unexpected destination on the trip. The radiation field of the destinations is distributed as followed in Fig. 5.

Supposing the experimental situation, the driver worked at the destination. His original route is going back home after work. On the way back home, the unexpected goal appeared which is going to buy vegetables in the market.

Because of the limited space, only one driver was chosen to be analyzed on his selected trip route. The information were showed as followed (Table 4):

In order to make a better study of the behaviors of the drivers' trip route decisions when they faced the unexpected goal, we first applied the way of the radiation field and the discrete dynamic Bayesian network to make a simulation under the supposed experimental situation, which acquired the drivers' trip routes as routes. Then supposing the unexpected goal as the original goal, on the process of the drivers' trip there existing two original goals could be simulated, which could acquire the drivers' trip route as the route ②, as Fig. 6 showed.

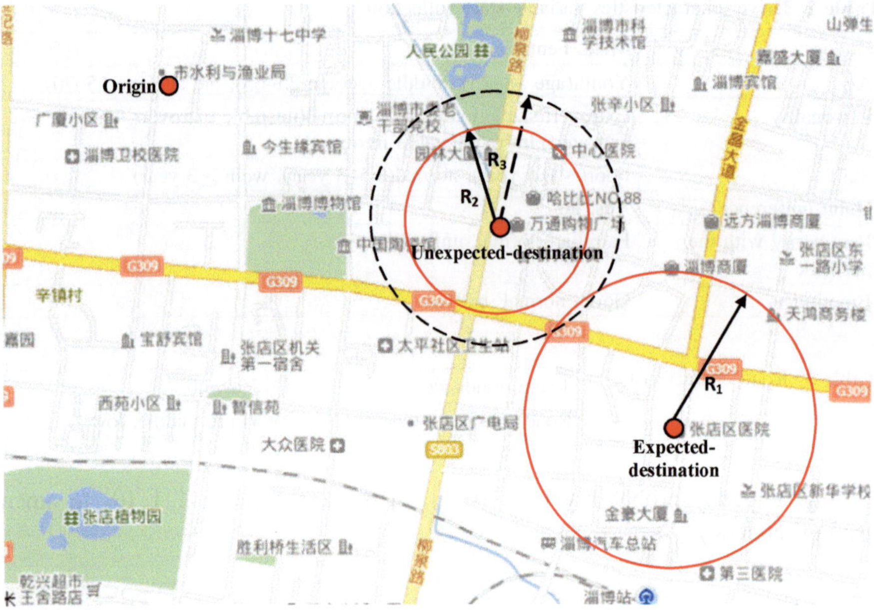

Fig. 5 Drivers' trip radiation field distribution

Table 4 Characteristics of selected target each time

	Gender (male, female)	Age (youth, middle age, old age)	Habit preference (yes, no)	Skill (poor, normal, well)	Character (introvert, normal, extrovert)
Time 1	(1.0, 0.0)	(1.0, 0.0, 0.0)	(0.7, 0.3)	(0.1, 0.2, 0.7)	(0.3, 0.5, 0.2)
Time 2	(1.0, 0.0)	(1.0, 0.0, 0.0)	(0.4, 0.6)	(0.1, 0.2, 0.7)	(0.3, 0.5, 0.2)
Time 3	(1.0, 0.0)	(1.0, 0.0, 0.0)	(0.3, 0.7)	(0.1, 0.2, 0.7)	(0.3, 0.5, 0.2)
Time 4	(1.0, 0.0)	(1.0, 0.0, 0.0)	(0.5, 0.5)	(0.1, 0.2, 0.7)	(0.3, 0.5, 0.2)
Time 5	(1.0, 0.0)	(1.0, 0.0, 0.0)	(0.8, 0.2)	(0.1, 0.2, 0.7)	(0.3, 0.5, 0.2)
Time 6	(1.0, 0.0)	(1.0, 0.0, 0.0)	(0.7, 0.3)	(0.1, 0.2, 0.7)	(0.3, 0.5, 0.2)
Time 7	(1.0, 0.0)	(1.0, 0.0, 0.0)	(0.1, 0.9)	(0.1, 0.2, 0.7)	(0.3, 0.5, 0.2)
Time 8	(1.0, 0.0)	(1.0, 0.0, 0.0)	(0.6, 0.4)	(0.1, 0.2, 0.7)	(0.3, 0.5, 0.2)
Time 9	(1.0, 0.0)	(1.0, 0.0, 0.0)	(0.4, 0.6)	(0.1, 0.2, 0.7)	(0.3, 0.5, 0.2)
Time 10	(1.0, 0.0)	(1.0, 0.0, 0.0)	(0.3, 0.7)	(0.1, 0.2, 0.7)	(0.3, 0.5, 0.2)
…	…	…	…	…	…

	Familiarity with the network (bad, normal, good)	Emotion (bad, normal, good)	Road rank (low, middle, high)	Generalized trip cost (low, middle, high)	vehicle characteristics (good, normal, bad)
Time 1	(0.0, 0.2, 0.8)	(0.2, 0.1, 0.7)	(0.0, 1.0, 0.0)	(0.3, 0.5, 0.2)	(0.0, 1.0, 0.0)
Time 2	(0.0, 0.2, 0.8)	(0.1, 0.4, 0.5)	(0.0, 1.0, 0.0)	(0.4, 0.4, 0.2)	(0.0, 1.0, 0.0)
Time 3	(0.0, 0.2, 0.8)	(0.4, 0.4, 0.2)	(0.0, 1.0, 0.0)	(0.6, 0.2, 0.2)	(0.0, 1.0, 0.0)
Time 4	(0.0, 0.2, 0.8)	(0.4, 0.4, 0.2)	(0.0, 1.0, 0.0)	(0.7, 0.2, 0.1)	(0.0, 1.0, 0.0)
Time 5	(0.0, 0.2, 0.8)	(0.7, 0.2, 0.1)	(0.0, 1.0, 0.0)	(0.6, 0.2, 0.2)	(0.0, 1.0, 0.0)

(continued)

Table 4 (continued)

	Familiarity with the network (bad, normal, good)	Emotion (bad, normal, good)	Road rank (low, middle, high)	Generalized trip cost (low, middle, high)	vehicle characteristics (good, normal, bad)
Time 6	(0.0, 0.2, 0.8)	(07, 0.2, 0.1)	(0.0, 0.0, 1.0)	(0.2, 0.5, 0.3)	(0.0, 1.0, 0.0)
Time 7	(0.0, 0.2, 0.8)	(0.8, 0.1, 0.1)	(0.0, 1.0, 1.0)	(0.5, 0.4, 0.1)	(0.0, 1.0, 0.0)
Time 8	(0.0, 0.2, 0.8)	(0.3, 0.5, 0.2)	(0.0, 1.0, 0.0)	(0.4, 0.5, 0.1)	(0.0, 1.0, 0.0)
Time 9	(0.0, 0.2, 0.8)	(0.3, 0.4, 0.3)	(0.0, 0.0, 1.0)	(0.4, 0.5, 0.1)	(0.0, 1.0, 0.0)
Time 10	(0.0, 0.2, 0.8)	(0.2, 0.4, 0.4)	(0.0, 1.0, 0.0)	(0.7, 0.2, 0.1)	(0.0, 1.0, 0.0)
…	…	…	…	…	…

Notes Numbers in the diagram, such as (1.0, 0.0) in the second line and second row means male; Habit preference means drivers prefer to choose which road in all the optional roads

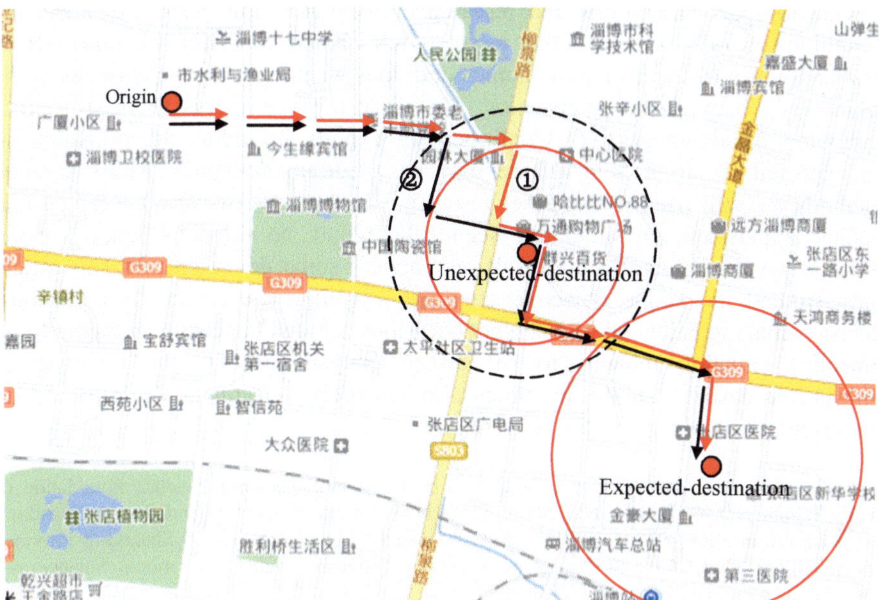

Fig. 6 Route-choice condition of drivers' trip

The results of the situational experiment turned out that the chosen route results of 87.5 % drivers attending the experiments are consistent with the simulated results. Thus, it explained that the radiation field and the discrete dynamic Bayesian network have the good applicability on the study of the choice route of the drivers under the unexpected goal, which also explained the existence of "false radiation field".

7 Conclusion

On the uncertain conditions, drivers may encounter a lot of unexpected-destination situations. As a result, "unexpected radiation field" is put forward and introduced, at the same time, the characteristic of it is also studied. On the basis of this, route-choice behavior of unexpected-destination trip under random dynamic conditions is researched with discrete dynamic Bayesian network. Studies display that the location and important of the unexpected destination affect the route choice behavior closely besides the effects of people, vehicle and environment. The simulation results show that, the combination of "radiation field method" and discrete dynamic Bayesian network is well applicable in route-choice behavior of unexpected-destination trip under random dynamic conditions, and enabled to simulate and reproduce the decision-making process more appropriately.

Due to space limitations and research convenience, this paper considers the condition of only one original target. route-choice behavior of unexpected-destination trip under random dynamic conditions when only using the forces of the relationship between the intersection of a circle to experiment. There still exists following disadvantages. (1) Confirmed by experts in radius determination of "unexpected radiation field", there may be a certain deviation. More accurate methods are still needed to further research. (2) As the experiment is taken real vehicle experiment on simulative scene, there may be some influences on the driver's psychological. So the result may appear some deviation. (3) This article only studied the single objective after the traveling, the driver's route choice decision-making behavior on emergent occasions, sudden targets under the multiple objective situation remains to be further analysis. (4) On the traveling of unexpected-destination, the importance of unexpected-destination's influence on the original one is needed further research.

Acknowledgments This study was supported by the National Natural Science Foundation of China (Grant NO. 61074140, 61573009, 51508315), Natural Science Foundation of Shandong Province (Grant NO. ZR2014FM027), Social Science Planning Project of Shandong Province (Grant NO. 14CGLJ27), and Project of Shandong Province Higher Educational Science and Technology Program (Grant NO. J15LB07).

References

1. Zhang, Y.: Study on Vehicle Routing Choice in Urban Traffic under Uncertainty. Southwest Jiaotong University, Chengdu (2006)
2. Wang, Z.-W., Luo, D.-Y., Hung, Z.-X., Wang, I.-J.: Multi-criteria multi-route choice under uncertainty. J. Syst. Eng. **24**(3), 355–359 (2009)
3. Tang, L.-S., Liang, J.: Comprehensive literature review on vehicle routing under emergencies. Railway Transp. Econ. **30**(12), 56–60 (2008)
4. Guo. H.-Y., Shi, H.-G.: Study on network travel time reliability under contigent events. J. Highw. Transp. Res. Dev. **22**(8), 102–105 (2005)

5. Feng, P.-Y.: Research on Traffic Evacuation under the Emergency Incidents. Southwest Jiaotong University, Chengdu (2008)
6. Murphy, K.P.: Dynamic Bayesian Networks: Representation, Inference and Learning. Computer Science Division, UC Berkeley, USA (2002)
7. Yu, Z.-Y., Chen, Z.-J., Zhou, R.: Study on algorithm of threat level assessment based on bayesian network. J. Syst. Simul. **7**(3), 555–558 (2005)
8. Guo, X.-B., Wang, Z., Hu, W.-D.: Information fusion with bayesian networks for target recognition. J. Syst. Simul. **17**(11), 2713–2716 (2005)
9. Liu, D.-Y., Wang, F.: Rearch on learning bayesian network structure based on genetic algorithms. J. Comput. Res. Dev. **38**(8), 916–922 (2001)
10. Zhang, Y., Ji, Q.: Active and dynamic information fusion for multisensor systems with dynamic bayesian networks. IEEE Trans. Syst. Man Cybern. Part B Cybern. **36**(2), 467–472 (2006)
11. Rose, C., Smaili, C.: A dynamic bayesian network for handling uncertainty in a decision support system adapted to the monitoring of patients treated by hemodialysis. In: Proceedings of the 17th IEEE International Conference on Tools with Artificial Intelligence. IEEE, USA (2005)
12. Tang, Z., Gao, X.-G.: Research on radiant point identification based on discrete dynamic bayesian network. J. Syst. Simul. **21**(1), 117–120 (2009)
13. Ye Y, Zhu B.: 《Psychology》. East China Normal University Press, Shanghai (2008)
14. Zhang, Y.-Y.: The verification method of vehicle driving tendency based on dynamic driver-vehicle-environment data. Shandong University of Technology, Shandong (2010)

OD Matrix Estimation Based on Mobile Navigation Technology

Tangjian Wei, Xiaosong Wang, Yuping Xv and Yuanyuan Feng

Abstract In traditional OD survey, too many manpower and material resources are consumed, but the life cycle of the data is shortening and survey cost is increasing. In tackling these questions, the paper put forward the theory of calculating flow of road segments by mobile navigation software, get all the traffic link flow in current city after fuzzy process, finally obtain estimate OD data. Analysis Nanchang main city zone, adopt STOCH algorithm, stochastic user equilibrium assignment and system optimum realize OD matrix estimation. The results of the example show that OD matrix estimation based on mobile navigation data has feasibility, which also large scale reduce the OD investigation cost at the same time improve the effectiveness of the OD data.

Keywords OD matrix estimation · Mobile navigation technology · TransCAD · Example analysis

1 Introduction

Compared to conducting an investigation to get the OD matrix of traffic zones, getting link traffic volumes to complete costs less distinctly in all respects [1–3], but it's still a troublesome work when we need to get the link flow in a large area. Cheng et al. [4] set the smallest square of deviance between link flow and actual measured flow as target function, bi-level programming model simplified to single level programming by deleting the function of discriminant error in OD matrix, then amount of calculation can be reduce greatly. Chen et al. [5] put forward the OD desire line intercept method develop for the choice of tested road sections based on initial OD matrix; however, adopting artificial field observation requires lots of labor, the observation data tend to be more subjective randomness, and it's difficult

T. Wei (✉) · X. Wang · Y. Xv · Y. Feng
School of Railway Tracks and Transportation East China Jiaotong University, Nanchang 330013, Jinagxi, China
e-mail: weitangjian@qq.com

© Springer Science+Business Media Singapore 2017
H. Lu (ed.), *Proceedings of the Second International Conference on Intelligent Transportation*, Smart Innovation, Systems and Technologies 53,
DOI 10.1007/978-981-10-2398-9_21

to get continuous observation data of related link. Exact and continuous data can get from series of professional equipment, but it costs giant cost investment which also result in huge difference between input and output in series of systemic traffic planning research. This paper put forward the theory of calculating flow of road segments by mobile navigation, this theory can work out the problem about mass tested road and huge amount of work, solve the OD estimate fast and effective.

2 Theory of Calculating Flow of Road Segments by Mobile Navigation Technology

Nowadays, with the rapid development of smart phone which brings great convenience to citizens, mobile phone has become one of the necessities for daily life. Facing with the increasingly serious traffic jams, some companies develop a serious of navigational software, providing not only navigation service but also real-time road conditions information. Using this kind of mobile software to get real-time road conditions information has become the main means by which citizens plan their going out. Combine relevant index with the road conditions information provided by mobile phone navigational software can estimate link flow easily, and get reasonable data to estimate OD matrix (Fig. 1).

The picture shows the whole traffic net and the traffic condition of each road segment in the city. Traffic conditions of different links segment are shown by lines in different colors. Green indicates an unblocked link, yellow an ambling link, light red a crowded link, dark red a serious traffic jam or a road having vehicles beyond its capacity. Conclusion can be made with reference of the fuzzy evaluation about related crowding road that the percentage traffic flow takes up discharge capacity in different degree of crowded links, the form can be show as below (Table 1):

Fig. 1 The real-time road conditions acquired through navigational software

Table 1 Fuzzy evaluate degree of link crowdedness

Link color	*Traffic condition*
Green	Unblocked
Yellow	Move slowly
Light red	Crowded
Dark red	Close-packed
Link color	*Link traffic flow take up capacity*
Green	Under 30 %
Yellow	3, 0–60 %
Light red	60–80 %
Dark red	More than 80 %

Table 2 The design speed of every road grade

Road grade	*Expressway*		
Design speed (km/h)	100	80	60
Road grade	*Arterial road*		
Design speed (km/h)	60	50	40
Road grade	*Minor arterial road*		
Design speed (km/h)	50	40	30

Table 3 Capacity of lane in expressway

Design speed (km/h)	100	80	60
Basic capacity (pcu/h)	2200	2100	1800
Design capacity (pcu/h)	2000	1750	1400

Table 4 Capacity of lane in other road grades

Design speed (km/h)	60	50	40	30
Basic capacity (pcu/h)	1800	1700	1650	1600
Design capacity (pcu/h)	1400	1350	1300	1100

Refer to the related rules in <CJJ 37 - 90—Design Standard for the industry standard of the people's Republic of China>, urban road mainly includes expressway, arterial road, minor arterial road, branch amount to five road grades, the design speed of every road grade show in Table 2.

In order to satisfy urban traffic demand, different road in different grades has its own capacity owing to design speed, for a lane belong to expressway, there exists relation between basic capacity and traffic speed (Table 3):

For the lane belong to other road grades, there exist relation between basic capacity and traffic speed (Table 4):

Using mobile navigational software can get basic information and real-time construction status of every link on the panoramic view, main includes unidirectional and bi directional road, the number of lane in every direction, such as Fig. 2.

With the basic statistics above, we can estimate the flow of each road segment. First, by referring data, learn the level of each road segment in the area, design

Fig. 2 Road pinworm

statistics such as the speed of vehicles, and with the standard of urban design, we can calculate out the fundamental traffic capacity of each road segment. Second, with the help of mobile navigation software we can perform a statistical analysis on the real-time road condition of road that relevant. At last, with the percentage of traffic flow from the roads capacity in different levels (chart 1), we can calculate out the traffic volume of each road segment. Above all, using mobile navigation software can simplify the workload of calculating the flow of road segments in the research area and get a relatively specific statistic of road segment flow.

3 Essential Process and Principle About OD Estimation Based on Link Flow

This paper processes OD estimation with transCAD, the workflow show as Fig. 3. The essential process of OD estimation in transCAD as follows: taking traffic flow in the investigative section of road as basic data. Before the OD estimation, refer to the divide of traffic zone edit an initial OD matrix as the formwork for output OD matrix. Choosing a kind of traffic distribute method, calculative starting point is the initial OD matrix, after multiple iteration and transition between traffic distribution and estimated matrix, repeat the process of traffic distribution and OD estimation, until the final result content the setting convergence condition, obtained the accurate OD matrix.

Multipath OD estimation is the promotion of unity path, which OD matrix come from link flow is more accurate. Only when the traffic distributable method is user equilibrium or stochastic user equilibrium, use multipath OD estimation. Compared to unity path, there are many best routes over traffic zones, compare traffic distributable flow with actual investigated flow, exacted traffic flow of observed link can be calculated as below:

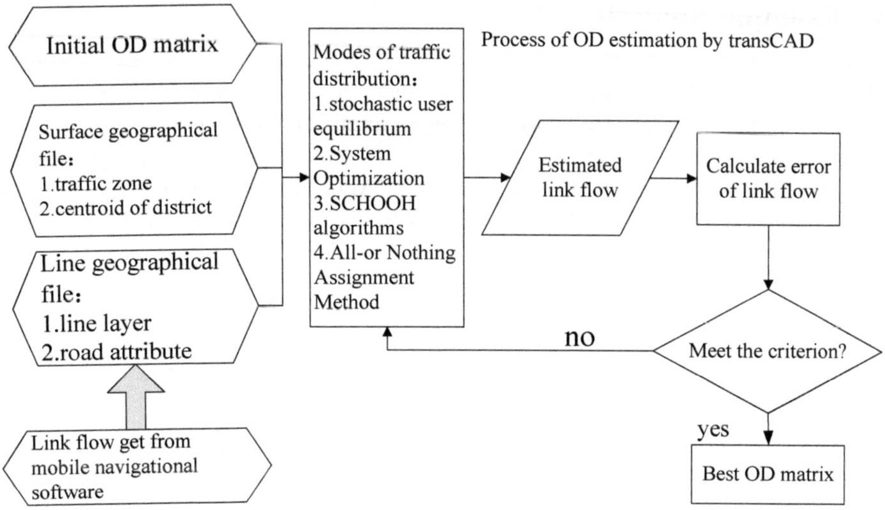

Fig. 3 Process of OD estimation by transCAD

$$T_{(E)ijr} = \frac{1}{N_{a\in(\tau,r)}}\left[\sum_{a\in(\tau,r)} T_{(E)ija}\right] \tag{1}$$

$$T_{ij(n)} = \sum_{r}(T_{(E)ijr} * P_{ijr}) \tag{2}$$

$$T_{(E)ija} = \frac{v_a}{T_{a(n-1)}} T_{ij(n-1)} \tag{3}$$

where:

$T_{(E)ija}$ exacted traffic flow of link a from traffic zone i to j
τ the set of traffic links
P_{ijr} probability of choose link r from traffic zone I to j
n iteration number

There are numerous optimal routes over traffic zones, so we need to distribute twice in every OD circulation update: the first distribution calculate flow, the second update OD flow.

4 Example Analysis

This paper uses the central areas in Nanchang as test region, amount to about 300 km including DongHu District, QingYunpu District, and some areas of QingShanhu District. By means of mobile navigational software can get the link flow of every road in study area, utilize TransCAD estimate OD matrix over traffic zones based on known link flow, which can be the basis for city road planning to study the travel distribution (Fig. 4).

4.1 *The Division of Traffic Zones*

In order to describe travel distribution of NanChang resident, combine the distribution of facilities such as business, travel, entertainment with actual situations, divided the research area to 18 traffic zones (Fig. 5). The specific result is as below:

4.2 *Input of Fundamental Road Network Information*

At present, the urban roads in NanChang have already reached 850, the total length of road is about 919.3 km. This paper considers expressway, arterial road, minor arterial road, some branches of higher vehicle flow. Unite TransCAD with schematic diagram of division in research area, then draw the traffic network, create a geography file of lane types. The traffic network of NanChang can show in Fig. 6:

Fig. 4 The link flow of research area

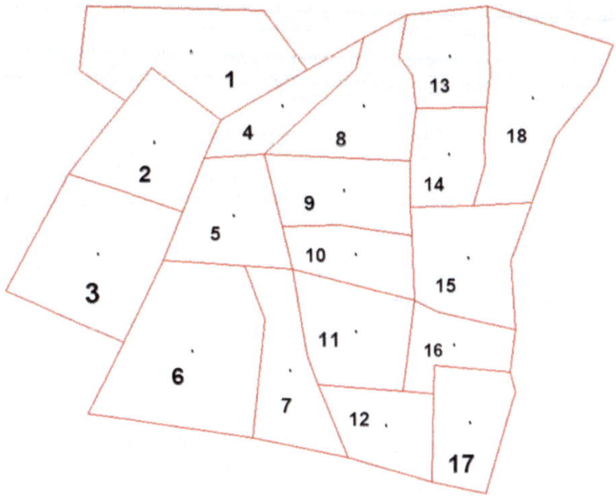

Fig. 5 Schematic diagram of division in research area

Fig. 6 The traffic network of research area

Before finish the chart of road section layer, input the own attribute of roads. Compared with geography file of surface types which has more attributes, line layer has fields when estimates OD matrix (Table 5):

Combine the demand of standard with a real situation of roads and then get the capacity of every link and design speed. Impedance function chooses BPR

Table 5 Attribute of roads

Fields	Content	Type
name	Road name	character
Time	Form time	Real number
Type	Road grade	Integer
AB-count	Vehicle flow in one direction	Real number
BA-count	Vehicle flow in another direction	Real number
capacity	Design capacity	Real number
speed	Design speed	Real number
Length	Length of road	Real number

impedance function $t_a(v_a) = t_a(0)\left[1 + \alpha\left[\frac{v_a}{c_a}\right]^{\beta}\right]$,$\alpha$ and β are undetermined parameters, according to the previous study $\alpha = 0.15, \beta = 4$ therein, over the above attributes, length of road can rough estimate from measurement tools inside the software of map.

An important datum of OD estimate this paper used is the vehicle flow in every link, in order to guarantee the accuracy of the OD matrix which was obtained from estimate, choose those links of high flow to get more comprehensive travel information over traffic zones from survey data. This paper chooses the main arteries in research area, the flows are relatively large, so that satisfy above demands; at the same time, every link is independent, there not exists linear relations, so that satisfy the principle of independence (Figs. 7, 8 and 9).

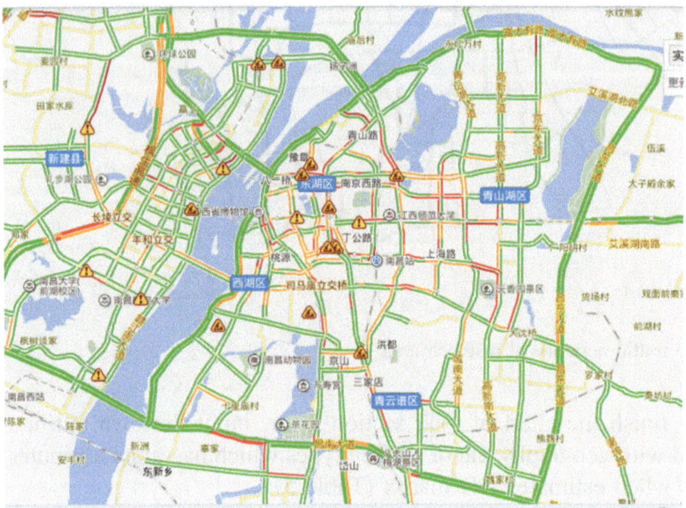

Fig. 7 The first picture of rush hour in NanChang

Fig. 8 The second picture of rush hour in NanChang

Fig. 9 The third picture of rush hour in NanChang

Owing to the statistics about each road traffic condition from mobile navigational software, we can have a whole understand about traffic conditions about the link flow. Generally speaking, the crowding phenomenon is kind of serious in NanChang, the statistical result shows in Table 6, crowding roads concentrate on most parts of the Honggutan New District, central Castle Peak Lakes, station in the East Lake area, roads near square of Bayi.

Edit the attribute of every link with the statistical data what is said above (Fig. 10).

Table 6 Statistical table about road traffic condition

Link color	Traffic condition	Related links	Proportion (%)
Dark red	Close-packed	Honggutanzhong street, Fenghezhong Street, Bayi Street, Lushannan Street (linking to the Bayi Bridge), Bayi Bridge, Taohua Road, Beijingxi Road, Zhanqianxi Road, Changdongnan Street, Ximazhuang Street, Yongshu Road, Shanghai Road, Huizhan Road, Zhongshan Road, etc.	8
Light red	Crowded	Ruzi Road, Yiyuan Road, Luyin Road, Chunhui Road, Hongcheng Road, Minfeng road, Xiangshanbei Road, Dieshan Road, Yangming Road, Qingshanhuzhong street, Jingdongnan Street, Hongdubei Street, Xiangshanbei Road, Shunwai Road, Shanghaibei Road, Jiefangxi Road, Beijingdong Road, Yangmingdong Road, Supu Road, Erqibei Road, etc.	15
Yellow	Move slowly	Jinggangshan Street, Taoyuan Road, Jiefangdong Road, Changnanxi Street, Nanjingdong Road, Fuhebei Road, Gaoxinzhong Street, Gaoxinnan Street, Yanjiangnan Street, Minqiang Road, Huoju Road, Yingbinbei Street, Xiangshan Road, Erqinan Road, etc.	30
Green	Unblocked	The rest of the road	47

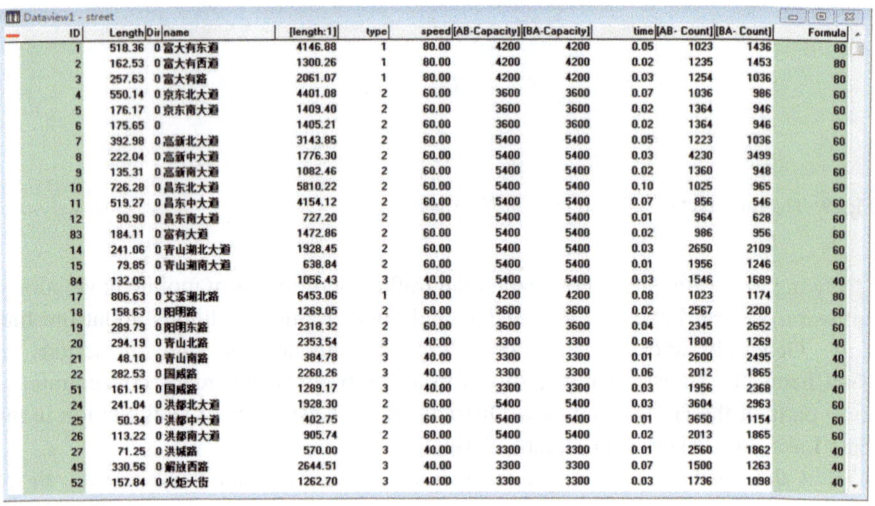
Dataview1 - street

ID	Length	Dir	name	[length:1]	type	speed	[AB-Capacity]	[BA-Capacity]	time	[AB- Count]	[BA- Count]	Formula
1	518.36	0	富大有东道	4146.88	1	80.00	4200	4200	0.05	1023	1436	80
2	162.53	0	富大有西道	1300.26	1	80.00	4200	4200	0.02	1235	1453	80
3	257.63	0	富大有路	2061.07	1	80.00	4200	4200	0.03	1254	1036	80
4	550.14	0	京东北大道	4401.08	2	60.00	3600	3600	0.07	1036	986	60
5	176.17	0	京东南大道	1409.40	2	60.00	3600	3600	0.02	1364	946	60
6	175.65	0		1405.21	2	60.00	3600	3600	0.02	1364	946	60
7	392.98	0	高新北大道	3143.85	2	60.00	5400	5400	0.05	1223	1036	60
8	222.04	0	高新中大道	1776.30	2	60.00	5400	5400	0.03	4230	3499	60
9	135.31	0	高新南大道	1082.46	2	60.00	5400	5400	0.02	1360	948	60
10	726.28	0	昌东北大道	5810.22	2	60.00	5400	5400	0.10	1025	965	60
11	519.27	0	昌东中大道	4154.12	2	60.00	5400	5400	0.07	856	546	60
12	90.90	0	昌东南大道	727.20	2	60.00	5400	5400	0.01	964	628	60
83	184.11	0	富有大道	1472.86	2	60.00	5400	5400	0.02	986	956	60
14	241.06	0	青山湖北大道	1928.45	2	60.00	5400	5400	0.03	2650	2109	60
15	79.85	0	青山湖南大道	638.84	2	60.00	5400	5400	0.01	1956	1246	60
84	132.05	0		1056.43	3	40.00	5400	5400	0.03	1546	1689	40
17	806.63	0	艾溪湖北路	6453.06	1	80.00	4200	4200	0.08	1023	1174	80
18	158.63	0	阳明路	1269.05	2	60.00	3600	3600	0.02	2567	2200	60
19	289.79	0	阳明东路	2318.32	2	60.00	3600	3600	0.04	2345	2652	60
20	294.19	0	青山北路	2353.54	3	40.00	3300	3300	0.06	1800	1269	40
21	48.10	0	青山南路	384.78	3	40.00	3300	3300	0.01	2600	2495	40
22	282.53	0	国威路	2260.26	3	40.00	3300	3300	0.06	2012	1865	40
51	161.15	0	国威路	1289.17	3	40.00	3300	3300	0.03	1956	2368	40
24	241.04	0	洪都北大道	1928.30	2	60.00	5400	5400	0.03	3604	2963	60
25	50.34	0	洪都中大道	402.75	2	60.00	5400	5400	0.01	3650	1154	60
26	113.22	0	洪都南大道	905.74	2	60.00	5400	5400	0.02	2013	1865	60
27	71.25	0	洪城路	570.00	3	40.00	3300	3300	0.01	2560	1862	40
49	330.56	0	解放西路	2644.51	3	40.00	3300	3300	0.07	1500	1263	40
52	157.84	0	火炬大街	1262.70	3	40.00	3300	3300	0.03	1736	1098	40

Fig. 10 The attribute of some links

4.3 OD Estimation in Different Method of Traffic Assignment

In order to estimate OD matrix, not only must finish the above work, but also prepare initial OD matrix. The ID of lane and line in the initial OD matrix must match centroid of district. There exist two methods to get initial OD matrix when the travel information of traffic zones cannot acquire:

Every corresponding element in the initial OD matrix assigned to 1 which is minimum.

Travelers' trips always influenced by the travel impedance between traffic zones, therefore calculate initial OD matrix through travel impedance between traffic zones:

$$t_{ij} = 1 - \frac{f_{ij}}{\sum_i \sum_j f_{ij}}$$

f_{ij} is the impedance value between the traffic zone I and j, use TransCAD screen out nodes and then get the centroid of district, therefore impedance matrix can easily get.

The initial OD matrix in this paper is the matrix whose elements are all 1, OD estimation considers travel circumstance not including the circumstance inside zones, set the diagonal elements as zero.

1. Estimate OD by stochastic user equilibrium assignment

User equilibrium assignment supposes all of the travelers master traffic information comprehensively, then choose the route which spend time least. However, travelers always calculate wrong about road impedance because they lack of comprehensive traffic information. Error of road impedance can regard as a random variable, this kind of circumstance is stochastic user equilibrium assignment, when travelers adopt this method they can't reduce the time by changing their own choice, so this method is more close to reality. Use successive algorithm to distribute traffic in TransCAD need large iterative loop, the picture below is the result of estimate by this software (Table 7):

2. Estimate OD by STOCH Algorithm

STOCH Algorithm applies logit model to traffic assignment. This algorithm doesn't calculate selection probability, only calculate the probability of connecting efficient paths. Now described STOCH Algorithm in detail:

First step: initial step

Defined $r(i)$, $s(i)$ as time when following the shortest path move from starting point r (node *i*); $down(i)$ as the set of all the downstream nodes related with *i*, $up(i)$ as the set of all the upstream nodes related with *i*.

Table 7 The result of estimate

	1	2	*3*	*4*	*5*	*6*	*7*	*8*	*9*	*10*	*11*	*12*	*13*	*14*	*15*	*16*	*17*	*18*
1	0	163.9	35.54	88.81	81.43	21.29	64.6	42.31	154.43	33. 66	184.4	54.24	188	51.57	55.65	117.6	79.19	83.86
2	355.98	0	23.82	82.09	45.04	92.66	30.71	32.21	95.12	74.92	251.6	32.08	59	32.31	46.41	156.2	63.35	72.13
3	34.15	19.93	0	336	24.05	39.71	13.23	22.19	68.04	50.48	37.98	35.27	41.2	14.5	51.88	70.45	55.36	68.37
4	81.59	68.8	276.1	0	312.5	261.2	72.51	379.6	316.48	300.07	950.5	81.65	254	121.3	383.7	464.9	146	140.1
5	115.54	49.55	28.24	298.5	0	141.6	38.7	292.6	127.54	82.11	277.8	30.25	61.8	37.04	403.9	269.4	75.86	80.38
6	17.65	62.26	37.37	216.1	70.39	0	51.44	179.4	106.25	64.64	11.97	14.72	43.5	47.18	87.8	63.16	74.13	18.23
7	119.87	52.66	21.16	101.1	48.61	79.53	0	32.98	71.04	71.51	64.95	52.64	58.1	34.08	52.58	75.66	98.42	101.5
8	54.42	38.39	28.28	512.6	704.9	136.6	35.03	0	154.65	58.01	273.4	91.73	75.3	34.64	600.9	236.7	60.3	67.19
9	192.01	67.02	62.64	244.6	75.05	89.5	83.73	212.5	0	35.2	201.6	20.61	49.3	82.18	82.78	39.91	70.47	108.47
10	29.64	64.85	48.64	345.8	67.96	67.13	58.58	63.48	35.79	0	23.88	32.3	130	54.58	46.34	36.28	49.32	28.43
11	182.23	170.9	47.19	634.4	373.7	14.88	69.53	293.1	150.1	27.5	0	22.22	189	60.63	353.4	108.8	92.48	51.74
12	24.72	26.14	34.52	70.83	23.36	13.43	41.32	49.69	19.64	28.92	6.94	0	380	40.11	32.96	21.77	55.29	23.44
13	140.27	49.49	39.16	273.5	49.49	42.12	46.46	81.61	49.19	110.04	134.9	418.99	0	44.82	58.95	47.13	59.54	102.05
14	47.62	23.68	13.84	134.2	28.94	50.29	12.2	24.91	90.14	56.64	45.86	40.82	47.1	0	77.03	98.85	434.6	206.46
15	67.08	51.8	53.99	493.5	503.9	109.6	52.38	373.7	117.55	50.25	293.8	37. 66	66	83	0	179.5	61.74	69.93
16	150.88	144.4	67.5	373.2	226	70.81	68.52	158.9	56.81	37.39	147.9	23.89	50	94.72	127	0	80.55	84.23
17	67.2	44.29	44.89	146.3	70.89	74.89	40.73	50.44	73.08	49.75	79.28	63.31	60.7	241.5	49.88	77.92	0	110.43
18	80.37	60.04	68.25	151.7	81.97	20.21	62.89	63.15	86.57	30.33	41.97	33.56	118	212.7	66.22	88.86	189.6	0

Calculate the least travel time from point r to node i and least travel time from node i to end point s, get the $r(i)$ and $s(i)$ of every node.

Calculate all of the link weight $L(i \rightarrow j)$ in effective link $i \rightarrow j$:

$$L(i \rightarrow j) = \begin{cases} e^{\theta[r(j)-r(i)-ti\rightarrow j)]}, \text{for effective links } (r(i) < r(j) \text{ and } s(i) > s(j)) \\ 0, \text{other links} \end{cases}$$

$t(i \rightarrow j)$ is the real traffic time of link $(i \rightarrow j)$. For point j, when $r(i) > r(j)$, i is the upstream node of j, otherwise, i is the downstream node of j

Second step: forward estimate

Start from starting point r, arrange nodes according to the value of $r(i)$, calculate path weight $w(i \rightarrow j)$ for every node $i,j \in down(i)$,

$$w(i \rightarrow j) = \begin{cases} L(i \rightarrow j), & i = r \\ L(i \rightarrow j) \sum\limits_{m \in up(i)} w(i \rightarrow j), & i \neq r \end{cases}$$

If narrate this formula to words:

The path weight of every lane equal to product of link weight multiplies the sum of all over the path weight upstream;

For the lane which includes starting point, path weight equal to link weight.

Third step: estimate

Start from final point s, arrange nodes according to the value of $s(j)$, then distribute the link flow $x(i \rightarrow j)$ to $i \in f(i)$

$$x(i \rightarrow j) = \begin{cases} q_{rs} \frac{w(i \rightarrow j)}{\sum_{m \in up(i)} w(m \rightarrow j)}, & j = s \\ \left[\sum\limits_{m \in down(j)} x(j \rightarrow m) \right] \frac{w(i \rightarrow j)}{\sum_{m \in up(j)} w(m \rightarrow j)}, & j \neq s \end{cases}$$

The formula looks complex, but it can be summarized with two sentences.

for every node, the sum of flow which come from upstream lane equal to the sum of flow which come from downstream lane;

For every node, there exists the direct proportion of all the upstream lane flow with their path weight.

This is the result of estimation through link flow use STOCH Algorithm in TransCAD (Table 8):

3. Estimate OD by System Optimum Traffic Assignment

The optimal system actually is an ideal mode of distribution, stochastic user equilibrium assignment what mentioned before is according to travelers, while system optimum traffic assignment is according to planners. In fact, crowded link flow usually influence a lot to the travel time of travelers, planners offer a kind of distributable scheme, add up to mutual cooperation of travelers therefore result to least total time in network (Table 9).

Table 8 Result of STOCH algorithm

	1	2	*3*	*4*	*5*	*6*	*7*	*8*	*9*	*10*	*11*	*12*	*13*	*14*	*15*	*16*	*17*	*18*
1	0	163.9	35.54	88.81	81.43	21.29	64.6	42.31	154.43	33. 66	184.4	54.24	188	51.57	55.65	117.6	79.19	83.86
2	355.98	0	23.82	82.09	45.04	92.66	30.71	32.21	95.12	74.92	251.6	32.08	59	32.31	46.41	156.2	63.35	72.13
3	34.15	19.93	0	336	24.05	39.71	13.23	22.19	68.04	50.48	37.98	35.27	41.2	14.5	51.88	70.45	55.36	68.37
4	81.59	68.8	276.1	0	312.5	261.2	72.51	379.6	316.48	300.07	950.5	81.65	254	121.3	383.7	464.9	146	140.1
5	115.54	49.55	28.24	298.5	0	141.6	38.7	292.6	127.54	82.11	277.8	30.25	61.8	37.04	403.9	269.4	75.86	80.38
6	17.65	62.26	37.37	216.1	70.39	0	51.44	179.4	106.25	64.64	11.97	14.72	43.5	47.18	87.8	63.16	74.13	18.23
7	119.87	52.66	21.16	101.1	48.61	79.53	0	32.98	71.04	71.51	64.95	52.64	58.1	34.08	52.58	75.66	98.42	101.5
8	54.42	38.39	28.28	512.6	704.9	136.6	35.03	0	154.65	58.01	273.4	91.73	75.3	34.64	600.9	236.7	60.3	67.19
9	192.01	67.02	62.64	244.6	75.05	89.5	83.73	212.5	0	35.2	201.6	20.61	49.3	82.18	82.78	39.91	70.47	108.47
10	29.64	64.85	48.64	345.8	67.96	67.13	58.58	63.48	35.79	0	23.88	32.3	130	54.58	46.34	36.28	49.32	28.43
11	182.23	170.9	47.19	634.4	373.7	14.88	69.53	293.1	150.1	27.5	0	22.22	189	60.63	353.4	108.8	92.48	51.74
12	24.72	26.14	34.52	70.83	23.36	13.43	41.32	49.69	19.64	28.92	6.94	0	380	40.11	32.96	21.77	55.29	23.44
13	140.27	49.49	39.16	273.5	49.49	42.12	46.46	81.61	49.19	110.04	134.9	418.99	0	44.82	58.95	47.13	59.54	102.05
14	47.62	23.68	13.84	134.2	28.94	50.29	12.2	24.91	90.14	56.64	45.86	40.82	47.1	0	77.03	98.85	434.6	206.46
15	67.08	51.8	53.99	493.5	503.9	109.6	52.38	373.7	117.55	50.25	293.8	37. 66	66	83	0	179.5	61.74	69.93
16	150.88	144.4	67.5	373.2	226	70.81	68.52	158.9	56.81	37.39	147.9	23.89	50	94.72	127	0	80.55	84.23
17	67.2	44.29	44.89	146.3	70.89	74.89	40.73	50.44	73.08	49.75	79.28	63.31	60.7	241.5	49.88	77.92	0	110.43
18	80.37	60.04	68.25	151.7	81.97	20.21	62.89	63.15	86.57	30.33	41.97	33.56	118	212.7	66.22	88.86	189.6	0

Table 9 Result of System Optimum traffic assignment

	1	2	3	4	5	6	7	8	9	*10*	*11*	12	*13*	*14*	*15*	*16*	*17*	*18*
1	0	197.5	24.02	82.25	81.22	16.88	67.31	40.99	131.9B	31.5	223.3	39.55	181	51.26	54.81	3.99	93.72	88.41
2	362.93	0	13.56	73.25	37.71	40.4	28.38	29.8	55.15	61.96	279.5	18.4	31.8	29.65	44.01	282.70	75.7	73.83
3	21.87	11.14	0	431.5	14.8	19.85	4.74	13.8	69.4	36.44	27.51	24.16	22.5	4.38	40.21	60.94	40.2	48.49
4	73.16	60.76	354.7	0	179.7	402.3	58.61	219.9	335.29	305.52	1334	98.54	266	112.8	196.2	421.7	154.3	138.62
5	91.76	31.83	16.57	263.5	0	44.09	29.75	264.6	64.63	67.6	280.8	16.12	31	30.12	389.9	658.41	85.05	82.57
6	11.6	48.51	24.82	193.8	60.69	0	41.78	227.2	150.4	89.97	6.4	10.95	27.5	39.18	96.19	55.7	77.67	12.86
7	148	81.04	9.58	89.57	54.74	46.83	0	30.3	68.54	63.82	63.1	45.3	35.5	31	49.68	75.31	132.4	107.5
8	53.03	37.35	18.42	466.5	677	192.4	29.25	0	154.87	53.62	300	76.51	152	29.67	589	274.6	73.47	72.33
9	201.44	83.15	54.83	210.6	118.3	605.3	86.68	260.7	0	30.34	182.1	23.59	176	88.81	83.25	22.73	69.88	115.34
10	26.48	76.63	41.85	336.9	89.53	111.6	67.11	62.93	29.95	0	19.44	35.53	132	52.45	41.23	28.73	45	25.83
11	315.18	245.8	36.11	844.8	401.9	10.76	65.81	348.3	116.22	24.88	0	10.22	175	57.35	400.4	79.91	111.7	66.16
12	13.78	21.26	22.57	64.37	19.92	11.46	30.67	33.07	19.28	30.46	1.54	0	428	30.65	32.73	19.25	51.53	15.78
13	128.18	40.47	28.65	288.6	42.19	35.92	38.58	88.73	46.58	150.96	111.5	494.19	0	37.6	56.89	40.91	57.44	92.75
14	40.18	18.54	3.37	118.7	23.13	27.89	5.22	18.86	113.01	44.52	38.9	31.73	27.7	0	77.81	104.7	911.7	238.95
15	67.49	52.24	43.2	447.4	507.9	512.2	47.16	409.1	197.32	52.13	326.5	59.17	88	93.44	0	209.9	59.12	61.85
16	157.59	104.6	62.7	349.4	168.6	316.1	70.77	212.3	183.44	37.99	176.2	33.28	173	114	173.2	0	89.14	86
17	68.22	45.39	27.49	146	77.58	132	34.39	59.01	84.09	48.87	90.49	63.32	68.6	450.4	40.45	72.61	0	74.64
18	81.86	61.61	49.46	152	88.67	17.14	56.65	71.82	80.32	29.26	49.48	28.4	114	293.4	57.12	84.4	195.2	0

Table 10 Statistic table of OD estimation

Method of traffic assignment	*Serial number*	*1*	2
Stochastic user equilibrium assignment	Origin	1387.12	1468.9
	Destination	1843.16	1054.04
STOCH algorithm	Origin	1466.74	1545.59
	Destination	1761.22	1158.09
System optimum traffic assignment	Origin	1277.61	1182.27
	Destination	1862.75	1157.04
Sum of OD		9599.6	7567.93
Ave of OD		3199.86	2522.64
Method of traffic assignment	*Serial number*	*3*	*4*
Stochastic user equilibrium assignment	Origin	872.48	5182.22
	Destination	803.98	5058.91
STOCH algorithm	Origin	982.73	4610.42
	Destination	931.3	4503.2
System optimum traffic assignment	Origin	982.03	4173.95
	Destination	831.85	4559.03
Sum of OD		5407.37	28,091.73
Ave of OD		1802.45	9363.91
Method of traffic assignment	*Serial number*	*5*	*6*
Stochastic user equilibrium assignment	Origin	2193.17	1554.77
	Destination	2772.77	1904.91
STOCH algorithm	Origin	2410.76	1103.06
	Destination	2788.09	1325.32
System optimum traffic assignment	Origin	1720.15	1269.83
	Destination	2643.55	2543.2
Sum of OD		14,533.49	9707.09
Ave of OD		4844.49	3235.69
Method of traffic assignment	*Serial number*	*7*	*8*
Stochastic user equilibrium assignment	Origin	1240.12	3379.62
	Destination	765.34	2776.69
STOCH algorithm	Origin	1136.39	3102.74
	Destination	842.56	2352.58
System optimum traffic assignment	Origin	1096.98	3292.81
	Destination	762.86	2391.41
Sum of OD		5851.25	17,303.85
Ave of OD		1950.41	5767.95
Method of traffic assignment	*Serial number*	*9*	*10*
Stochastic user equilibrium assignment	Origin	2159.32	1295.64
	Destination	1996.66	1170.45

(continued)

Table 10 (continued)

STOCH algorithm	Origin	1717.58	1182.75
	Destination	1772.42	1127.76
System optimum traffic assignment	Origin	2323.13	1263.65
	Destination	1768.49	1159.84
Sum of OD		11,746.6	7210.09
Ave of OD		3915.53	2403.36
Method of traffic assignment	*Serial number*	*11*	*12*
Stochastic user equilibrium assignment	Origin	3332.52	624.37
	Destination	3827.63	813.62
STOCH algorithm	Origin	2481.5	892.89
	Destination	3028.7	1048.28
System optimum traffic assignment	Origin	3260.06	903.58
	Destination	3510.58	1108.96
Sum of OD		19,451.99	5403.7
Ave of OD		6483.99	1801.23
Method of traffic assignment	*Serial number*	*13*	*14*
Stochastic user equilibrium assignment	Origin	1780.36	1783.39
	Destination	1950.63	1541.67
STOCH algorithm	Origin	1747.71	1433.17
	Destination	1809.59	1286.79
System optimum traffic assignment	Origin	1608.8	1844.79
	destination	2129.19	1508.63
Sum of OD		11,039.3	9412.44
Ave of OD		3679.76	3137.48
Method of traffic assignment	*Serial number*	*15*	*16*
Stochastic user equilibrium assignment	Origin	3458.37	2003.5
	Destination	2529.8	2247.58
STOCH algorithm	Origin	2627.67	1962.56
	Destination	2577.21	2089.88
System optimum traffic assignment	Origin	3024.18	2508.7
	Destination	2423.13	863.55
Sum of OD		16,655.36	11,691.8
Ave of OD		5551.78	3897.26
Method of traffic assignment	*Serial number*	*17*	*18*
Stochastic user equilibrium assignment	Origin	1582	1429.59
	Destination	2299.2	1370.92
STOCH algorithm	Origin	1284.78	1456.56
	Destination	1685.84	1416.94
System optimum traffic assignment	Origin	1583.63	1510.61
	Destination	2323.11	1279.59
Sum of OD		10,775.56	8482.21
Ave of OD		3591.85	2827.4

4.4 *Data Analysis*

4.4.1 Data Analysis Based on the Number of Traffic Zones

Before estimate OD in three different method of traffic assignment, this form is an OD summary of traffic zones in different method of traffic assignment (Table 10).

Figure 10 is the OD estimation summary of 18 traffic zones according to population density. Related to above data therefore get OD matrix, from which can acknowledge the travel circumstance of NanChang residents. Large travel areas mainly include Donghu District, middle of Qingshanhu District, Gaoxin District and Honggutan New District (Figs. 11 and 12).

4.4.2 Data Analysis Based on Different Method of Traffic Assignment

Observe statistical data will find out that results from three different method of traffic assignment are nearly same in magnitude, although there still exist difference, this software not only estimate OD but also assign traffic in the same assignment (Fig. 13).

Analyze the data in Table 11, three traffic assignment methods exist error in different degrees, but for most of path, the error rate between repeat traffic assignment and original link traffic assignment is less than 30 %, accuracy meet the requirement. Compare three traffic assignment, the error of STOCH algorithm is a

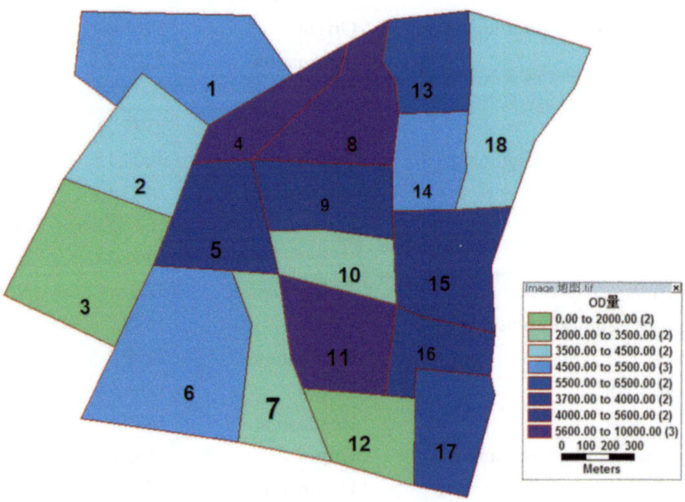

Fig. 11 OD of traffic zones

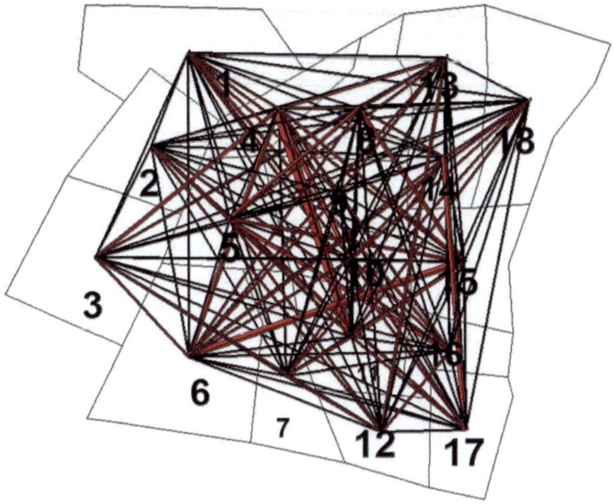

Fig. 12 Flow of traffic zones

[AB- Count]	[BA- Count]	Formula	ID1	AB_Flow	BA_Flow	TOT_Flow
1023	1436	80	1	1077.0527	1240.4122	2317.4649
1235	1453	80	2	1378.6731	1208.8547	2587.5278
1254	1036	80	3	1114.8701	969.7275	2084.5976
1036	986	60	4	825.5631	840.4380	1666.0010
1364	946	60	5	1498.5922	1424.7718	2923.3641
1364	946	60	6	1323.3095	838.4292	2161.7388
1223	1036	60	7	1066.4795	714.0967	1780.5761
2230	2499	60	8	3068.5940	2997.9097	6066.5036
1360	948	60	9	551.8744	422.2966	974.1710
1025	965	60	10	965.9916	1119.6574	2085.6490
856	546	60	11	806.0065	830.8841	1636.8906
964	628	60	12	878.0192	801.8264	1679.8456
986	956	60	83	1018.0414	967.0098	1985.0511
2650	2109	60	14	2795.2279	2157.3376	4952.5655
1956	1246	60	15	2592.0945	2181.7497	4773.8441
1546	1689	40	84	1676.6463	1493.3118	3169.9580
1023	1174	80	17	1044.5252	1275.4685	2319.9937
2567	2200	60	18	3168.7017	2896.1271	6064.8287
2345	2652	60	19	3168.7017	1865.8911	5034.5928

Fig. 13 Comparison diagram of link flow before distribution

little larger, the error of single road section also larger than others, so it's best not choose STOCH algorithm. In contrast, average error of both stochastic user equilibrium assignment and System Optimum Traffic Assignment are less, results

Table 11 Comparison diagram of the error in link flow before distribution

Best path	Direction AB (%)	Direction BA (%)
	STOCH algorithm	
1	12.6	8.5
2	−24.5	−16.4
3	−8.4	−13.9
4	17.9	16.1
5	12.5	30.8
6	−14.3	24.6
7	−22.1	−12.4
8	11.8	22.1
9	−7.4	−6.4
10	21.3	−19.1
11	−15.1	22.4
12	24.3	−13.7
13	15.6	9.8
14	−16.2	−29.5
15	−17.2	−13.8
16	12.1	−15.4
17	15.4	18.3
18	38.4	33.3
Error-sum	56.7	45.3
Average error	3.2	2.5
	Stochastic user equilibrium assignment	
1	5.3	13.2
2	8.6	6.4
3	−11.1	−12.3
4	−6.3	−8.4
5	9.8	5.6
6	−12.9	−8.8
7	−12.7	−10.3
8	7.6	5.8
9	8.9	6.7
10	−8.4	−4.9
11	−8.1	−10.9
12	−13.3	−12.9
13	11.2	15.8
14	−24.1	−14.4
15	15.9	14.3
16	−16.4	−8.1
17	10.4	−6.7
18	9.7	6.4

(continued)

Table 11 (continued)

Best path	Direction AB (%)	Direction BA (%)
Error-sum	−25.9	−23.5
Average error	−1.4	−1.3
	System optimum traffic assignment	
1	−20.3	−12.4
2	−8.6	2.6
3	−6.6	4.2
4	33.2	−16.2
5	−18.4	−5.7
6	15.2	21.4
7	−9.4	−13.5
8	4.3	33.1
9	−6.4	−12.6
10	11.3	5.8
11	14.6	−6.1
12	13.5	−1.6
13	−24.1	−12.4
14	18.3	−14.7
15	−8.6	−12.3
16	19.4	20.4
17	−14.7	−19.4
18	12.6	11.3
Error-sum	25.3	−28.1
Average error	1.4	−1.6

more actualize. The error of stochastic user equilibrium assignment is least. So stochastic user equilibrium assignment is the best traffic assignment in OD estimation based on link flow.

5 Conclusion

This paper put forward theory of calculating flow of road segments by mobile navigation software, maximal improve the fussy condition when obtaining link flow. Based on the link flow, estimate OD through the traffic planning software TransCAD, then verify the feasibility of this method in example analysis.

Example analysis based on Nanchang central urban area, adopt STOCH algorithm, stochastic user equilibrium assignment and system optimum, and analyze around 200 link flow statistical data which assigned by software, compare with origin link flow and error analyzing, in a conclusion, stochastic user equilibrium assignment is the best traffic assignment in OD estimation, system optimum secondly, STOCH algorithm was not often recommended because the result of it has larger difference.

References

1. Lu, H.: Transport Management and Planning. Tsinghua University Press (2006) (In Chinese)
2. Shao, C., Gu, Y.: Transportation Planning. Beijing Jiaotong University Press (2012) (In Chinese)
3. Lu, H., Huang, H.: Transportation Planning Theory Research Front. Tsinghua University Press (2007) (In Chinese)
4. Wei, C., Yueming, C., Yupei, D.: OD matrix estimation algorithm with consideration to uncertainty of link flows. Urban Trans China **3**, 79–81 (2008) (In Chinese)
5. Chen, Z., Li, K., Sun, J.: Count location selection for OD matrix estimation. Trans. Comput. China **1**, 104–106 (2007) (In Chinese)

Weight-Dependent Equilibrium Solution for Weighted-Sum Multiobjective Optimization

Yang Wu, Zhiyong Zhang, Jianhua Yuan and Qing Ma

Abstract As a popular approach to solve Multiobjective Optimization Problem (MOP), weighted-sum (WS) method obtains a series of weight-dependent Pareto Optimalities (i.e. multi-objective global optimums) forming Pateto Front. Each priori (preset) combination of single-objective (SO) weights stands for a certain way to compromise all of SO, e.g. a popular opinion is "Balanced weights lead to the equilibrium solution". To verify this notion, this paper proposes a method to obtain adaptive posteriori weights derived from heuristic search rather than human-judged priori weights, so as to generate an unique Equilibrium Pateto Optimality (Equi-PO) out of the Pareto Front of multiobjective-function (MOFunc), where mutual interest of every single-objective- function (SOFunc) is achieved to a certain "equal" extent. The numerical example reveal that an unique Equi-PO is obtainable with adaptive weights converging towards an unique end, and: (1) For and only for the WS-MOP whose Pareto Front is symmetric to the Equiangular Utopia Ray, "balanced weights" results in "equilibrium solution"; (2) For other conditions, "balanced weights" can't.

Keywords Weighted-Sum multiobjective optimization problem (WS-MOP) · Adaptive weight · Equilibrium pare to optimality

1 Introduction

As a popular scalarized modeling, Weighted-sum (WS) method converts Multi-objective Optimization Problem (MOP) into a Single-objective Optimization Problem (SOP), using a certain combination of weights to compose all of

Y. Wu (✉) · Z. Zhang · J. Yuan · Q. Ma
Traffic Management Research Institute of Ministry of Public Security of China, Wuxi, China
e-mail: wuyang8848@qq.com

Y. Wu · Z. Zhang · J. Yuan · Q. Ma
Collaborative Innovation Center of Modern Urban Traffic Technologies, Nanjing, China

© Springer Science+Business Media Singapore 2017
H. Lu (ed.), *Proceedings of the Second International Conference on Intelligent Transportation*, Smart Innovation, Systems and Technologies 53,
DOI 10.1007/978-981-10-2398-9_22

Single-Objective-Functions (SOFunc) into a Multi-Objective-Function (MOFunc). A generalized Weighted-Sum MOP (WS-MOP) is formulated as:

$$\min_{X} J(\mathrm{W}, X) = \min_{X} W \cdot F(X) = \min_{X} \sum_{i=1}^{m} \omega_i f_i(\mathrm{X}) \tag{1.1}$$

$$s.t\ c_k(X) = 0,\ k = 1,\ 2,\ \ldots,\ k_E \tag{1.2}$$

$$c_k(X) \leq 0,\ k = k_E, k_E + 1, \ldots,\ k_\Omega \tag{1.3}$$

$$X \in R^n \tag{1.4}$$

Where: $J(\mathrm{W}, X)$ the scalarized weighted-sum MOFunc; $F(X) = [f_i(X)]_{m\times 1} = [f_1(\mathrm{X}), f_2(\mathrm{X}), \ldots, f_m(\mathrm{X})]^{\mathrm{T}} \in R^m$ and $W = (\omega_i)_{1\times m} = (\omega_1, \omega_2, \ldots, \omega_m) \in R^m$ respectively the objective function vector and the the weight vector, with $f_i(X)$ and ω_i respectively the ith SOFunc and its weight for i = 1, 2, …, m with $\sum_{i=1}^{m} \omega_i 1$; $X = (x_j)_{n\times 1} = (x_1, x_2, \ldots, x_n)^{\mathrm{T}} \in R^n$ the decision-variable point (also known as design point) for j = 1, …, n; $c_k(X)$ for $k = 1, 2, \ldots, k_E$ and $c_k(X)$ for $k = k_E, k_E + 1, \ldots, k_\Omega$ respectively the Equality and the Inequality constraint (linear/non-linear), with $k_\Omega = 0$ and $k_\Omega{>}0$ respectively representing an Unconstrained and Constrained WS-MOP.

Given one combination of weights W*, diversified methods including Mathematical Programming of Operational Research and Evolutionary Computation (e.g. Genetic Algorithm or Particle Swarm Optimization) [1], can generate one "trade-off" global-minimum at $X = X^{\mathrm{global}} = (x_j^{\mathrm{global}})_{n\times 1} = (x_1^{\mathrm{global}}, x_2^{\mathrm{global}}, \ldots, x_n^{\mathrm{global}})^{\mathrm{T}}$, as: $J(\mathrm{W}^*, X^{\mathrm{global}}) = \min_{X} \sum_{i=1}^{m} \omega_i^* f_i(X) = \sum_{i=1}^{m} \omega_i^* f_i(X^{\mathrm{global}})$, i.e. so-called "Pareto-Optimality/Pareto-Efficiency (also known as Non-dominated/Non-inferior solution)". Given diversified W*, a number of Pareto-Optimalities can be obtained for decision maker to select from, all of which construct "Pareto Front" (which is curve/surface/hyper-surface in 2/3/multi-dimension objective-space).

In form of line/plane/hyper-plane in 2/3/multi-dimension objective-space, weighted-sum MOFunc can't approach possible non-convex part of Pareto Front. Therefore, in order to generate the entire Pareto Front, scalarized WS-modeling and its Mathematical-programming or Evolutionary-computation methods (or quasi-scalarized "Stratified/Sequential" modeling and methods such as "Lexicographic/Multi-layer" or "Goal" Programming) was dominated by non-scalarization model-free methods such as Evolutional Computing Approaches, for solving each and every SOFunc simultaneously by decomposing MOP into a number of sub-MOP within fragmented feasible region in differential-geometry level [1–3]. Until adaptive Weighted-Sum methods (AWS) was proposed [4, 5]) as a kind of Pareto-based posteriori methods, combining the easy-to-interpret of scalarization and the fragmentation of feasible region in differential-geometry level, both convex and non-convex part of Pareto Front for WS-MOP can be generated, then Weight-sum is led to a new life.

Based on abovementioned abundant methods producing Pareto Optimalities for WS-MOP with a certain combinations of weights, the relation between the ways to combine weights and the meaning of the trade-off among SOFuncs, becomes an issue worth digging. Each priori (preset) combination of weights stands for a certain way to compromise all of SOFunc. E.g. 1st/2nd/… largest weights 0.5/0.3/… are respectively assigned to 1st/2nd/… most valued SOFunc by empirical means e.g. Expert Knowledge, without sufficient mathematical grounds. Another example is the popular opinion as "Balanced weights ($\forall$i: ω_i = 1/m) lead to the equilibrium solution", which is prevalently considered as the sole prerequisite to obtain a "win-win" solution for a check- and -balance WS-MOP with each SOFunc standing for a stakeholder. It also needs to be verified.

To figure out whether "balanced weights" contribute to "equilibrium solution" or not, this paper propose a method to seek a certain combination of weights and Pareto Optimality, where mutual interest of every SOFunc is achieved to an equal extent, in a evolutionary computation way rather than human-judged way.

2 Understand the Equilibrium Solution for WS-MOP Using Adaptive Weights

"What an equilibrium solution (Equilibrium Pareto Optimality, Equi-PO) means for a WS-MOP" and "How to combine weights to realize the Equi-PO" are analyzed in this section.

2.1 The Concept of Equilibrium Solution for WS-MOP

At the $X = X^{\text{global}}$, ith SOFunc value is $f_i(X^{\text{global}})$ which in most cases can NOT achieve ith SOFunc's own ideal local-minimum $f_i(Xi^{\text{local}})$ generated by Xi^{local}. The deviation $dev_i = \left|f_i(Xi^{\text{local}}) - f_i(Xi^{\text{global}})\right|$ is chosen to stand for the extent to which ith SOFunc's local-minimum is achieved: Smaller dev_i represents higher optimization of $f_i(Xi^{\text{local}})$, down to dev_i=0 representing accomplishing the utmost ideal optimization of $f_i(Xi^{\text{local}})$. The concept of "Equilibrium Pareto Optimality" of this paper on check- and -balance basis, means an unique Pareto Optimality at which every SOFunc's deviation dev_i equals each other.

Within the 2-dimension objective-space of WS-MOP As shown in Fig. 1, $(dev_i)_{m\times 1}$ (for i = 1, 2, …, m) represents a vector originating form Utopia Point and pointed at a global minimum point $[f_i(X^{\text{global}})]_{m\times 1}$ (with a length called "Utopia Distance" in this paper), further dev_i=represents Utopia-Distance's projection on ith axis. If Utopia Point acts as the origin of coordinates, then Equi-PO $J(\mathrm{W}^{\text{equi}}, X^{\text{global}})$

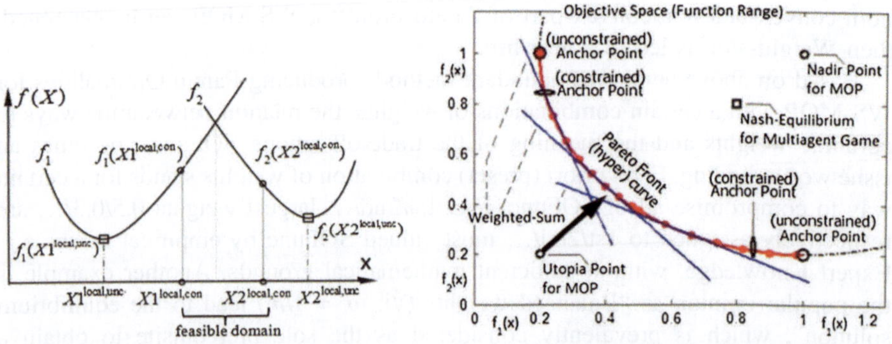

(a) Local extremes of $f_i(Xi^{local})$ (i=1,2)

(b) Utopia Point, Equiangular Utopia Ray, Pareto Front in 2-D objective space

Fig. 1 Extremes of $X = (x_j)_{n\times 1}$ (n = 1) and $[f_i(X)]_{m\times 1}$ (m = 2) for (un)constrained WS-MOP *Note* Constrained Utopia/Anchor/Nadir points distribute closer to each other, Constrained Pareto Front is more narrow, than unconstrained

in geometric terms is the intersection point between: The (hyper-)curve/surface of Pareto Front, and such a (hyper-)ray originating from Utopia Point that the included angles between this ray and every axis are equal (called "Equiangular Utopia Ray" in this paper). Consequently, Equi-PO signifies an unique Utopia-Distance vector with equal projections on every axis ∀i:dev_i, in contrast to another unique Utopia-Distance vector with the shortest Utopia-Distance as in the related study [6]. Our studies indicate 2 kinds of vector abovementioned equal each other for and only for symmetrically-structured WS-MOP.

Where: The sole (only one) Utopia Point is $[f_1(X1^{\text{local,unc}}), f_2(X2^{\text{local,unc}}), \ldots, f_m(Xm^{\text{local,unc}})]^T$, ith out of m Anchor Points is $[f_1(Xi^{\text{local,unc}}), f_2(Xi^{\text{local,unc}}), \ldots, f_m(Xi^{\text{local,unc}})]^T$.

2.2 The Weights Self-adapted to Equilibrium Solution

Whether balanced $(\omega_i)_{1\times m} = (\frac{1}{m})_{1\times m}$ can lead to the Equi-PO or not? If not, what kind of weights can? To answer the questions, we design a kind of adaptive posteriori weights derived from heuristic search rather than human-judged priori weights coming with disputes between conflicting values.

Initiated by a random weights' combination leading to a certain Pareto Optimality, the weights W = $(\omega_i)_{1\times m}$ is iterated as (and normalized in the range of [0,1]):

$$\omega_i = \omega_i + \lambda_i s^{\text{weight}}. \tag{2.1}$$

Where: the iteration step $s^{\text{weight}} \in [0, 0.1]$ is pre-set as a constant depending on weights' converging speed, the iteration speed λ_i of ith SOFunc is defined by $dev_i = \left|f_i(Xi^{\text{local}}) - f_i(X^{\text{global}})\right|$ and $dev_{\text{sum}} = \sum_{i=1}^{m} \left|f_i(Xi^{\text{local}}) - f_i(X^{\text{global}})\right|$, in the way as follow. During each iteration step, according to a new combination of iterated weights dependent on the CURRENT Pareto Optimality and Utopia Distance, the NEXT Pareto Optimality and Utopia Distance are solved by mathematical planning until the approximation of Equi-PO is obtained.

Design λ_i: static $\lambda_i = \pm 1$ (with a faster iteration speed), as follows: All of dev_i^{con} are sorted into a array $\{dev_i^{\text{con}}\}_{\text{descend}}$ in descending order; the number of the weights to be increased/decreased by s^{weight} is respectively M^{incre} and M^{decre}, where $M^{\text{incre}} = M^{\text{decre}} = floor(\frac{m}{2})$ (equals $\frac{m}{2}$ if m is even number, and $\frac{m}{2}$-1 if m is odd number); ω_i(for i = 1,2,…, M^{incre}) matched to 1st/2nd/3rd/…/and M^{incre}th element of $\{dev_i\}_{\text{descend}}$ designed to be increased as $\omega_i = \omega_i + s^{\text{weight}}$ (for i = 1, 2, …, M^{incre}), whereas ω_i(for i = m − M^{decre} + 1, …, m − 1, m) matched to corresponding ith element of $\{dev_i^{\text{con}}\}_{\text{descend}}$ designed to be decreased as $\omega_i = \omega_i - s^{\text{weight}}$ (for i = m-M^{decre} +1, …, m-1, m); therefore, with half of weights increased and another half decreased, the sum of weights stays as $\sum_{i=1}^{m} \omega_i$=1.

The Xi^{local} at which SOFunc obtains local-minimum $f_i(Xi^{\text{local}})$, exist in two ways as follow:

(1) For and only for constrained WS-MOP, at certain $Xi^{\text{local}} = Xi^{\text{local,con}} = (Xi_j^{\text{local,con}})_{n\times 1}$ located within or at the boundary of a constrained feasible domain (closed space), $f_i(X)$ obtain its local minimum $f_i(Xi^{\text{local,con}}) = \left[\sum_{j=1}^{n} (xi_j^{\text{local,con}} - xi_j^{\text{target}})^2\right]^{1/2}$ subject to constraints e.g. formula (1.1–1.2) and (1.1–1.3). In this condition of Constrained Utopia Point, $dev_i^{\text{con}} = \left|f_i(Xi^{\text{local,con}}) - f_i(X^{\text{global}})\right|$ or $\sum_{i=1}^{m} f_i(Xi^{\text{local,con}}) - f_i(X^{\text{global}})$.

(2) For unconstrained or constrained WS-MOP, there always exist a certain $Xi^{\text{local}} = Xi^{\text{local,unc}} = (xi_j^{\text{local,unc}})_{n\times 1}$ located within an unconstrained feasible domain (open space), where $f_i(X)$ obtain its local minimum $f_i(Xi^{\text{local,unc}}) = \left[\sum_{j=1}^{n} (xi_j^{\text{local,unc}} - xi_j^{\text{target}})^2\right]^{1/2}$ despite any constraint. In this condition of a Unconstrained Utopia Point, $dev_i^{\text{unc}} = \left|f_i(Xi^{\text{local,unc}}) - f_i(X^{\text{global}})\right|$ or $\sum_{i=1}^{m} \left|f_i(Xi^{\text{local,unc}}) - f_i(X^{\text{global}})\right|$.

3 Exemplify the Equilibrium Solution for WS-MOP Using Adaptive Weights

Assuming $f_i(X) = \|X - Xi^{\text{target}}\| = \left[\sum_{j=1}^{2}(x_j - xi_j^{\text{target}})^2\right]^{1/2}$ the Euclidean Distance between the decision-variable point $X = (x_j)_{2\times1} = (x_1, x_2)^{\mathrm{T}} \in R^2$ and ith target point (fixed-point) $Xi^{\text{target}} = (xi_j^{\text{target}})_{2\times1} = (xi_1^{\text{target}}, xi_2^{\text{target}})^{\mathrm{T}} \in R^n$ for j = 1 and 2, then a constrained WS-MOP is described as:

$$\min_X \sum_{i=1}^{m} \omega_i f_i(\mathrm{X}) = \min_X \sum_{i=1}^{m} \omega_i \|X - Xi^{\text{target}}\| =$$

$$\min_X \sum_{i=1}^{m} \omega_i \left[\sum_{i=1}^{2}(x_j - xi^{\text{target}})^2\right]^{1/2} \quad m = 4 \tag{3.1}$$

$$s.t.\ c_1(X) = x_1^2 + x_2^2 - 2^2 \le 0 \tag{3.2}$$

$$c_2(X) = (x_1 - 1)^2 + x_2^2 - 2^2 \le 0 \tag{3.3}$$

$$X = (x_j)_{2\times1} \in R^2. \tag{3.4}$$

The heuristic search of adaptive weights approaching the Equi-PO is applied for WS-MOP with (a)symmetric structure, using two kinds of local optimum i.e. constrained $f_i(Xi^{\text{local,con}})$ and unconstrained $f_i(Xi^{\text{local,unc}})$ for a contrast, to calculate dev_i as $dev_i^{\text{con}} = \left|f_i(X_i^{\text{local,con}}) - f_i(X^{\text{global}})\right|$ and $dev_i^{\text{unc}} = \left|f_i(X_i^{\text{local,unc}}) - f_i(X^{\text{global}})\right|$.

3.1 *Example 1: Symmetrically Structured WS-MOP*

It can be verified that the Eqi-PO for a symmetrically structured WS-MOP results from balanced weights i.e. (0.25, 0. 25, 0. 25, 0.25), e.g. assuming $[X1^{\text{target}}, X2^{\text{target}}, X3^{\text{target}}, X4^{\text{target}}]^{\mathrm{T}} = [(-1,-2), (-1,2), (2,2), (2,-2)]^{\mathrm{T}}$ symmetric to the point (0.5, 0) in the 2-dimension variable-space, no matter the iterative initial-weights are balanced (0.25, 0. 25, 0. 25, 0.25) or unbalanced (0.1, 0.1, 0.3, 0.5).

3.2 Example 2: Asymmetrically Structured WS-MOP

Assuming $[X1^{\text{target}}, X2^{\text{target}}, X3^{\text{target}}, X4^{\text{target}}]^{\mathrm{T}} = [(-2.5, 0), (-1.5, 2.2), (2, 2.2), (2.5, -1)]^{\mathrm{T}}$ asymmetric to any point in the 2-dimension variable-space, this WS-MOP can be considered asymmetrically-structured compared to example 1.

No matter weights' iteration starts from balanced initials (as in Fig. 2) or unbalanced initials (as in Fig. 3), finally adaptive weights converge at unbalanced (0.40, 0.28, 0.20, 0.12) corresponding to Equi-PO; and Equi-PO results from an unbalanced combination of weights, because Pareto Front (including Anchor Points) stays asymmetric with respect to the "Equiangular Utopia Ray" abovementioned.

Whereas decision-variable points converge towards different destinations, respectively nearby at $X^{\text{global, unc}} = (0.1,\ 0.3)$ and $X^{\text{global, con}} = (0.6,\ 0.2)$. Where: The difference of X^{global} as between Figs. 2 and 3, reflects the distinction between the two designs of Utopia Point/Distance on which the heuristic measure dev_i is based.

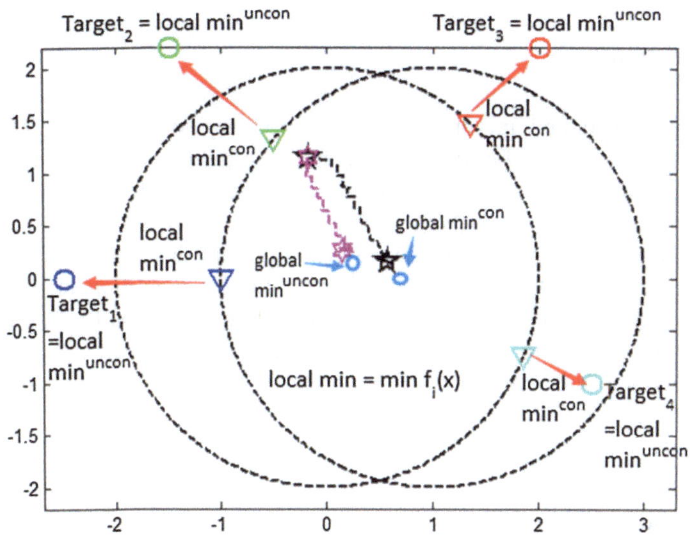

Fig. 2 Asymmetric WS-MOP: the evolution and convergence of the X obtaining Equi-PO, given balanced initial weights

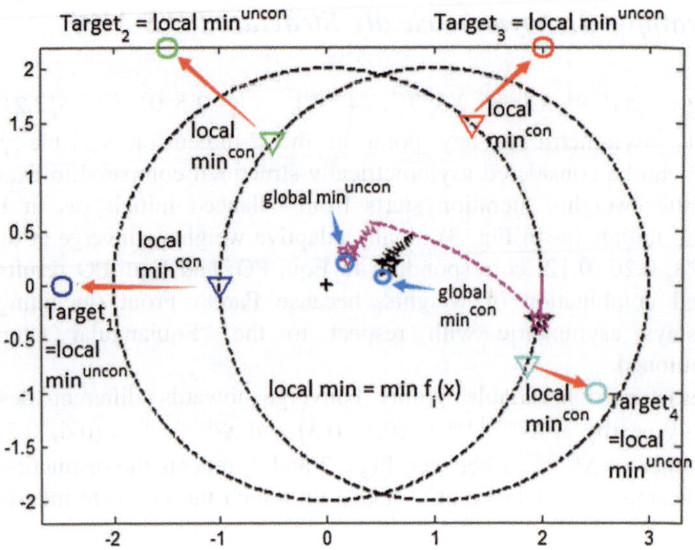

Fig. 3 Asymmetric WS-MOP: The evolution and convergence of the X obtainning Equi-PO, given Unbalanced initial weights

4 Conclusions

(1) Despite diversified initial-weights, the convergence of Pareto Optimalities towards an unique Equilibrium Pareto Optimality driven by the heuristic Utopia Point/Distance, reveals that an equilibrium solution within Pareto Front is obtainable for a given WS-MOP.

(2) The adaptive weights converge towards different distributions contingent on the structure of WS-MOP: For and only for the condition in which Pareto Front stays symmetric with respect to "Equiangular Utopia Ray", the balanced weights can lead to the Equi-PO; otherwise, the balanced weights show no connection with the Equi-PO. Therefore the so-called "Equilibrium Pareto-solutions" consist not in the balanced quantities of weights, but in the symmetric structure of WS-MOP.

Acknowledgments Supported by National Key Technology R&D Program of China "Key Technologies and System Integration of Network-based Coordinated Control of Freeway Traffic Safety (Project No.: 2014BAG01B04)", Key Laboratory of Road Traffic Safety of Ministry of Public Security of China.

References

1. de Weck, O.: Multiobjective optimization: history and promise. Lecture note at department of aeronautics-&-astronautics and engineering systems division, MIT. http://www.learningace.com/doc/2255311/f1295672b2a379be18867783de471afc/3_46_cjk-osm3-keynote (2004)
2. Pil'at, M.: Evolutionary multiobjective optimization: a short survey of the state-of-the-art. In: Proceedings of WDS 2010 Contributed Papers, Part I, pp. 13–18, (2010). http://www.mff.cuni.cz/veda/%20konference/wds/proc/pdf10/WDS10_102_i1_Pilat.pdf
3. Coello, C.A.C: An updated survey of evolutionary multiobjective optimization techniques: state of the art and future trends. Laboratorio Nacional de Informatica Avanzada. ftp://dca.fee.unicamp.br/pub/docs/vonzuben/%20ia707_1s06/textos/coello_multiobjetivo2.pdf (2010)
4. Kin, I.Y., de Weck, O.: Adaptive weighted-sum method for bi-objective optimization: Pareto front generation. Structural Multi. Optimi. **29**, 149–158 (2005)
5. Kin, I.Y., de Weck, O.: Adaptive weighted-sum method for multiobjective optimization: a new method for Pareto front generation. Structural Multi. Optim. **31**, 105–116 (2006)
6. Zavala, Victor M., Tlacuahuac, Antonio Flores: Stability of multiobjective predictive control: a utopia-tracking approach. Automatica **48**, 2627–2632 (2012)

Research on Urban Spatial Structure of Nanchang City Based on Mobile Communication Data

Yu-ping Xu, Zheng Zhang and Tian-tian Wu

Abstract With the advent of the age of social mobile and big data, communication devices not only become the necessity of modern life, but also permeate almost every facet of people's lives. The arrival and development of the information era has brought a series of revolution to people's special activities, thus causing the change in the cities' special structure. In recent years, the studies on urban internal spatial structure have always been the hot field of architecture, economics, geography and etc. This paper first explains the internal space structure of the city. According to the situation of limitations of traditional researching methods, the article based on the big data of mobile location-based service that mesh divides the cities and defines the regions where the density of mobile usage is higher than the average as mobile hotspot. Then through the observation of grid mobile hotspot, we can acquire the recognition of urban public system and internal functional divisions under different space and time law. Taking Nanchang City for example, through mobile hotspot method, we can come to the conclusion that the CBD business area in Honggutan region, industrial area in Xiaolan region, and most living area in the eastern part of River Fu qualifies the current city planning of Nanchang, which could provide reference for the planning department.

Keywords Urban spatial structure · Big data · Mobile location-based service

1 Introduction

Urban space structure is the expression of the relationship between the material environment, the function and the cultural value of the city. Some experts also call it "urban internal space". Urban spatial structure can affect the time and space habits of the residents, and the time and space habits of the residents can also change the structure of urban space. Including architecture, economy, geography and other

Y. Xu (✉) · Z. Zhang · T. Wu
East China JiaoTong University, Nanchang, China
e-mail: 1423907384@qq.com

© Springer Science+Business Media Singapore 2017

H. Lu (ed.), *Proceedings of the Second International Conference on Intelligent Transportation*, Smart Innovation, Systems and Technologies 53,
DOI 10.1007/978-981-10-2398-9_23

disciplines are hot fields of the subject. However, most of the scholars of our country start to study the urban space structure from the macro level at present. Take Nanchang City as a example, from the study of the spatial structure of Nanchang City, Zhongrong MIN and Xianfang YANG started with the dynamic mechanism of urban spatial structure evolution to study the optimization of urban spatial structure in 2011 [1]; Xiujuan YI [2]; researched on the mechanism of interaction between urban spatial structure and urban road traffic development from road traffic in 2007 Jiang LV studied the urban spatial structure formation mechanism and the relevant laws of the various elements from the change of the land use in 2013 [3]. Qin GUO [4] analyzed the spatial organization relationship between the various elements of the tourism system in Nanchang in 2007. But employment and living population, land use are both the macro level of the residents of the time and space.

In this paper, a new method based on mobile location data is proposed, which is based on the analysis of the characteristics of urban residents' behavior, and through the analysis of data, exploring the scale of urban spatial structure and urban function zoning. With the gradual improvement of the mobile communication network construction, the proportion of mobile phone has reached a very high proportion of urban and rural residents. Take Nanchang City as an example, as of the end of 2013, the number of mobile phone users has reached 4,900,000, and in Nanchang City, the resident population is 5,180,000 [5], mobile phone penetration rate has reached 95 %. Due to the high rate of mobile phone use, and mobile positioning is the most powerful and the most widely used data in mobile information technology, so that the social and economic attributes of the mobile phone positioning observation can be reflected in the urban space structure.

2 Traditional Methods of Urban Space Structure

From the perspective of the behavior of the residents, traditional studies of the internal structure of the city are mostly methods for the census, transportation, and remote sensing data and other related technologies. So the study is based on the small sample size of the survey or census and other statistical data, compared with the new mobile positioning technology, the traditional technology is an extensive data, that is, through the input of production factors to achieve the accumulation of data, the way to collect data consumption, higher cost, product quality is difficult to improve, the efficiency is low.

On the other hand, because of the low availability of the traditional data format, the data cannot be transmitted synchronously, so the traditional method is mostly static data, But we want to explore the law of urban internal structure, we must grasp the dynamic population distribution law, real-time reflect the mobile phone holder's dynamic activity space position, which provides the possibility to describe the holders of employment, residence, entertainment and other activities of the time and space.

3 Mobile Hot Spot Identification Method

In recent years, according to the analysis of large scale mobile phone positioning data to get residents' activity characteristics has been widely representative, which has become the international forefront of research hotspot. According to Nielsen's statistical report ≪The mobile consumer≫ in 2012, the United States, Britain, China's 16 year old people's use of mobile phone has reached 94, 97, and 89 %. So mobile phones become the most populous, the use of the most widely used, more convenient to carry, and the impact of the most widely used mobile communication equipment terminal. And mobile positioning data coverage of a wide range of data, real-time data generated, the sampling period is long, can be a large scale of the objective and long-term records of urban residents' activities of the time and space characteristics, and will not cause a burden to be investigated. So the academic community that mobile positioning can continue to effectively analyze the support of large-scale resident activities; The use of large scale mobile positioning data access to the distribution of residents, on the one hand, it can support the large-scale sample of job balance and other related empirical studies, which increases the completeness and diversity of the study population and reduce the cognitive bias caused by the sample size and sample deviation; On the other hand, real time data can also be updated to support continuous observation and rapid update. So it can provide a mature technology base for the research of the urban internal spatial structure by using the data of the mobile phone.

In view of the shortcomings of the traditional method, this paper based on Louail Thomas of the method, mobile communication based on big data, and it is first defined, then set the threshold value, when the user density is greater than a set value, it can be called hot spot. The number of hot spots can determine the size of the city, while the time distribution of regional hot spots, from which we can identify the function of the city, to identify the industrial area, commercial area, residential area, etc.

3.1 *Definition of Mobile Hot Spot Identification Method*

Under the background of big data, the data scale is huge, huge number, variety. It can be said that big data for data extraction and access to open up a new way of thinking, which is bringing the third industrial revolution. However, in the face of such a large database, the selection of effective information and the way of data acquisition has brought about a certain degree of confusions, only the fast, accurate and efficient processing of information, which can ensure the value of information, making the data play a greater value and function.

Using the mobile data to survey the city space structure, the analysis takes the city as the main body, and divides each city through the grid, and then analyzes the whole characteristic of the city by the weighted distance. On the micro level, the city area is divided into several grids, and the distribution of the important position of the city is obtained by analyzing the information of the mobile phone users.

As shown in the Fig. 1, the analysis is based on the summary of the 11 Jiangxi provincial cities in the area covered by the city of the weekend in the mobile phone mobile data, these cities on the geographical features, the number and density of urban population are not the same, the blue area indicates that the city's hot spots. From the map is not difficult to analyze, Nanchang City is the most active mobile hot city in Jiangxi province, and its central economic zone in Jiangxi Province as a comprehensive economic center, transportation, logistics, research and development, consumption and the center of the high-tech industry is not unrelated.

Table 1 shows the relationship between the major cities in Jiangxi province and the city population, the map can be found in Nanchang City, a substantial increase in the area of the city, the most popular, the most active area of mobile hotspot. Ganzhou ranks second because of the big population. We can understand the size of the city by a mobile hotspot. We can understand the size of the city by a mobile hotspot.

Fig. 1 Hotspots for 11 cities in Jiangxi Province

Table 1 The major city hot spots science and technology in Jiangxi province

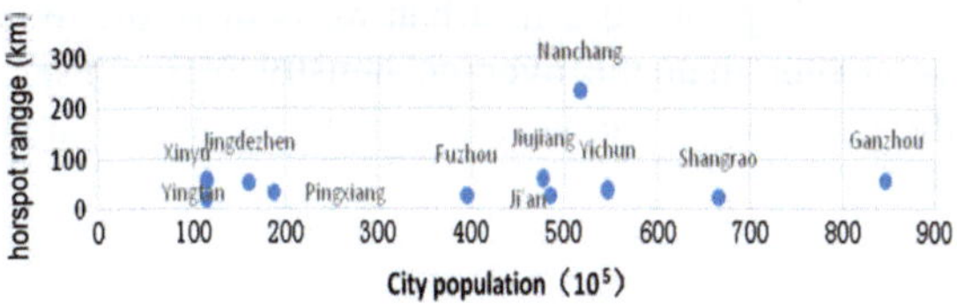

3.2 *The Model of City Hot Spot Analysis on Macro Level*

On the macro level, we mainly focus on the specific regional structure system of the urban development and the surrounding area. The development of any city, in addition to its own specific conditions (such as the location of urban location, historical basis, construction conditions and the city and adjacent areas of resources conditions, etc.), and its regional economic basis is closely related to the region. Through the observation of the hot spot in the research, we can find the relationship among the city. Here we introduce two concepts [6].

1. Average distance between individual and dilatation index:

$$v = s_i s_j d_{ij} \tag{3.1}$$

   With $s_i(t) = n_i(t)/N(t)$ the share of individuals present in cell i at time t, and d_{ij} the distance between i and j. When all activity is concentrated in one spatial unit only, the minimum value zero of v is reached. An important point of this dilatation index is that one doesn't need to determine hotspots to compute it.
2. Weight coefficient of urban area:

$$D_{(v)}t = \frac{\sum s_i(t)s_j(t)d_{ij}}{\sum s_i(t)s_j(t)} \tag{3.2}$$

   With $s_i(t) = n_i(t)/N(t)$ the share of individuals present in cell i at time t. In order to compare the value of $D_{(y)}$ across cities, we compute $D_y(t)/\sqrt{A}$ with A the area of the city. $D_y(t)$ signals how much the important places of the city at time t are distant from each other.

3.3 *The Model of City Hot Spot Analysis on Micro Level*

Most hot spot analysis of the micro level is the study of the distribution and combination of different functional areas within the city. We observe the grid by dividing the number of cities by the number of cities. Figure out the weighted average density of mobile phone use in urban resident ρ_{AVG}. When the density of the grid user ρ is greater than the average density of the user ρ_{AVG}, it can be determined that the grid range is more active than the active area.

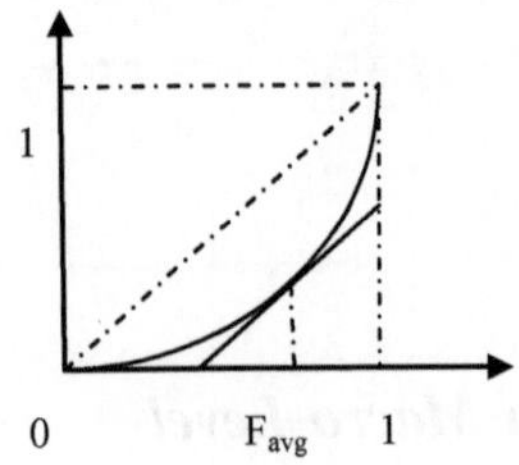

$$L(F) = \frac{1}{m}\int_{0}^{F} \rho(F)\, dF \tag{3.3}$$

$$m = \rho\,(F_{AVG}) \tag{3.4}$$

$$\frac{dL}{dF} = 1 \tag{3.5}$$

The research shows that grid cell phone user density is distributed in the Lorenz curve and to observe the grid mobile hotspot activity by the tangent of F value, according to the different time under the condition of hot spots of active rules can to the urban center and urban functional zoning recognition function.

4 Application of Mobile Hotspot Analysis Method in Nanchang City

According to ≪Nanchang overall planning(2001–2020)≫, Nanchang city will build to the urban center of the city, the Changleng area and periphery of the five towns group (Liantang, Wangcheng, Wanli, Lehua, Maqiu) constitute the metropolitan area of Nanchang. About 510 square kilometers and 3,500,000 people.

Public center main function includes commercial and business office, we choose the average density at working days 10:00 to identify business office activities agglomeration situation, choose the average density at weekend 15:00 days to identify commercial activities agglomeration status. Areas with high average density at working day 10:00 usually would be the administrative area; Areas with high average density at weekend 15:00 usually would be the commercial area. By observing the average density distribution of multiple mobile phones at working day 10:00 and weekend 15:00, we found that the Bayi square is the center of Nanchang city public system, because of being as the center of Nanchang City and main business district.

As for the identification of urban functional zoning of Nanchang city, we use the density ratio of day and night and population density at night to reflect the distribution of urban employment area, recreation area and residential area. Calculate the mean ratio of user density at working day 10:00 23:00, and weekend 10:00 23:00. If density ratio is higher at working day 10:00 23:00, it shows that the employment of this area is more than the resident population: the higher the ratio, the more the employment function of area is. If density ratio is higher at weekend 10:00 23:00, it shows that Recreation or shopping consumption is more than the area of the population: the higher the ratio, the more recreation functional. Through comparative analysis we found that Honggutan region is the CBD business district, xiaolan dominated industrial zone, area east of the Fu river consist of residential district. The research is in accordance with the existing urban structure in Nanchang City, and it is verified that the mobile hotspot analysis method is a new way of thinking to the traditional method of investigation.

5 Conclusion

This article only raised a new thought for studying urban special structure. Through mobile bid data and mobile hotspot method, the usage density of average user and the characteristics of regions under different time and space law, we could effectively recognize the urban public system and urban internal spatial structure. From the current results, we can see that mobile big data can be a good substitute of the traditional method whose renew cycle is longer and sampling costs is also high. In contrast, mobile hotspot analyzing method has the advantage with low data sampling costs and high reliability of data acquiring. Also it can master the law regularity dynamically. The structure of informational society will change the rule of traditional planning, thus making the planning more fit into human needs.

Acknowledgments This work has been supported by Youth Foundation of China Ministry of Education and Social Science Project (**grant** 15YJAZH091) and Social science planning project of Jiangxi province (**grant** 14YJ24) and 2014 Jiangxi Provincial Department of education scientific and technological research project (**grant** GJJ14355) and 2015 Jiangxi provincial science and Technology Department of foreign cooperation research project (**grant** 20151BDH80015)

References

1. Ming, Z., Yang, X.: Optimization of urban spatial structure and control division of urban space: a case study of Nanchang, pp. 43–47 Modern Urban Research, Nanjing (2011)
2. Yi, X.: Study on the interactive influence between urban spatial structure and road traffic—a case study of Nanchang. Jiangxi Teacher College, Nanchang (2007)
3. Lv, G.: The urban spatial structure based on land use research of Nanchang perspective. Jiangxi Teacher College, Nanchang (2013)

4. Guo, Q., Huang, P.: On the spatial structure of Nanchang's tourism. J. Jiangxi Agric. Univ. Nanchang **9**(2) (2010)
5. All data showed on this paper comes from the Statistical Yearbook of Jiangxi Province
6. Louail, T., Lenormand, M., Ros, O.G.C.: From mobile phone data to the spatial structure of cities. Sci Rep. **4** (2014)

Deterrent Effect of Fixed-Site Speed Enforcement in Freeways

Mengdie Yang, Jun Ma, Qiang Chen, Yichi Yang and Ning Shen

Abstract Speeding is common in freeways. Speed cameras have limited coverage and the effect of them on speed behavior is unclear. This may lead to irrational decision making on the investment and configuration of enforcement facilities. In this study, required sample speed data were collected in five cross-sections around a speed camera in G45 (Daguang Freeway) of China. Statistics showed that the distance effect model of the speed camera is a concave parabola function. Speeding drivers begin to slow down to speed limits when they approach advance warning signs and then immediately speed up after passing the enforcement site. The deterrent effect of fixed-site speed enforcement is limited to a 300-m range. It was concluded that there is an obvious distance halo effect of fixed-site speed enforcement by deterrence theory. The findings imply that the deterrent effect of fixed-site speed enforcement is weak due to its short effective distance and overt form of control. Reasonable configuration of enforcement facilities and combination with mobile speed enforcement will be appropriate to reduce speeding.

Keywords Fixed-site speed enforcement · Deterrent effect · Distance halo effect

1 Introduction

Speed enforcement technology, which refer to off-site enforcement in China, consists of fix-site speed enforcement and point to point speed enforcement. Fix-site speed enforcement is used more commonly than point to point speed enforcement to target speeding violation in freeways in China. The coverage of speed camera in freeways is limited so their effect on speed behavior is unclear. The use and effect of

M. Yang (✉) · J. Ma · Y. Yang
Department of Traffic Management Engineering, People's Public Security University of China, 102614 Beijing, China
e-mail: 459505432@qq.com

Q. Chen · N. Shen
Tianjin Traffic Police Branch of Freeway, 300402 Tianjin, China

© Springer Science+Business Media Singapore 2017

H. Lu (ed.), *Proceedings of the Second International Conference on Intelligent Transportation*, Smart Innovation, Systems and Technologies 53,
DOI 10.1007/978-981-10-2398-9_24

this technology is argumentative. Zhang [1], Zhang [2], Li [3] and Tian [4] discuss, in law perspective, problems of obtaining evidence, the procedure and the value that off-site speed enforcement have on law-executor. Wang [5] conducted a questionary survey on drivers' attitude to off-site enforcement technology, and they conclude the importance of effectiveness in enforcement in off-site enforcement. Hess [6] studied the number of traffic accidents by quantitative analysis, and proved that the effective range of a camera can make the accident quantity less than before. But in terms of evaluating the effect of law enforcement based on reducing accidents is not convincing.

This study mainly evaluates the deterrence effect of fixed-site enforcement has on drivers and, by analyzing the speed data from the five sections around the enforcement site, five sections around The purpose is to provide a quantitative basis of the development of fix-site enforcement.

2 Deterrent Effect of Traffic Enforcement Technology

Fixed-site speed enforcement is usually being set up in hot spots or special places in freeways, to monitor passing vehicles' instantaneous speeds and records the speeds of illegal speeding vehicles.

Velocity measurement detectors always sets both radar detectors and camera. Since a freeway is always closed and very long, it is not economical to set detecting points to realize covering needs in freeways. In order to expand enforcement scope, it is necessary to emphasize the deterrence effect of automatic speeding enforcement technology.

Deterrence effect is a kind of psychological effect. Its cause is based on human's psychological change of relevant driving behavior punishments, which would make people actively take measures to avoid punishments.

Chen [7], by integrating deterrence effect's development process, estimated its foreground. An [8] studied the deterrence effect from criminal law perspective and found the relationship between them. Jiang [9] built a cloud model of deterrence effect of automated enforcement on drivers in signalized intersections, and concludes that automated traffic enforcement can effectively decrease the numbers of drivers' traffic law violations.

Fixed-site speed enforcement will have deterrence effect on drivers. When drivers see a warning sign, they will decelerate gradually in order to have enough time to control vehicle velocity in limit speed before arriving at velocity enforcement sites, and avoid punishment of speeding.

The objective of speed enforcement is to decrease probability and risk level of traffic accidents. In order to avoid drivers' violation behavior, deterrence effect should be used in these aspects:

1. Drivers' safe-driving behavior habits can be redressed by detecting vehicle speed;
2. Make drivers understand and actively obey to laws and regulations by punishing and warning them;
3. With punishments being used in regulating drive behavior, deterrence effect about laws and regulations on drivers will be more effective.

In a word, deterrence effect of fixed-site speed enforcement guarantees traffic safety.

3 Method

Fixed-site speed enforcement to produce a deterrent effect's manifestation is the "distance halo effect". "Distance halo effect" can be explained as a multicyclic model, around enforcement site. Vehicle velocity which is firstly high and then decreases and at last gradually increases along distance. This model starts from the warning sign and ends at some distance after enforcement site, and it can influence driver's behavior.

3.1 Background

The objective of speed: the selected location is the confluent area of Niutuo service area and Daguang freeway from 1378 km + 500 m to 1379 km + 300 m. These five speed detection positions are 1378 km + 500 m (A1), 1379 km (A2), 1379 km + 100 m (A3), 1379 km + 200 m (A4) and 1379 km + 300 m (A5). Warning sign is located in 1378 km + 500 m (A1) and velocity measurement instrument are set in 1379 km (A2). This section's maximum speed limit is 120 km/h and survey sites are in Fig. 1.

1. The time of survey: May 13, 2015.
2. Instruments: Hand-held radar velocimeter. This instrument needs to be calibrated before using in order to detect velocity accurately.

Sampling and sample size: since the purpose of the survey is to detect the deterrent effect about fixed-site speed enforcement, so the samples are selected in cars which drive in the free flow state and at a speed of 100 km/h or higher.

During the survey, each velocity measurement point's sample size needs to be greater than the minimum number of samples required. Therefore the number of samples need to be tested sufficient or not by the theory of samples survey. The minimum number of samples formula is as follows:

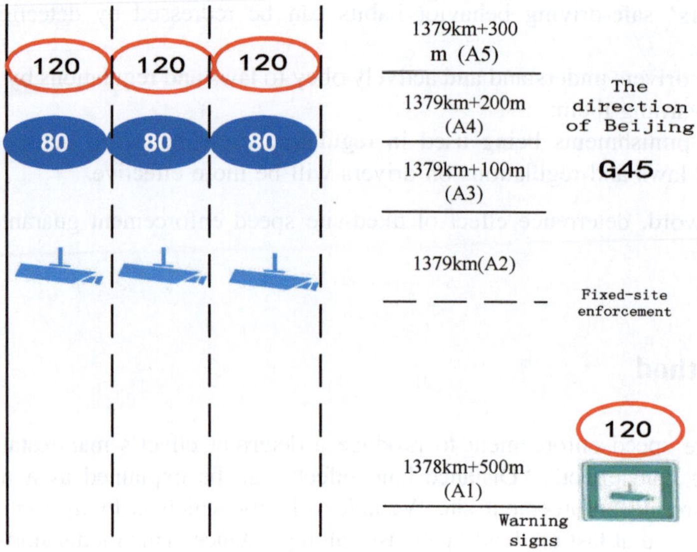

Fig. 1 The survey schematic figure

Table 1 Measured sample size and the minimum sample size

	Measured	Std	Required
A1	333	9.15	322
A2	170	6.46	161
A3	250	7.75	231
A4	250	7.64	225
A5	250	6.82	179

$$N_{min} = xS^2/e^2$$

where:

N_{min} minimum sample size of survey;

S sample standard deviation;

e tolerance.

Takes a confidence level of 95 %, x = 3.84, e = 1.0, then the required minimum sample size determined in Table 1.

Table 1 shows that the numbers of samples in every location meets the need measured sample volume.

3.2 *The Distance Halo Effect*

The analysis of these data obtained in each detect point accordance with the following data comparison above Table 2.

Table 2 Speed data statistics

	Mean (km/h)	Max (km/h)	Min (km/h)	Speeding (%)
A1	113.46	147	100	23.00
A2	109.21	134	100	5.59
A3	111.18	156	101	12.64
A4	111.78	141	101	16.89
A5	112.27	144	101	12.00

Table 2 shown:

1. From the point A1 to A2, average speed and proportion of speeding show a decline trend, which shows the deterrent effect is good. It can achieve the purpose of reduce the vehicle speed effectively.
2. Seen the detect results from A3 to A4, average speed and proportion of speeding shows an upward trend. The results also show that the maximum speed at detect point A3 has exceeded the speed A1, the speed of A4 overdrive ratio points. It is close to the level of detect point A1.
3. In A3 the maximum speed beyond the speed of the remaining instrument position, indicating that some of the drivers at detect point immediately acceleration after this point, speed points for such drivers slow down the speed just to avoid speeding violation.
4. Above these analysis shows that, A4 point after 300 m, the driver of the speeding proportion will be close to the value of A1; A3 point after 300 m, the driver's average speed is almost the same speed as the A1 point.

4 Conclusion

This research verifies the enforcement technology's deterrence effect on drivers by analyzing data of five detect point vehicle speed samples in five detect points. The deterrence effect's manifestation is the distance halo effect that happens in the area from notice sign to some distance the detect point. The research shows if a faster vehicle meets a slower one, it will overtake the slower one. And vehicles surrounded phenomenon will happen if these faster vehicles do overtaking one by one. In this situation, velocity detectors can hardly measure an overtaking vehicle's speed, what's more, collected data is influenced by the surrounding environments, velocity measurement detector and anonymous vehicles, sampling data analysis result is not ideal. But deterrence effect is still obvious based on these data and, combined with the distance halo effect, explains positive effect that fixed-site speed enforcement has on speed law enforcement.

This researches results would provide references for traffic management departments to make appropriate decision about automatic speeding enforcement.

Acknowledgments Supported by the Project of Ministry of Public Security Research Technology (NO. 2014JSYJB024).

Supported by the Project of National Science & Technology Support Program (No.2014BAG01B0501).

References

1. Zhang, W.: The development of off-site law enforcement and forensics direction of law enforcement. J. Zhejiang Police **6**, 93–96 (2008). (in Chinese)
2. Zhang, H.: Off-site law enforcement of road traffic safety management. Master Thesis, Soochow University, 2009 (in Chinese)
3. Li, S.: The exploration of off-site law enforcement in law. Master Thesis, Chinese University of Political Science and Law, 2013 (in Chinese)
4. Tian, D.: The problem and countermeasure of off-site law enforcement in public security and traffic management. Master Thesis, Guizhou University, 2011 (in Chinese)
5. Wang, S., Jin, B., Shi, J.: Analysis the off-site law enforcement effectiveness. Traffic Inf. Saf. **32**(3), 62–68 (2014). (In Chinese)
6. Hess, S.: An analysis of the effects of speed limit enforcement cameras with differentiation by road type and catchment area. Transp. Res. Rec. **1865**, 28–34 (2004)
7. Chen, Y., Chen, G.: Theoretical and empirical research of deterrent effect: past, present and future. Study Inst. Econ. (03) (2009) (in Chinese)
8. An, C.: Deterrent effect of punishment from the "strike hard". J. Shanghai Univ. 11(4) (2004) (in Chinese)
9. Jiang, X., Huang, K., Wang, B.: Quantitative analysis the deterrent effect in automated traffic enforcement. J. Harbin Inst. Technol. 46(12) (2014) (in Chinese)

Weather-Responsive Freeway Speed-Limits Using Approximated Friction Coefficient of Road Surface

Jianhua Yuan, Yang Wu, Zhiyong Zhang and Qing Ma

Abstract The regular weather-responsive design of variable speed limit (VSL, as a subsystem of Intelligent Transport System) of freeway relys on expensive realtime detection of road surface friction coefficient, not available for economy-underdeveloped regions. A low-cost approach is proposed to design 2 types of weather-responsive VSL (straight-running VSL subject to visible distance, and curve-running VSL aimed at skid-avoidance) in the absence of friction coefficient, using equivalently-approximated discrete friction coefficient (thresholds) depending on the types and intensities of adverse weather elements (i.e. the road surface conditions of dry or wet, further of water or snow/ice covering adhered on road surface), in stead of expensive realtime-detected continuous friction coefficient.

Keywords Variable speed limit (VSL) · Adverse weather · Visible distance · Radius of curve · Friction coefficient

1 Introduction

Under circumstances of adverse weather, declined visibility caused by fog/haze or rain/snow, along with decreased friction coefficient of road surface due to rain/snow/ice, result in lower speed limit (i.e. permitted ceiling speed) of vehicle on freeway than normal weather, thus necessitate the weather-responsive design of variable speed limit (VSL, as a subsystem of Intelligent Transport System).

Prevailing weather-responsive design of VSL uses approximately 2 methods: (1) To suit different thresholds of VSL to 85 % of the statistical mean of speed distribution (in unlimited regions) corresponding to different adverse weather

J. Yuan · Y. Wu (✉) · Z. Zhang · Q. Ma
Traffic Management Research Institute of Ministry of Public Security of China, Wuxi, China
e-mail: wuyang8848@qq.com

J. Yuan · Y. Wu · Z. Zhang · Q. Ma
Collaborative Innovation Center of Modern Urban Traffic Technologies, Nanjing, China

© Springer Science+Business Media Singapore 2017

H. Lu (ed.), *Proceedings of the Second International Conference on Intelligent Transportation*, Smart Innovation, Systems and Technologies 53,
DOI 10.1007/978-981-10-2398-9_25

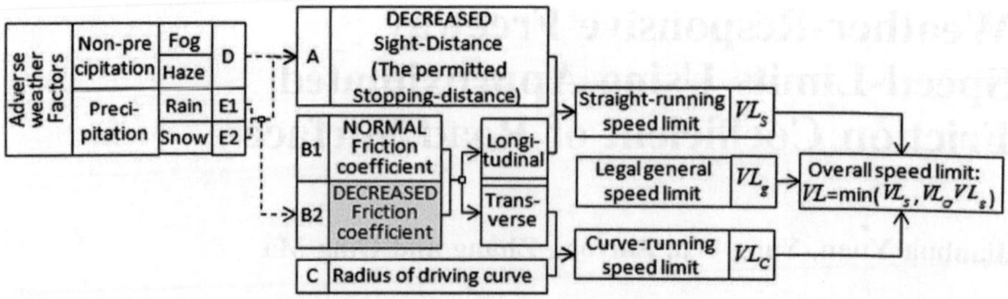

Fig. 1 The cause-effect (*dotted arrow*) and computating logic (*solid arrow*) between adverse weather and speed limit using directly realtime-detected friction coefficient

intensities [1], if large-scale statistical data available; (2) To analyze the permitted ceiling speeds at beginning of braking subject to declined stopping-sight-distances and decreased friction-coefficients [2] (as in Fig. 1), if statistic not available.

Besides directly realtime-detected stopping-sight-distance (factor A as in Fig. 1), the 2nd method requires another fundamental factor: friction coefficient (factor B1/B2 as in Fig. 1), the realtime-detection [2, 3] technologies of which are expensive for economy-underdeveloped urban/rural areas—for example, the prices of Finland Vaisala™ Remote Road Surface State Sensor (DSC) is around 40- to 80-times of the expense of regular meteorological sensors (of water/snow/ice thickness or temperature/humidity of road surface).

A low-cost approach is proposed by this paper to design 2 types of adverse-weather-responsive VSL when friction coefficient not available, using equivalently-approximated friction coefficient depending on the types and intensities of adverse weather elements, in stead of expensive realtime-detected friction coefficient.

2 Normal Friction Coefficient Dependent VSL Under Dry Weather

Combining the "straight-running VSL" subject to visible distance, and the "curve-running VSL" aimed at skid-avoidance, the overall (comprehensive) VSL is determined by the minimum amongst 3 speed-limits as in Fig. 1: "straight-running VSL", "curve-running VSL", "general legal VSL".

2.1 Straight-Running VSL Subject to Visible Distance

Vehicle's stopping sight distance from a certain speed v_0 to full halt, shall be no longer than visible distance $d_{visible}$, as shown in Eq. 1 and Fig. 2:

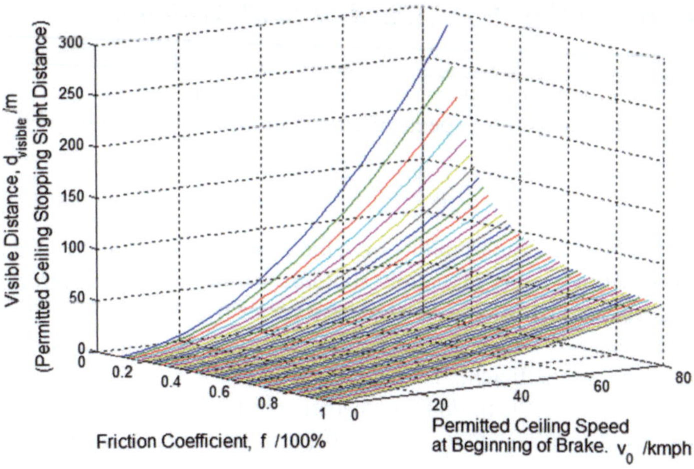

Fig. 2 Correlation between visible distance and *f* and the permitted ceiling speed at beginning of braking

$$\frac{v_0}{3.6}t_r + \left(\frac{v_0}{3.6}\right)^2 \frac{1}{2g(f+i)} + d_{safe} \le d_{visible}. \tag{1}$$

Where: $d_{visible}$ (m) realtime-detected visible distance, i.e. permitted ceiling stopping sight distance; v_0 (km/h) permitted ceiling speed at beginning of braking; t_r = 1.5 (s) driver response time(normal value = 1–2 s in free/steady traffic flow condition); d_{safe} = 5–10 (m) safe space between vehicles; g = 9.8 m/s^2; $f = \frac{\text{vehicle weight}}{\text{friction force}}(100\,\%)$, the longitudinal friction coefficient between road surface and vehicle tire; *i* (100 %) the field-measured longitudinal gradient (i > 0 if uphill, i = 0 hereinafter).

Under adverse dry (e.g. fog/haze) weather (i.e. no rain/snow/ice detected, meanwhile visible distance is detected reduced down to or below a normal standard, e.g. 500 m hereinafter), the longitudinal friction coefficient *f* can be considered as a constant of 0.8–1.0 (proposed = 0.8), showing no difference from the case of dry road surface in normal weather as Factor B1 in Fig. 1.

Therefore, straight-running speed limit VL_S (kmph), as the combination of Eqs. 1 and 2, depends on v_0 resulting form the visible distance $d_{visible}$ (m) detectable by a variety of economic means.

$$VL_S = VL_{interval} * \text{int}_{floor}(v_0/VL_{interval}) \tag{2}$$

where: $VL_{interval}$ a multiple of 10 (kmph), the interval between each and every speed-limit grade, hereinafter = 20; int_{floor} the operator of integer-conversion at the floor value.

2.2 Curve-Running VSL Aimed at Skid-Avoidance

The critical condition of vehicle' skidding [2] (as shown in Eq. 3 and Fig. 3), confines the permitted ceiling speed of curve-running v_{curve} (kmph):

$$v_{curve}^2 \leq 127 \cdot r(\varphi + \varepsilon) \tag{3}$$

where: r(m) the field-measured radius of vehicle path conforming to road curvature; ε the field-measured super-elevation (default value = 0 representing the most dangerous horizontal surface); φ the transverse friction coefficient (in lateral direction) between road surface and vehicle tire (can be considered equal to the longitudinal friction coefficient *f*).

Under adverse dry (e.g. fog/haze) weather, the transverse friction coefficient φ can be considered as a constant of 0.8–1.0 (proposed = 0.8), identical to longitudinal friction coefficient *f* of dry road surface in normal weather as Factor B1 in Fig. 1.

Therefore, curve-running speed limit VL_C (kmph), as the combination of Eqs. 3 and 4, depends on v_{curve} resulting form the radius of vehicle path r (m) field-measured.

$$VL_C = VL_{interval} * \text{int}_{floor}(v_{curve}/VL_{interval}) \tag{4}$$

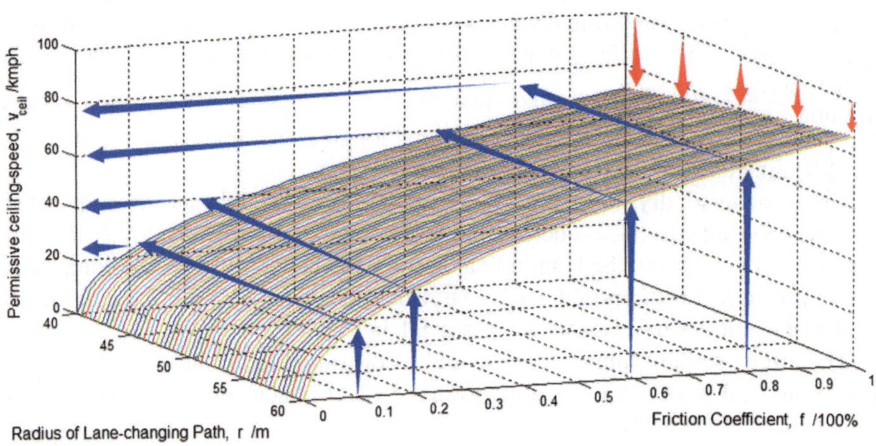

Fig. 3 Correlation between radius of vehicle path and φ and the permitted ceiling speed of curve-running

2.3 Overall VSL

The overall speed limit *VL* is determined by the minimal amongst 3 speed-limits: longitudinal VL_S, transverse VL_C, and general legal VL_g, as Eq. 5.

$$VL = \min(VL_S, VL_C, VL_g) \tag{5}$$

3 Decreased Friction Coefficient Dependent VSL Under Wet Weather

Under adverse wet (e.g. rain/snow) weather (i.e. any of rain/snow or water/snow/ice adhered onto road surface detected), the varying longitudinal/transverse friction coefficient $f(\varphi)$ is significantly lower than the case of normal dry weather. Given a certain DECREASED *f*, Eqs. 1 and 3 can calculate longitudinal v_0 (kmph) and transverse v_{curve} (kmph) subject to adverse wet weather.

In the absence of realtime-detection of Friction Coefficient (Factor B2 as in Fig. 4), we require: ① The low-expense realtime-detection of the thickness/depth of Water/Snow/Ice adhered onto road surface, i.e. Factor F as in Fig. 4; and ② The equivalent-approximation (by Factor F) of Factor B2 (Friction Coefficient).

3.1 Approximation of Friction Coefficient in the Absence of Real-Time Detection

3.1.1 The Mapping of "Rainfall-Related Water Thickness" to "Friction Coefficient"

Railfall in different intensities give rise to Water Film with different Thickness adhered onto road surface (detected by regular meteorological sensor with a typical

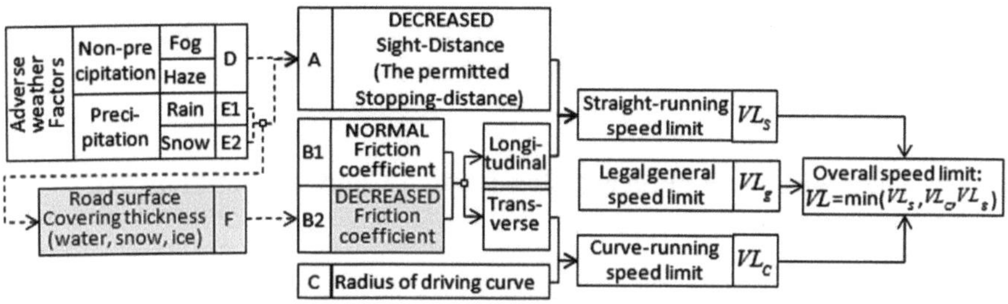

Fig. 4 The cause-effect and computating logic between adverse weather and speed limit in the absence of realtime-detected friction coefficient

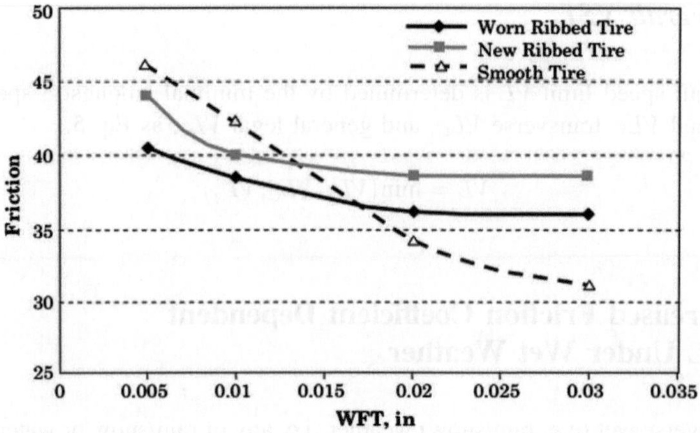

Fig. 5 The correlation between WFT (in./25.4 mm) and friction coefficient (100 %) from Ref. [2]

range of 0–6 mm and the precision of 0.5 mm), reducing the road surface friction coefficient *f*, unbalancing vehicle's equilibrium between transverse friction and centrifugal force, causing skid of vehicle to different extents.

The research results from Ref. [2] indicates that: (i) a Water Film Thickness (WFT) of at least 0.025–0.05 mm can reduce *f* by 20–30 % up to 60 % of the dry friction coefficient; (ii) *f* decreases EXPONENTIALLY as WFT increases, furthermore, the rate at which *f* decreases generally becomes smaller as WFT increases; (iii) When WFT ≧ 0.5 mm, Friction coefficient *f* remains around 0.40 with slight volatilities less than 5–10 % subject to pavement micro/macro/mega-texture and tire tread depth, etc., as in Fig. 5; (iv) As WFT increases further up to the range of 2.5–5.0 mm, Aquaplaning or so-called Hydroplaning phenomenon (i.e. a vehicle tire is separated from the pavement surface by the water pressure building up at the pavement–tire interface) can occur, causing friction coefficient drop to a near-zero level around 0.15–0.25, on incorporated grounds of the research results of Air Accidents Investigation Branch UK and NASA US [4–6].

In contrast to the realtime-detected continuous spectrum of friction coefficient, the equivalently-approximated discrete thresholds of friction coefficient resulting form realtime-detected WFT can be concluded as Table 1, applied in Eq. 1/3 to design the speed limit under rainfall-related adverse weather.

3.1.2 The Mapping of "Precipitation-Related Conditions of Snow/Ice" to "Friction Coefficient"

As long as ice, caused by snowfall and/or rainfall, is detected (by regular meteorological ice and temperature/humidity sensor, mechanically-vibrational or thermo-dynamics or photoelectrical or ultrared, with a typical range of 0–6 mm and

Table 1 The discrete thresholds of Friction coefficient f subject to WFT

Realtime-detected ceiling of WFT (mm)	Aquaplaning occurs?	Threshold partitioned friction coefficient f (100 %)
<0.5	Negative	Exponential decrease from the value of dry condition (i.e. 0.8–1.0)
0.5–2.5	Negative	0.4
$\geq$2.5	Yes	0.2

the precision of 0.5 mm, plus the function to recognize the featured types between "wet/loose ice including the complex of ice and snow" and "dry/compacted ice"), no matter in form of the wet/loose or the dry/compacted, the decreased friction coefficient ranges from 25 to 50 % of the dry condition, with a floor value around 0.1–0.2 [6]—this principle learned by field studies is officially applied by Finland Vaisala™ friction coefficient sensor.

Using the minimal floor value of friction coefficient 0.1 and Eq. 1/3, the speed limit subject to the conditions of rain/snow/ice covering under precipitation-related adverse weather can be calculated.

3.2 VSL Based on Approximated Friction Coefficient

Depending on the types and intensities of adverse weather (i.e. the road surface conditions of dry or wet, further of water or snow/ice covering adhered onto road surface), the threshold-partitioned discrete friction coefficient are concluded as in Table 2, applied in Eq. 1/3 to design the speed limit under adverse weather in the absence of realtime-detected continuous friction coefficient.

Table 2 The threshold-partitioned discrete thresholds of Friction coefficient f under adverse weather

Rainfall detected?	Water covering detected?	Ice covering detected?	Realtime-detected ceiling of WFT/mm	Aquaplaning occurs?	Threshold-partitioned friction coefficient f (100 %)
N	N	N	–	–	0.8 (normal dry condition)
N/Y	Y	N	0.5–2.5	Negative	0.4
N/Y	Y	N	$\geq$2.5	Yes	0.2
N/Y	N/Y	Y	–	–	0.1

4 Conclusion

1. A low-cost method is proposed to design 2 types of weather-responsive VSL (straight-running VSL subject to visible distance, and curve-running VSL aimed at skid-avoidance), using threshold-partitioned discrete friction coefficient depending on the types and intensities of adverse weather elements (i.e. the road surface conditions of dry or wet, further of water or snow/ice covering adhered onto road surface), in stead of expensive realtime-detected continuous friction coefficient.
2. A series of supplements to the method is in progress, adaptive to more varieties of contaminant e.g. dust and/or sand (along with wind, etc.) on road surface.

Acknowledgments This paper is supported by National Key Technology R&D Program of China "Key Technologies and System Integration of Network-based Coordinated Control of Freeway Traffic Safety (Project No.: 2014BAG01B04)", Key Laboratory of Road Traffic Safety of Ministry of Public Security.

References

1. Watanatada, T., Harral, C.G., Paterson, W.D.O., Dhareshwar, A.M., Bhandari, A., Tsunokawa, K., et al.: The World Bank Publication—Highway Design Maintenance Standards Series 995 vol. 1: Description of the HDM-III Model. The Johns Hopkins University Press (1987)
2. Applied Research Associates Inc. Champaign IL, CDRM Inc. State College PA, NASA Langley Research Center Hampton VA, Penn. Transp. Institute, University Park, PA: NCHRP 01-43 Report: Guide for Pavement Friction. NCHRP web-only document No. 108, TRB (2011)
3. Andersson, M., Bruzelius, F., Casselgren, J., et al.: Road Friction Estimation. IVSS Final Report, Reference number 2004:17750 (2007)
4. Horne, W.B., Dreber, R.C.: Phenomena of pneumatic tire hydroplaning, NASA Tech Note, D-2056, p. 10 "depth of fluid" (1963). Accessed at www.hathitrust.org
5. Katz, B., O'Donnell, C., Donoughe, K., et al.: FHWA-SA-12-02 Final Report: Guidelines for the Use of Variable Speed Limit Systems in Wet Weather. FHWA, U.S. DOT (2012)
6. Haavasoja, T., Pilli-Sihvola, Y.: Friction as a measure of slippery road surfaces. In: Proceedings of the 15th Standing International Road Weather Commission (SIRWEC) Conference, University of Birmingham, UK (2010)

Algorithm of Speed-up Turnout Fault Intelligent Diagnosis Based on BP Neural Network

Kai Zhang, Yongfeng Ju, Kai Du and Xu Bao

Abstract Based on analysis of action current curves change law when the speed-up turnout is normal and fault, this paper summarized the current curve eigenvalues, proposed the turnout fault intelligent self-diagnostic algorithms based on change characteristics of the turnout action current curve. Then mapping sample set between action current curve eigenvalues and turnout fault types, and using BP neural network to establish speed-up turnout fault intelligent diagnosis algorithm. The results show that: the fault diagnosis algorithm of the speed-up turnout is high precision and adaptability.

Keywords Speed-up turnout · Fault diagnosis · Action current curve · Neural network

Railway signal and communication system is an important technical means to ensure traffic safety, improve transport efficiency and improve railway staff working conditions. Turnout, signal, track circuit is the 3 big items outside the signal and communication system, in which the turnout is conversion and locked, is critical equipment directly related to traffic safety, and is the equipment directly affect rail transport [1]. Statistical Report about National Railway signal and communication system fault on China Railway Transport Bureau of 2014 and 2015 displays that turnouts equipment failures in 2013 totaled 184, accounting for 18 % of the total number of signal and communication system failure, turnouts equipment failure in 2014 is 1124, accounting for 36.76 % in all the signal and communication system failure, which is the highest railway signal and communication system failure rate. Existing turnout fault diagnosis mainly depends on indoor and outdoor staff collaboration to develop manual troubleshooting process. The phenomenon of mis-

K. Zhang (✉) · Y. Ju · K. Du
Department of Electronic and Control Engineering, Chang'an University, Xi'an 710064, China

X. Bao
Key Laboratory for Traffic and Transportation Security of Jiangsu Province, Huai'an 223003, China

© Springer Science+Business Media Singapore 2017

H. Lu (ed.), *Proceedings of the Second International Conference on Intelligent Transportation*, Smart Innovation, Systems and Technologies 53,
DOI 10.1007/978-981-10-2398-9_26

carriage of justice and missing is higher by artificial troubleshooting, making troubleshooting occurred too long time which has a serious impact on the efficiency of rail transport. With the rapid development of science and technology and universal application of artificial intelligence, the existing turnout troubleshooting mode has not adapted to the development needs of the railway.

In recent years, domestic and foreign experts, scholars have begun to research turnout fault intelligent diagnosis. In 2009, Atamuradov et al. [2] proposed using expert system to diagnose turnout fault; 2012, Li and Wei [3] developed turnout fault diagnosis system research based on fuzzy neural network; 2012, Zhang and Ji [4] modeled and analysed fault diagnosis based on fuzzy probability Petri nets; 2012, Di [5] applied bayesian network to the turnout control circuit fault diagnosis; 2013, Li and Dong [6] used information fusion to research turnout fault diagnosis; 2015, Guan [7] proposed the high-speed railway turnout troubleshooting based on FOA-LSSVM; 2015, Tian [8] proposed high-speed railway turnout fault diagnosis based on Fuzzy Neural Network. In this paper, divided the speed-up turnout action current curve into N time zones by time, extracted the eigenvalues to form eigenvectors, then created the sample set which eigenvectors corresponding with the fault. Finally, using BP neural network algorithm designed speed-up turnout intelligent fault diagnosis algorithm. Algorithm can effectively intelligently diagnose speed-up turnout typical faults, greatly reduced fault delay, thus improved transport efficiency.

1 Action Current Curve Analysis of Speed-up Turnout

S700 K electric switch machine, ZYJ-7 type electro-hydraulic switch machines and ZDJ9 electric switch machine is suitable for speed-up turnout, this paper use the type S700K electric switch machine as an example to analyse. Type of S700K speed-up turnout, the motor adopts three-phase ac power supply, in the action current curve use red, green, blue three lines representing respectively ac 380 v power supply of A, B, C three-phase current. Normal turnout action current curve can be divided into: unlock area—action area—locking area—slowly release area. T_0–T_1 is the unlock area: turnout starting current is large, complete the unlock process. In this process, when the motor starts, there is a large starting current, accompany with produce larger torque at the same time. T_1–T_2 is the action area: which is the conversion process of turnout. If the current action is small, then show the resistance is small; if the current action is slightly larger, indicate convert resistance is large or jammed. T_2–T_3 is the locking area: the turnout turn into the locking process. The process for turnout point rail is driven to the other side and point rail post with stock rail tightly, locking block is pop-up, cut the action current. T_3–T_4 is the slow release area: The fourth session is stage of 1DQJ slow release. Turnout conversion is completed, the snap switch contact group convert to a predetermined position, disconnect the starting circuit, 1DQJ autistic circuit breaking into the slow release state. Within 1DQJ slow release time, the two-phase small current still exist in start-up circuit, which is due to automatic opening and closing contacts connecting a outdoor indication circuit when the turnout is in place, slow release area varies depending on 1DQJ slow discharge time, current value depends on a circuit resistance, usually is 0.5 A (Fig. 1).

After the actual site investigation and analysis, turnout typical faults of S700K type electric switch machine are summarized in Table 1 and turnout action current curves are shown in Fig. 2.

Fig. 1 The turnout action current curve decomposition

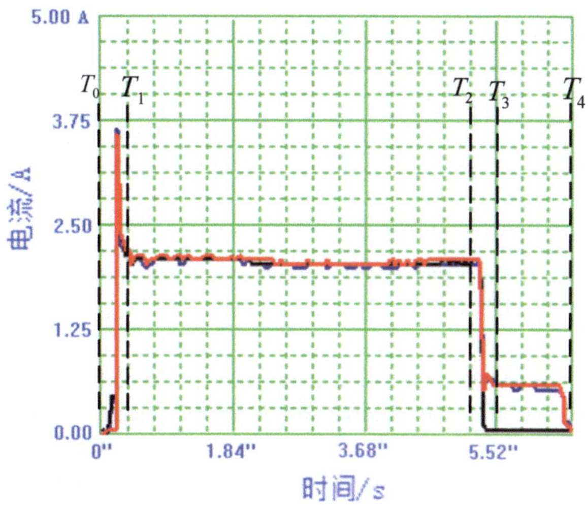

Table 1 Summary of turnout typical faults

Type code (figure no.)	Phenomenon	Fault reason
1(E_1)	Turnout action current curve is normal, but a turnout circuit fault	Indicates relay parallel branch open
2(E_2)	Action current curve is idling curve, idle time is about 13 s	Dense paste strength or card gap
3(E_3)	It represents three-phase current curves are zero value curve	Outdoor indication circuit open
4(E_4)	It represents two-phase current curves are slightly over switching current curve	Diode and a resistor shorted in outdoor indication circuit
5(E_5)	Action current curve is a zero value curve	Turnout DBQ are bad plug or a two-phase, three-phase power outage
6(E_6)	One phase current curve is zero value curve, the other two-phase current curve down quickly after rising rapidly	Power supply missing one phase

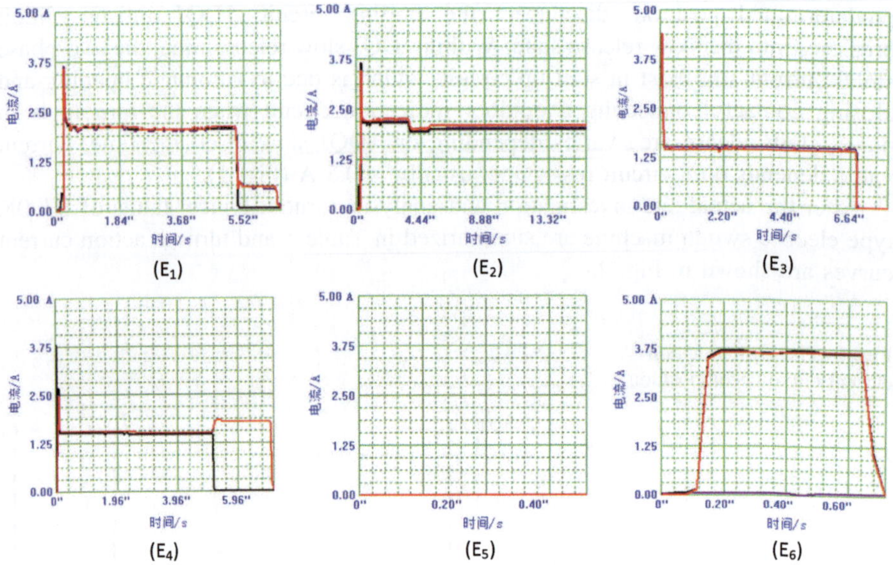

Fig. 2 The action current curve of turnout fault (the microcomputer monitor actually screenshot when turnout fault)

With the acceleration of railway, speed-up turnout will be the development direction of the future, which S700K electric switch machine is mainly used for speed lines or high-speed lines turnout, and ZYJ-7-type electro-hydraulic switch machine action current curve is basically the same with S700K type electric switch machine action current curve, this study also apply for turnouts fault diagnoses of ZYJ-7-type electro-hydraulic switch machine. Also series of ZD6 electric switch machine action current curve and S700K electric switch machine has some similarities, so this study also has reference value for series of ZD6 electric turnout switch machine fault diagnosis.

2 Algorithm of Speed-up Turnout Fault Intelligent Diagnosis Based on BP Neural Network

2.1 Selection of Eigenvectors

S700K type action current curve is constituted of a three-phase current curve. The raw data is divided into N time segment according to the time sequence, so that each time zone can be extracted fault characteristics difference relatively weaker compared with the whole sample. Then extract three-phase current value when the current is maximum, minimum, mean for every time segment, There totaled $6 * N$ parameter statistical characteristics, and then add the actual turnout conversion time into eigenvectors, so each curve corresponds to a length of $6 * N + 1$ eigenvectors. Extracting a feature for each time zone, save the details of the characteristic curve of every section.

Set actual conversion time is T, then the time length of each segment was $t = T/N$, N takes 2^n, where $n = 1, 2, 3, \ldots, n$. If the value of n is too large, resulting in an increase in the time zones N, the length of time segment t reduced, along with the feature vector will be excessive, causing the input eigenvectors dimension algorithm increases, prolonged training, and important eigenvectors of action current curve are ignored. Turnout action current set I is divided into N matrixes with three rows and M = ceil(t/0.04) columns, which is shown in equation.

$$I_N = \begin{bmatrix} I_{A1} & I_{A2} & \cdots & I_{AM} \\ I_{B1} & I_{B2} & \cdots & I_{BM} \\ I_{C1} & I_{C2} & \cdots & I_{CM} \end{bmatrix}, \quad N = 1, 2, 3, \ldots, 2^n \tag{1}$$

Calculate the sample eigenvector within I_N matrix according to the following formula:

1. Maximum of time zones: in the Nth time zone, when $max(I_K)$ is maximum, collection time is P, select the three-phase phase current value of the same time P.

$$I_{max(K)} = max(I_K) \text{ where } K = 1, 2, 3, \ldots, N$$

$$I_{max(A)} = I_{AP}$$

$$I_{max(B)} = I_{BP}$$

$$I_{max(C)} = I_{CP}$$

The method matlab program to calculate P values are as follows:

$$[3, k] = \text{size}(I_k);$$

$$\text{P} = ceil(find(I_K == max(max(I_k)))/3)$$

2. Minimum of time zones: in the Nth time zone, when $min(I_K)$ is maximum, collection time is p, select the three-phase phase current value of the same time p.

$$I_{min(K)} = min(I_K) \text{ where } K = 1, 2, 3, \ldots, N$$

$$I_{min(A)} = I_{Ap}$$

$$I_{min(B)} = I_{Bp}$$

$$I_{min(C)} = I_{Cp}$$

The method matlab program to calculate p values are as follows:

$$[3, k] = \text{size}(I_k);$$

$$\text{p} = ceil(find(I_K == min(min(I_k)))/3)$$

3. Mean of time zones:

$$I_{mean(A)} = (I_{A1} + I_{A2} + I_{A3} + \cdots + I_{AM})/M$$

$$I_{mean(B)} = (I_{B1} + I_{B2} + I_{B3} + \cdots + I_{BM})/M$$

$$I_{mean(C)} = (I_{C1} + I_{C2} + I_{C3} + \cdots + I_{CM})/M$$

When turnout actual conversion time is T, n takes 2, action current curve is divided into four sections with the time period, extract the current value when the current is maximum, minimum, mean for every time segment, establish speed-up turnout eigenvectors, constitute the sample set with fault type.

2.2 Intelligent Diagnosis Algorithm Implementation

Step 1: Set samples of (X, Y), where X is a speed-up turnout fault feature vector, Y is the fault type code.

Step 2: When the system starts to run, θ^1, θ^2, w^1, w^2 are designed to be a random number in [−0.5,0.5].

Step 3: Calculate the output of the hidden layer Q_j.

$$Q_j = f\left(\sum_{i=1}^{m} w_{ji}^1 x_i - \theta_j^1\right) \tag{2}$$

Then afferent the output layer with the hidden layer neuron output Q_j, calculate the output layer Z_k.

$$z_k = g\left(\sum_{j=1}^{l} w_{kj}^i Q_J - \theta_K^2\right) \tag{3}$$

where *the* hidden layer function $f(\cdot)$ is *tansig*, and output layer function $g(\cdot)$ is *purelin*.

Step 4: According to the error of the network output Z_k and the desired output Y_k, constantly adjusting θ^1, θ^2, w^1, w^2, until the error meets the requirement of precision. The error between the network output and the desired output is E [9, 10].

$$E = \frac{1}{2}\sum_{k=1}^{n}(y_k - z_k)^2 = \frac{1}{2}\sum_{k=1}^{n}\left\{y_k - g\left[\sum_{j=1}^{l} w_{kj}^2 f\left(\sum_{i=1}^{m} w_{ij}^1 - \theta_j^1\right) - \theta_k^2\right]\right\}^2 \tag{4}$$

$$w_{ji}^1(t+1) = w_{ji}^1(t) + \eta^1 \delta_j^1 x_i \tag{5}$$

$$w_{kj}^2(t+1) = w_{ki}^2(t) + \eta^2 \delta_j^2 Q_j \tag{6}$$

where η_1 and η_2 are the learning step length of hidden layer and output layer respectively.

$$\theta_j^1(t+1) = \theta_j^1(t) + \eta^1 \delta_j^1 \tag{7}$$

$$\theta_k^2(t+1) = \theta_k^2(t) + \eta^2 \delta_k^2 \tag{8}$$

3 Experiment and Analysis

The collection of turnout fault action current curves is very difficult, but in order to ensure the validity of the sample diversity and verification algorithm, the research requires a lot of samples. This article is divided into the training sample and test sample sets with random method. Use the MATLAB toolbox to achieve the establishment of BP neural network model algorithm. As shown in Fig. 3, training process of the algorithm is convergence quickly, the error is small, which indicating

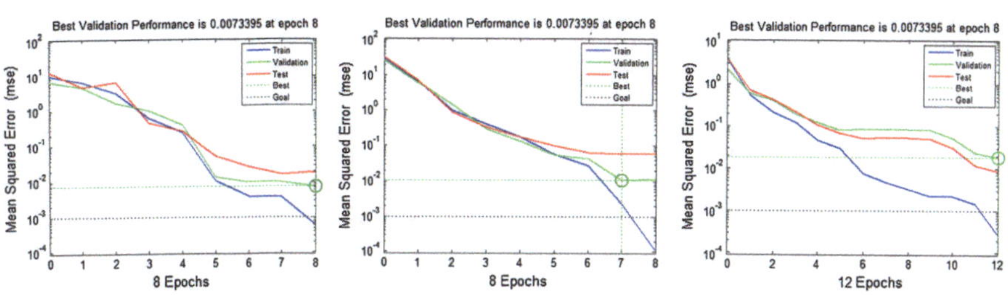

Fig. 3 Three times in the training process of the algorithm

good performance established by BP neural network. This paper unfolds three tests, three total correct rate are 100 %. Experiments show that algorithm of speed-up turnout indication circuit fault diagnosis based on BP neural network of S700K electric switch machine has better fault recognition (Figs. 4, 5 and 6).

Fig. 4 First test results

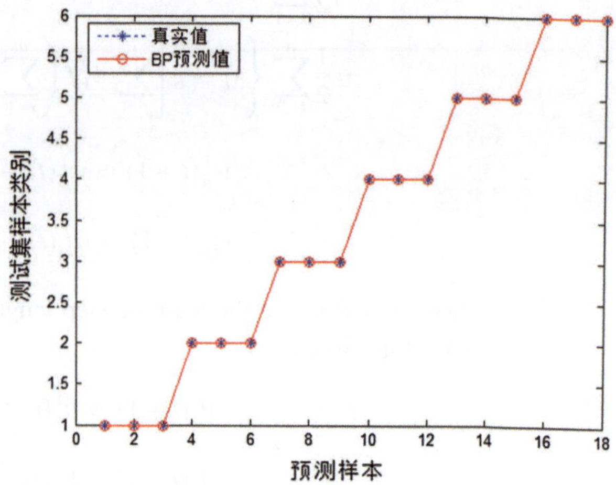

Fig. 5 Second test results

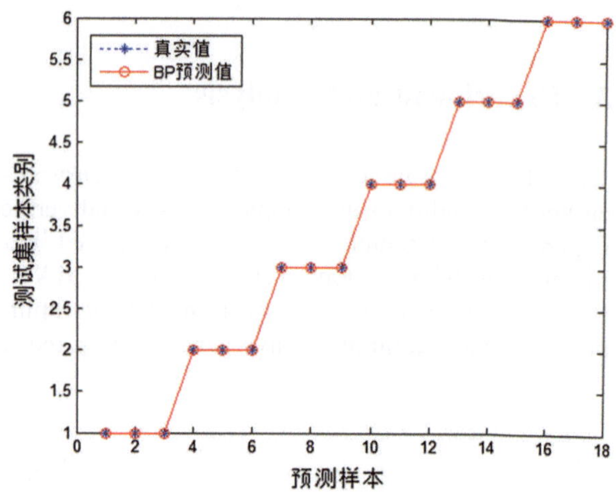

Fig. 6 Third test results

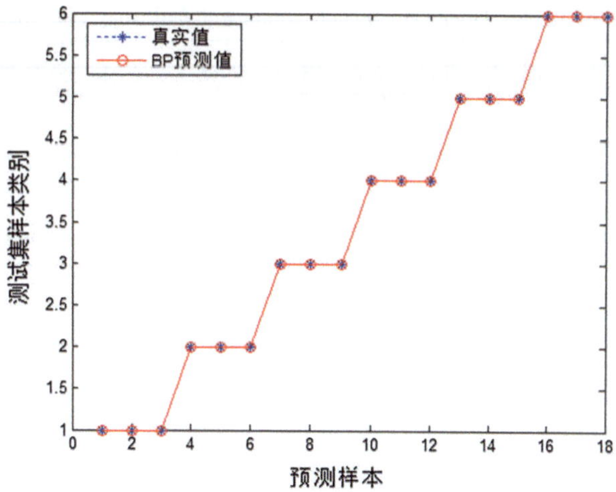

4 Conclusion

This article summarizes under typical turnout fault the S700K type electric switch machine action current curve which is divided into N time zone, extract eigenvectors in each time zone, establish the sample set of a eigenvectors corresponding to the type of fault, using BP neural network to intelligently diagnose turnout faults. Experiments show that the algorithm has high accuracy, and can meet the actual needs of railway, the algorithm is suitable for ZYJ-7 electro-hydraulic switch machines and ZDJ9 electric switch machine turnout, also has reference to the use of electric switch machine ZD6 turnout.

Acknowledgments Supported by the Science and Technology Industrial Research Project of Shaanxi(2015GY033), the Fundamental Research Funds for in the Central Universities (310832142008), the open fund for the Key Laboratory for Traffic and Transportation Security of Jiangsu Province (TTS2015-04).

References

1. Lin, Y.: Railway signal foundation. China Railway Press, Beijing (2006)
2. Atamuradov, V., Camci, F., Baskan, S.: Failure diagnostics for railway point machines using expert systems. In: SDEMPED 2009, pp. 1–5 (2009)
3. Li, Y., Wei, W.: Research of switch fault diagnosis system based on fuzzy neural network. Comput. Commun. Signal **21**(1), 35–39 (2012)
4. Zhang, W., Ji, J.: Modeling and analysis of fault diagnosis based on fuzzy probability petri net. J. Suzhou Univ. Sci. Technol. (Nat. Sci.) **29**(1), 57–60 (2012)
5. Di, Y.: Research on Bayesian networks in the application of rail switch control circuit fault diagnosis. Lanzhou Jiaotong University, Lanzhou (2012)

6. Li, N., Dong, H.: Switch equipment fault diagnosis method based on information fusion research. Railway Oper. Technol. **19**(2), 1–3 (2013)
7. Guan, Q.: High-speed railway switch failure diagnosis based on FOA-LSSVM. Bull. Sci. Technol. **31**(4), 230–232 (2015)
8. Tian, J.: Fault Diagnosis Method for Railway Switch Point Based on Fuzzy Neural Network. Beijing Jiaotong University, Beijing (2015)
9. Zhao X.: A Study on Techniques of Car License Plate Recognition Based on Neural Network. Xidian University, Xi'an (2005)
10. Shi, F., Wang, X., Yu, L., Li, Y.: MATLAB Neural Network Analysis of 30 Cases. Beijing University of Aeronautics and Astronautics Press, Beijing (2011)

Research and Application of Traffic Visualization Based on Vehicle GPS Big Data

Xin Wang, Shuxu Zhao and Liang Dong

Abstract GPS data of vehicle on the road can reflect the actual road status, its analysis can be help to urban road planning, but the multi-source, mass and high dimension features of GPS big data has restricted its application for road traffic. Aiming that, a GPS big data visual computation architecture has been design in this paper, speed attribute and two-pass corner detection has been introduce to improve map matching and clustering analysis methods. Vehicle GPS data of Zibo city has been selected as the case, the relevant result shows that the improved methods can be effective and visual, and can get a better effect than flow map method for vehicle GPS big data processing.

Keywords Visual analysis · GPS big data · Clustering analysis · Map-matching algorithm · Traffic data

1 Introduction

With the continuous development of urban traffic and positioning technology, the increasing of vehicle has leaded to a explosive growth of trajectory data, it is important that how to induce information and rules from these complex data in order to service urban transportation development [9]. Visualization method mentioned in this paper mainly is used to grasp the whole scene of the massive GPS trajectory data, help users to balance conflicts, and get the implicit knowledge data to improve road traffic planning.

Tobler [8] has studied the flow map, finished population migration map based on the population data from 1965 and 1970. Kapler has exploited the development of visualization software called GeoTme [4], which can show each attribute of the trajectory data and track incidents involving a particular attribute, this method not

X. Wang · S. Zhao (✉) · L. Dong
School of Electronic and Information Engineering, Lanzhou Jiaotong University, Lanzhou, China
e-mail: Zhaosx_2012@163.com

© Springer Science+Business Media Singapore 2017
H. Lu (ed.), *Proceedings of the Second International Conference on Intelligent Transportation*, Smart Innovation, Systems and Technologies 53,
DOI 10.1007/978-981-10-2398-9_27

only can not break the trajectory space attribute features but also introduce the time attribute, this method makes utilizing trajectory data fully to be possible.

2 Visualization Algorithm Architecture

2.1 Map Matching

Vehicle GPS (Global Position System) trajectory data needs to be rebuilt and calibrated before its visualization [10], and then its attributes, such as space, time, and others can be visualized. Map matching is a kind of technology which can make use of network information to modify the original track data, the basic idea of this method aims to serialize vehicle position by comparing with the electronic position data of road network, and acquire the related location in the road electronic map. The process includes the trajectory data rebuilding, data cleaning, data saving, and so on.

Generally vehicle GPS data acquired at first time have some features, such as poorer accuracy, serious loss signal, massive volume, and lower sampling rate. All these features makes it difficult to locate vehicle trajectory on the electronic map correctly. At the same time, lower sampling rate enlarges the distance of two adjacent GPS points, so the connection of two adjacent GPS points can not meet with the real electronic map, if both points have been connected with each other directly. Low sampling rate of GPS data matching algorithm [3] makes some improvements, which has introduced the network topology information to solve connection of two trajectory points for lower sampling rate. At the same time, because of the vehicle positioning error, the probability of GPS point located at adjacent sections is increasing, it is difficult to locate the accurate track of vehicle on the road, the algorithm shows a higher accuracy and shorter computing time in dealing with this problem. In this paper, the low sampling rate of GPS data matching algorithm has been used and improved.

Map matching including road network data and GPS data. Road network data including road network element and topology (node, link, connectivity and other information). Given the vehicle trajectory L: $p_1 \rightarrow p_2 \rightarrow \cdots \rightarrow p_n$ and candidate points set P related with each GPS point data: $c_1^{j_1} \rightarrow c_2^{j_2} \rightarrow \cdots \rightarrow c_n^{j_n}$, c_i^j is the jth candidate point of p_i. Firstly, retrieves the candidate road sections which is a circle range (for each point p_i, the circle's centre is p_i, its radius is r, $1 \le i \le n$).Then the candidate points can be calculated, which are projection of p_i to these road sections. Distance calculation is as formula (2.1).

$$d = 2\pi R\gamma/36 \tag{2.1}$$

$$\gamma = 2\arcsin\sqrt{\sin^2[(x - x_1)/2] + \sin^2[(y - y_1)/2]\cos x \cos x_1} \tag{2.2}$$

In the formula (2.2), x_1 and y_1 are coordinates of point $n_1(x_1, y_1)$ in the candidate road section, R is the earth radius. After retrieving the track all over the sampling points on the L and completing the candidate point set P calculation, the next step is how to choose a candidate point from each set to make $P: c_1^{j_1} \rightarrow c_2^{j_2} \rightarrow \cdots \rightarrow c_n^{j_n}$ match with $L: p_1 \rightarrow p_2 \rightarrow \cdots \rightarrow p_n$ perfectly.

There need to join in the road topology information to evaluate the candidate in the matching process. This paper, observation probability is used to improve the road topology constraints, it is described as the possibility of GPS sampling points p_i matching with the candidate point c_i^j, its value is based on the distance between the two points $dist(c_i^j, p_i)$. The observation probability of c_i^j to p_i is as (2.3).

$$N\left(c_i^j\right) = \frac{1}{\sqrt{2\pi}\sigma} e^{\frac{(x_i^j - \mu)^2}{2\sigma^2}} 0 \leq x_i^j \leq +\infty \tag{2.3}$$

$x_i^j = dist(c_i^j, p_i)$ is the distance between p_i and c_i^j. In this paper, the zero mean normal distribution is used ($\mu = 0$), the standard deviation σ is set as 50 based on experience. About the final path, the transmission probability described as (2.4) is combined with the observation probability to determine the combination of space and topological constraints of optional sections set $G'(V, E)$, V is candidate points set with each GPS data, E is the road edge candidate set.

$$\mathrm{V}\left(c_{i-1}^t \rightarrow c_i^s\right) = \frac{d_{i-1 \rightarrow i}}{w_{(i-1,t) \rightarrow (i,s)}} \tag{2.4}$$

$d_{i-1 \rightarrow i} = dist(p_{i-1}, p_i)$ is the Euclidean distance between p_{i-1} and p_i, $w_{(i-1,t) \rightarrow (i,s)}$ is the length of the shortest path between c_{i-1}^t and c_i^s. Based on two formulas, the definition of spatial analysis function $F_s\left(c_{i-1}^t \rightarrow c_i^s\right)$ can be described as (2.5).

$$F_s\left(c_{i-1}^t \rightarrow c_i^s\right) = N\left(c_i^s\right) * V\left(c_{i-1}^t \rightarrow c_i^s\right), \quad 2 \leq i \leq n \tag{2.5}$$

c_{i-1}^t and c_i^s are the candidate points of p_{i-1} and p_i respectively, in (2.5) two probability has been included, so the geometry and topology attributes has been shown in the model. But in fact, moving objects are unable to always follow the shortest path strictly, so the observation probability $N\left(c_i^j\right)$ can't be ignored in formula (2.5). Through spatial analysis, a set of candidate paths for the two arbitrary GPS points p_{i-1} and p_i can be got, and a spatial value can be obtained via formula (2.5). The shortest route from G' can be calculated from p_0 to p_n, and as the travel path matching results. If route is unreachable, the matching process fail, if many routes exist, the shortest route can be as the matching result. Algorithm process is as the following.

Input: Road network data G, GPS data $(p_1 \rightarrow p_2 \rightarrow \cdots \rightarrow p_n)$, Output: Matching sequence $G'(c_1^{j_1} \rightarrow c_2^{j_2} \rightarrow \cdots \rightarrow c_n^{j_n})$

1. Initialize candidate paths list.
2. Calculate the candidate route of $p_i (1 \le i \le m)$ based on formula mentioned above.
3. Add the satisfied points into $G'(V, E)$, then the candidate route set can be got.
4. Calculate the best matching route G' based on $N\left(c_i^j\right)$ and the shortest route constraints.

2.2 Trajectory Clustering Computation

Clustering analysis is an important method of data mining [6], mainly used to aggregate on the whole trajectory space. Visual analysis of paper mentioned mainly aim to reduce the vehicle trajectory data to road network and display the state of the vehicle through UI as far as possible, so that traffic state and driver behavior characteristics can be find out.

In this paper, DBSCAN [5] clustering algorithm based on density is referred and some improvement has been finished. Traditional DBSCAN clustering method mainly adopts the whole path of clustering, speed attribute of track section has been ignored. The improvement of paper has introduced speed attribute and two-pass corner detection [2].

Firstly, a trajectory is separated according to the following process. Supposed that in distance (D_{min}, D_{max}), if the included angle value of points (P^-, P, P^+) is less than or equal to a given value $\alpha_{\max}$, or the speed difference between any point and point P is equal to or greater than ε_2, the point P is named key point. It is described in Fig. 1. So if a point P is key point, it must meet requirements, such as $d_{min}^2 \le |p - p^-|^2 \le d_{max}^2$, $d_{min}^2 \le |p - p^+|^2 \le d_{max}^2$, $\alpha \le \alpha_{max}$ or $|V_p - V_{p^+}| \ge \varepsilon_2, |V_p - V_{p^-}| \ge \varepsilon_2$. The definition of parameters follows the literature [5].

And then to compare the distance within determined trajectories, Discrete' Freshet distance method has been used, the constraint conditions is $\delta_{df}(L_i, L_j) \le \varepsilon_1, |\overline{V_{L_i}} - \overline{V_{L_j}}| \le \varepsilon_2$ (Discrete' Freshet between P and Q is $\delta_{df}(P, Q) = \min_{coupling} \max_{C(p_i, q_j) \in C} |p_i - q_j|$, $N_\varepsilon(L_i) = \{L_j \in D | \delta_{df}(L_i, L_j) \le \varepsilon_1, |\overline{V_{L_i}} - \overline{V_{L_j}}| \le \varepsilon_2)$. Algorithm process is described as follows.

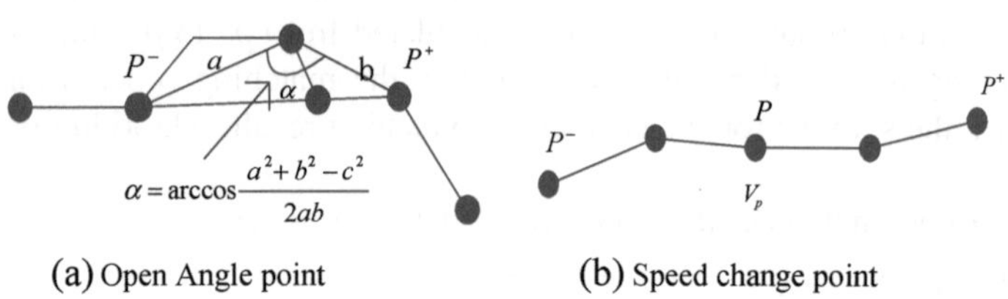

Fig. 1 Determine the key point

Input: divided trajectory set D={$L_1, L_2, \ldots, L_{num_{ln}}$, ε_1, ε_2, and MinLns.
Output: cluster set ()={$C_1, C_2, \ldots, C_{num_{clus}}$}
For each($L \in D$)do{
If(L is unclassified) then
Computc $N_{\varepsilon_1,\varepsilon_2}(L)$;
Assign clusterID to $\forall X \in N_{\varepsilon_1,\varepsilon_2}(L)$;
ExpandCluster(Q,clusterID,ε_1,ε_2,MinLns);
Increase clusterID by 1;
Else Mark L as noise;}
check the clusters cardinality.

3 Visualization Analysis Scheme

3.1 GPS Data Preprocessing

Vehicle GPS trajectory data and road network data is used as input to calculate and analyze traffic jams in this paper. GPS trajectory data includes massive track points. Each trace contains a series of sampling points. Each sampling point contains IP vehicle identification, location, record (longitude and latitude), time, speed, current instantaneous direction Angle, submitting time and a series of random attributes ($\alpha_0, \alpha_1 \ldots \alpha_{n-1}$. The sampling point is arranged by the time series, each region between two adjacent sampling points is called a track section. GPS abnormal data mainly includes anchor point drift error, lot of useless information, if the vehicle is at a gas station or parking, at the same time the transmission signal and the blind spots can cause lack of GPS data. This application is adopted the vehicle GPS data from Zibo city of Shandong province. Because of the GPS data are sampling data, the time interval of data submitting is about 1–2 min, so several groups of filter conditions (F1–F5) is set to eliminate the noise data in Fig. 2.

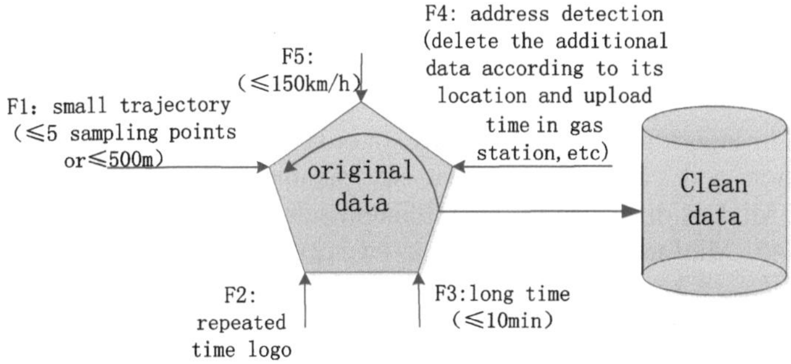

Fig. 2 Filter setting

3.2 *Road Network Data Preprocessing*

The OpenStreetMap [1] source map tool is used, it allows users via API interface to customize individualized maps. The road network data preprocessing has followed the following rules:

1. Filter the road which can not connected to the main road network and to ensure all roads connected together.
2. The relationship between two roads is clear and one-way.
3. In order to ensure the spatial resolution, the length of each road is less than 50 km.

3.3 *Algorithm Implementation*

3.3.1 Map Matching

According to the low sampling rate of GPS data. the following constraints has been defined:

1. GPS trajectory section L is composed of continuous GPS track points, and the sampling interval is not more than the $\Delta T(\ \Delta T \leq 10$ min), L: $p_1 \rightarrow p_2 \rightarrow \ldots \rightarrow p_n$, $p_i \in L$. In this paper, the time interval is 1–2 *min*.
2. Not only single GPS points and the distance between candidate section but also the road network topology information has been considered.
3. Considering the GPS error and the width of road network, the error range of the radius is set as 50 m.

Take No. 77110 vehicle matching process for example. $\mathrm{F}(C_1^1 \rightarrow C_1^{41}) = 0.034$, $\mathrm{F}(C_2^1 \rightarrow C_2^{33}) = 0.027$, $\mathrm{F}(C_3^1 \rightarrow C_3^{29}) = 0.013$, $\mathrm{F}(C_4^1 \rightarrow C_4^{36}) = 0.041$. In the experiment, when the buffer radius is 40 m, quantity of roads in the buffer becomes stable. If $\mu = 0$ and σ = 50 m, trajectory recovery rate of about 91 % vehicles is higher than 80 %, the GPS data in a straight road or cross the intersection have been matched successfully.

3.3.2 Vehicle GPS Data Clustering Analysis

Based on the algorithm mentioned above, in order to observe computing time, track segmentation and clustering number, six parameters, such as $d_{min}, d_{max}, \alpha_{max}, \varepsilon_1, \varepsilon_2$, and MinLns, has been set as 5 different sets of values. (1) $\alpha_{max} = 160$, $\varepsilon_1 = 4.4$, $\varepsilon_2 = 45$, MinLns $= 35$, $d_{max} = d_{min} + 1$; (2) $\varepsilon_1 = 4.4$, $\varepsilon = 45$, MinLns = 35, $\mathrm{d}_{max} = 6$, $d_{min} = 3$; (3) $\alpha_{max} = 160$, $\varepsilon_2 = 45$, MinLns $= 35, d_{max} = 6$, $d_{min} = 3$; (4) $\alpha_{max} = 160$, $\varepsilon_1 = 4.4$, MinLns $= 35, d_{max} = 6$, $d_{min} = 3$; (5) $\alpha_{max} = 160$, $\varepsilon_1 = 4.4$, $\varepsilon_2 = 45$, $d_{max} = 6$, $d_{min} = 3$

The results is as Table 1.

Table 1 Clustering results

Set (1)			
d_{min}	Time (s)	Num_{Traj}	Num_{clus}
1	389.783	2993	1
2	1223.67	2829	2
3	1976.33	2797	3
4	2368.07	2611	4
5	2487.63	2577	5
6	2314.51	2434	6
7	2765.97	2452	7
8	2787.77	2468	8
Set (2)			
α_{max}	Time (s)	Num_{Traj}	Num_{clus}
125	1945.90	2573	2
130	1991.23	2592	2
135	1945.11	2596	2
140	1886.78	2632	2
145	1812.05	2655	2
150	1876.03	2674	2
155	1844.57	2711	1
160	1465.21	2760	1
Set (3)			
ε_1	Time (s)	Num_{Traj}	Num_{clus}
4.1	1971.00	2774	9
4.4	1971.42	2774	9
4.5	1971.23	2774	5
4.6	1971.97	2774	4
4.9	1971.51	2774	2
5.0	1971.48	2774	2
5.2	1971.58	2774	1
5.4	1971.17	2774	1
Set (4)			
ε_2	Time (s)	Num_{Traj}	Num_{clus}
17	1903.337	2789	4
21	1991.500	2766	5
25	1936.740	2701	3
29	1912.830	2715	3
33	1974.621	2693	2
37	1909.520	2662	2
41	1884.010	2674	2
45	1755.255	2 638	3
Set (5)			
MinLns	Time (s)	Num_{Traj}	Num_{clus}
27	1867.06	2774	2
28	1867.55	2774	2
30	1867.12	2774	4
32	1867.33	2774	3
33	1867.41	2774	4
35	1867.88	2774	4
41	1867.75	2774	4
42	1867.27	2774	5

The experiment shows that according to d_{min} increasing, the clustering results has increased, and according to α_{max} increasing, the number of the trajectory has increased respectively, values of ε_1, and MinLns affect clustering result, according to ε_2 enlargement, the trajectory segment and the number of cluster have decreased. All above, set (2.4) can be as the input of algorithm parameters.

4 Result Analysis

4.1 Method in This Paper

In this paper, the experimental environment includes R i386 3.1.3, Windows 7, CPU (2.2 GHz the CORE 2 DUO), RAM 2 GB. The data quantity of GPS is about 21,000 items. Zibo city includes Zhangdian, Linzi, and Zhoucun 3 main districts, and about 453 roads, about 3000 points in the map is used to describe the average traffic congestion state. Figure 3 is the visual interface of average road traffic state from 1:30 to 24:00, which is based on the method mentioned in this paper, the color from blue to red presents the traffic state from clear to congestion. Figure 3 shows that traffic jam phenomenon is more serious along with Lutai avenue, Shiji road, Zhongrun avenue, and Jinjing avenue, vehicles' speed is about 10–15 km/h.

Figure 4 has shown the average traffic state in the whole time, in which morning and evening peaks was included. Road traffic state has become better from 0:00 to 6:00, become worse from 6:00 to 9:30 and gradually got ease until 1:00 p.m.

Finally, the traffic flow of Jiqing highway between Zhangdian and Linzi has increased obviously, the vehicles' speed keep the 60–100 km/h in this section, that shows the section of highway plays a very important role between the two districts,

Fig. 3 Visual interface based on method mentioned

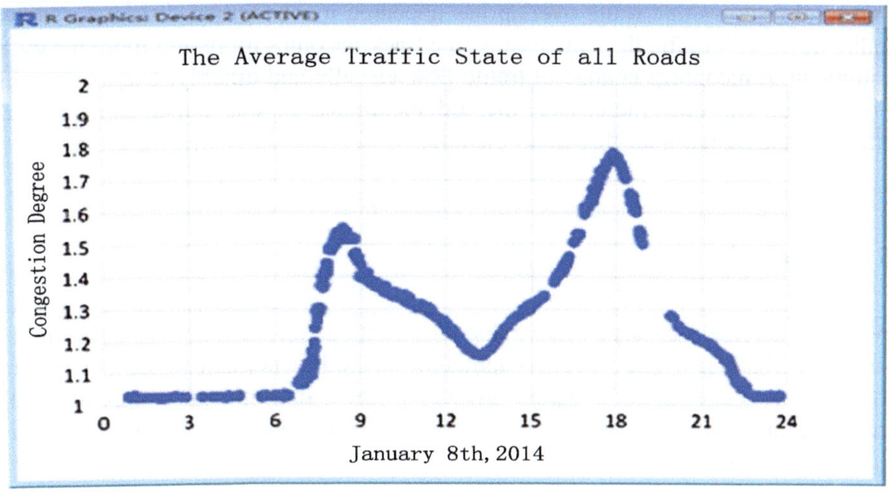

Fig. 4 Road traffic state in a whole day

if permitted, a new road between the two districts should be planned to meet the increasing urban traffic pressure.

4.2 Compared with Flow Map Method

Flow map method is the current mature data visualization method [7]. With the same GPS data, the visual interface based on flow map has been shown in Fig. 5.

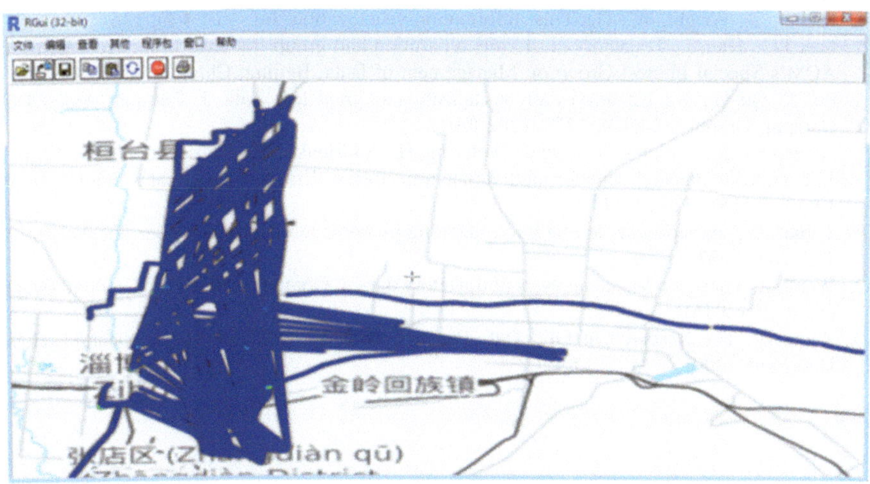

Fig. 5 Visual interface based on flow map method

The results shows that route overlap has appeared in Zhangdian district for heavy traffic flow, with traffic flow increasing, the track becomes more and more heavy, it is difficult to recognize change of traffic flow visually and quickly. Compared with Fig. 3, with the data increasing, method mentioned in this paper can be more effective and visual than flow map method.

5 Conclusions

With the increment of large scale traffic data (GPS data), the data visualization has become a challenge. The visual computation method mentioned in this paper aims to visualize large scale complex trajectory data, through the improved map matching and clustering analysis methods. It can be effective for traffic GPS big data visualization.

Acknowledgments This work of paper is supported by The Gansu province The National Key Technology R&D Program (1504GKCA018).

References

1. Chen, S.: A Web-based Accessibility Analysis Service Using OpenStreetMap Data. Shanghai normal university (2010)
2. Chetverikov, D., Szabo, Z.: A simple and efficient algorithm for detection of high curvature points in planar curves. In: Proceedings of the 10th International Conference on Computer Analysis of Images and Patterns. Groningen, The Netherlands [s. n.] (2003)
3. Enjian, Y.A.O.: Real-time map matching algorithm based on low-sampling-rate probe vehicle data. J. Beijing Univ. Technol. **39**(6), 2–4 (2013)
4. Kapler, T., Wright, W.: GeoTime information visualization. Inf. Vis. **4**(2), 136–146 (2005)
5. Lee, J.G., Han, J.: Trajectory clustering: a partition-and-group framework. In: Proceedings of ACM's Special Interest Group on Management of Data, Beijing, China [s. n.] (2007)
6. Pu, J., Qu H., Ni, L.: Survey on visualization of trajectory data. J. Comput. Aided Des. Comput. Graphics **24**(10), 1273–1282 (2012)
7. Scheepens, R., Willems, N., Van de Wetering, H., Andrienko, G., Andrienko, N., van Wijk, J. J.: Composite density maps for multivariate trajectories. IEEE Trans. Visual. Comput. Graph. **17**(12), 2518–2520 (2011)
8. Tobler, W.: Experiments in migration mapping by computer. Cartogr. Geogr. Inf. Sci. **14**(2), 155–163 (1987)
9. Wang, Z, Yuan, X.: Visual analysis of trajectory data. J. Comput. Aided Des. Comput. Graph. (1), 9–25 (2015)
10. Zhou, B.: The Research of GPS Data Pre-processing Methods and its Application. Hehai University (2005)

A CA Model with Variable Cell Size for Passengers Behavior in Subway

Yifan Zhuang, Yichen Zheng, Yi Zhang and Xudong Xie

Abstract The Cellular Automaton (CA) is a kind of discrete system whose dynamic performance depends on simple reactions among single cells. It has been widely applied in many fields nowadays because it can mimic some intricate situations. The present designs of cellular automaton, however, are mostly such systems with stationary cell size. In particular cases, there could be some huge differences between simulation and real statistics. The CA model with variable cell size (CA-VCS) then becomes significant, and the design for extended CA based on the existing model is proposed to simulate the movement of passengers in subway. In most cases, the cell size is referred to psychological size not actually physical body size. With the consideration of high density in subway in rush hours, Cells could alter their cell size based on the normal CA models to describe passengers much accurately according to the condition around. When meeting highly dense population, cells will be compressed and transformed into small size until they could find enough space to enlarge their size. The simulation based on the defined movement rules of cells shows the necessary processes for the cell transformation. The relative research and its results provide the proof to apply the proposed model in analysis of subway.

Keywords Cellular automaton · Variable cell size · Passenger behaviors

Y. Zhuang (✉) · Y. Zheng · Y. Zhang · X. Xie
Department of Automation, Tsinghua National Laboratory for Information Science and Technology (TNList), Tsinghua University, Beijing 100084, China
e-mail: zhuangyf12@mails.tsinghua.edu.cn

Y. Zheng
e-mail: zhengyc10@mails.tsinghua.edu.cn

Y. Zhang
e-mail: zhyi@mail.tsinghua.edu.cn

X. Xie
e-mail: xdxie@mail.tsinghua.edu.cn

© Springer Science+Business Media Singapore 2017

H. Lu (ed.), *Proceedings of the Second International Conference on Intelligent Transportation*, Smart Innovation, Systems and Technologies 53,
DOI 10.1007/978-981-10-2398-9_28

1 Introduction

Cellular Automaton (CA) is a sort of dynamic system, which is discrete both in time and space and widely applied recently in transportation systems to investigate the behaviors and movement of passengers. It divides the space into the regular network. Each cell corresponding to each element in this network has limited status and follows the same rules. Depending on those well-defined regulations, cells update their own status concurrently. It is well known that CA is not conducted by strict mathematical equations or functions but a series of certain rules to determine the next status at every certain moment. There are some mature designs in various fields like Game of Life [1]. To be more specific, CA model is also utilized in the transportation field to mimic or predict the population flow [2]. In order to better explain the complex pedestrian behavior, several theories are put forward to support the construction of CA model [3–6].

Kazuhiro and etc. proposed a real-coded cellular automata (RCA) model based on real-coded lattice gas (RLG) model, and cells in this model have continuous velocity distributions, which means the collision does not constrained by the explicit lattice structure [7]. What's more Song Weiguo and etc. did researches in the evacuation behaviors at exit [8]. Fang Weifeng and etc. did simulations of bi-direction pedestrian movement [9]. Thus, much work has been done on almost all aspects of pedestrian movement, and these results also prove that CA model is a very effective tool.

Although CA is widely employed and could simulate a large of practical events well, it is constrained in some particular situations due to the unchangeable cell size [10]. Take the passenger movement in subways as an example where population density will change a lot at different time. During the peak hour, the great population density will lead to a huge decrease of the area occupied by each passenger. In addition, passengers in different areas of a subway carriage would not possess the same area size. Obviously, the original CA model could not solve these conditions perfectly because of its fixed size and this feature is contradicted to the real fact [11–13]. Suppose our cell size matches the normal situation where there are only few people, but when the carriage becomes crowded each person will occupy less area soon. The area size discussed here is not physical area as usual but psychological area which is always larger than the former area. According to our life experience, most people would like to have more space and thus they will choose to stay a distance from others rather than close to each other. In addition, the degree of comforts is obviously different due to different distance from other passengers. We will feel more comfortable when having more space. Therefore the psychological distance is a very important index of estimating comfortable degrees. And the improvement on CA is expected to satisfy the demands above with the cell size for cells changeable during the simulation, and put forward the idea of CA-VCS.

In this paper, an available thought is proposed in Sect. 2 firstly. The simulations are then made to check whether this model coincide the real situation in Sect. 3. Finally, conclusions are drawn in Sect. 4.

2 System Model

2.1 General Parameters

Compared with the basic model whose behavior is only related to the surrounding area and fixed cell size, the pedestrian movement in subway is much more complicated. The latter question involves the destination of each cell and trend of others' movement. So, more parameters should be introduced to simulate passenger behavior suitably. The general parameters for basic models are shown in the Table 1.

In this model, a passenger can only move to one of eight cells around it at every step. After defining the passenger is currently at the position, $p_{0,0}$, it could move to eight cells around or just stand still, as shown in Fig. 1. The parameter $P_{i,j}$ depicts the probability of moving to the cell $p_{i,j}$. As a result, passengers would choose a cell with the maximum possibility to move in at the next step if there will not be collision happening for no more than one passenger is going to move to the same place.

The calculation of $P_{i,j}$ is shown as follows.

$$P_{i,j} = \alpha Dis_{i,j} + \beta E_{i,j} + \gamma F_{i,j}, \quad \alpha + \beta + \gamma = 1.$$

where $Dis_{i,j}$ indicates the distance of the possible cell (i,j) to the final target cell (i_T, j_T), $E_{i,j}$ defines whether the possible cell is occupied or not, and $F_{i,j}$ represents the density of occupied cells on the way to the target cell. They are described as:

$$Dis_{i,j} = |i - i_T| + |j - j_T|$$

$$E_{i,j} = \begin{cases} 1, & \textit{if cell } (i,j) \textit{ is occupied} \\ 0, & \textit{if cell } (i,j) \textit{ is empty} \end{cases}$$

$$F_{i,j} = \frac{m - n}{(|i - i_T| + 1) \cdot (|j - j_T| + 1)}$$

where m represents the total number of cells occupied while n stands for the number of those empty cells.

Where N represents the total number of passengers in a carriage. The maximum principle is that choose the cell that owns the maximum possibility to occupy the square.

The collision free situation is an ideal hypothesis, so we must consider the collision rules. The rule process is simply shown in the Fig. 2.

Table 1 General parameters

Symbol	Description
$P_{i,j}$	Evaluate the possibility of a passenger moving to cell (i,j) in next step
$Dis_{i,j}$	Contribution to approaching destination cell by moving to cell (i,j)
$E_{i,j}$	Symbol whether cell (i,j) is occupied
$F_{i,j}$	Congestion level of the area from cell (i,j) to destination

$P_{-1,1}$	$P_{0,1}$	$P_{1,1}$
$P_{-1,0}$	$P_{0,0}$	$P_{1,0}$
$P_{-1,-1}$	$P_{0,-1}$	$P_{1,-1}$

Fig. 1 All possible cells for the passenger at $p_{0,0}$

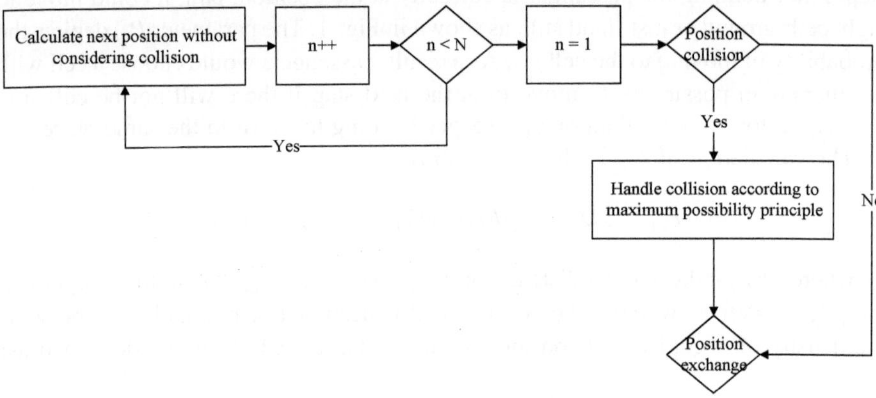

Fig. 2 Collision rules

In conclusion, the fundamental rules for movement of cells have been stated above. Furthermore, to change its cell size during the simulation properly, some special parameters should be introduced in this model. The quantitative indicators should be then adopted instead of qualitative description, which results in our models more accurate.

The status of congestion in subway can be defined when people in a subway carriage are exceeding a well-set threshold. It also can be easily described using mathematical descriptions as follows.

$$\begin{aligned} & if\ N_{pred} = N_{curr} - N_{off} + N_{on} > N_{thre} \\ & \quad the\ status = crowded \\ & or \\ & \quad the\ status = uncrowded \end{aligned}$$

where N_{pred} is the total number of passengers who are supposed to be able to appear in the subway, N_{curr} is the total number of passengers who are in the subway currently, N_{off} is the number of passengers who will get off the subway, N_{on} is the

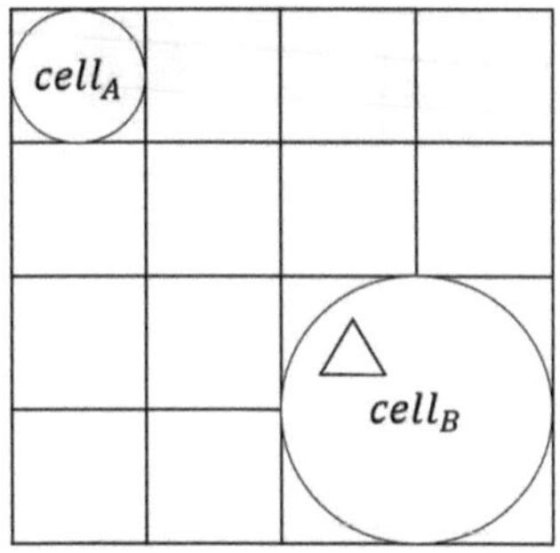

Fig. 3 Comparison of Two types of cells

number of passengers who will get on the subway, N_{thre} is the threshold used to evaluate the subway crowded or not.

The size of cells will change on the basis of their location in carriage and their moving speed while their step length would also be changed along with that at the same time. That is, one big square can be divided into a four-rectangle-grid if necessary. When the carriage has more free space to accommodate passengers, each cell occupies the whole square and its movement rules are kept the same. When the carriage has a pretty high population density, the cell size would probably be changed from four squares to one square. Their size status can be described using the parameter E_{size} to represent whether one cell is in small size. Take $E_{size} = 1$ when the cell has the small size while $E_{size} = 0$ when it indicates that the cell stay in large size.

The comparison for cells in different sizes is shown in Fig. 2. The different parts of the carriage have different possibilities to change the cell size, and detailed rules would described in the next part of this article.

The triangle in the large cell in Fig. 3 suggests the referring location of the big cell and it is always on the upper left corner of one big square.

2.2 *Movement Rules*

After the introduction of general status parameters and moving rules, the rules for size changing can be achieved, which represent the most particular features differing from the original models. These rules could guarantee the new model to be closer to our real situations.

Rules for cell size changing:

- The step length of each movement is determined by the size of cells. In order to make it more clearly and accurately, we suppose that the basic unit length of each step is equal to the diameter of the small cell. To be more specific, the large cell parallel to the area occupied by passengers when the subway is pretty in real

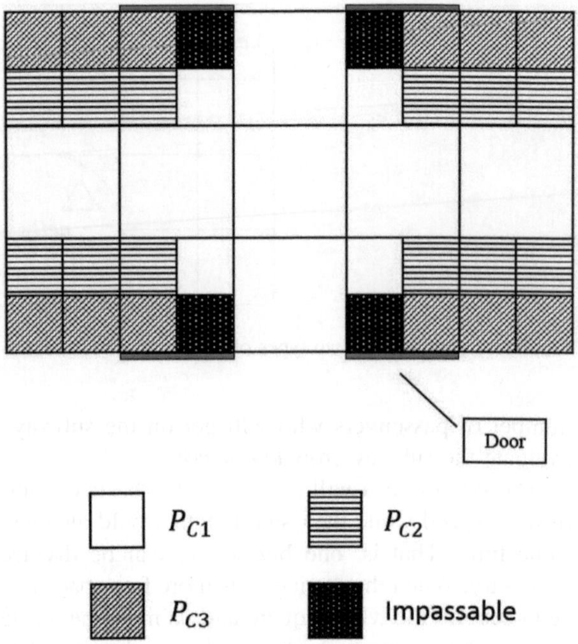

Fig. 4 Possibilities of different areas

moves forward two unit lengths at each stage. Meanwhile, the small cell which indicates the crowded conditions moves forward only one unit length.

- To make our description more clearly, we would firstly define a population threshold. Only when the total population is higher than T_{popu}, the cell could change its size. The different area in the carriage has different possibilities for cell size changing. We suppose that there are three kinds of possibility P_{C1}, P_{C2} and P_{C3} for three parts of a carriage. Their location distribution is shown in the Fig. 4.
  Where the value relationship among them is $P_{C1} > P_{C2} > P_{C3}$. And it also coincides well with our observations the population density on the passageway is much higher than it nears the seat.
- When the carriage is crowded, the situation that cells change from big size to small size is shown in the Fig. 5. In order to move forward without enough room to keep its original size, the cell has to change into small type. Unless it makes some changes, the only solution is to stand still and let others leave first. Where the part in gray in the figure stands that the cells are occupied, and $cell_A$ could not move to that section
- Each passenger who is represented by cell in the simulation has his or her rights to decide whether to stay close to each other in most cases. We use the possibility $P_{Ci}(i = 1, 2, 3)$ of changing size to mimic their decisions. However, in a few extremely crowded situations, they have no choice but to stay really close to other people due to social force. One of the great feature of this situation is that passengers move so slowly when they want to get off the subway. Therefore, we

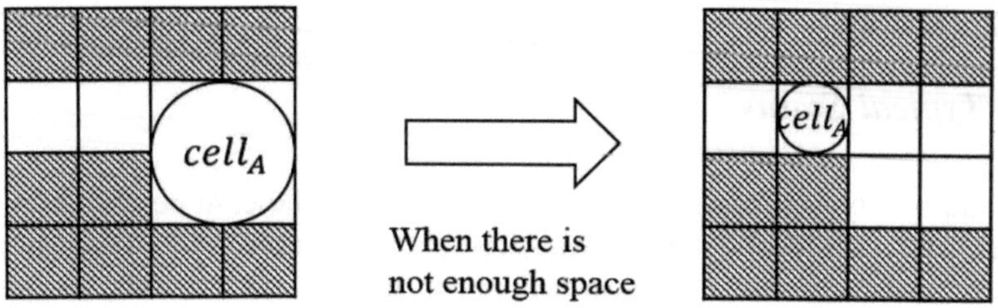

Fig. 5 Two kinds of status of cells

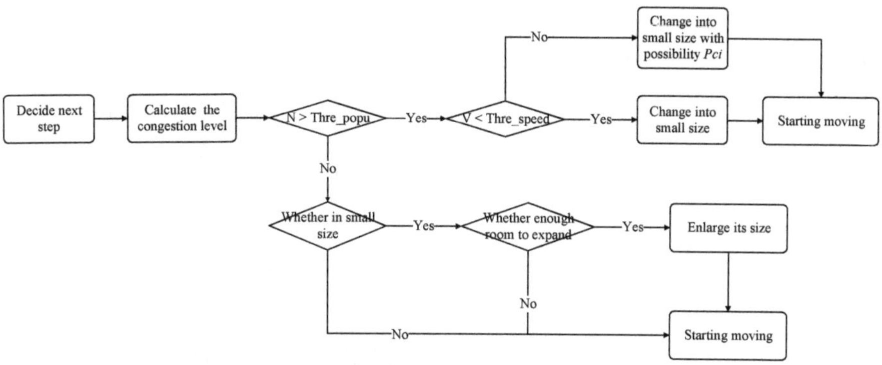

Fig. 6 Process of the size change

Table 2 Advantages and disadvantages

Advantages	Disadvantages
Simple movement rules	Hard to determine the cell status
Mimic the real situation well	Complex collision rules

could also define a speed threshold T_{speed}. When both the average speed of a single cell is lower than T_{speed} and the total population is higher than T_{popu}, the cell must compress its size.

- Only when the carriage becomes un-crowded, cells could transform from small one to big one. The size transformation of cell is obviously restricted by its surroundings. In specific, if no enough room for transforming, the cell would not change its size. Figure 6 shows all procedures happening for cell transformation, and comparison of advantages and disadvantages is listed in the following Table 2.

3 Simulation

3.1 Typical Status

There are two typical cases shown in Figs. 7 and 8. The case described in Fig. 7 indicates the most ideal condition where only a few people are in the carriage and then everyone can hold the maximum space, the big size in four squares. The second case shown in Fig. 8, however, indicates a totally opposite condition where there are too many passengers in carriage and each passenger hold the minimum space, the small size.

Fig. 7 Uncrowded situation

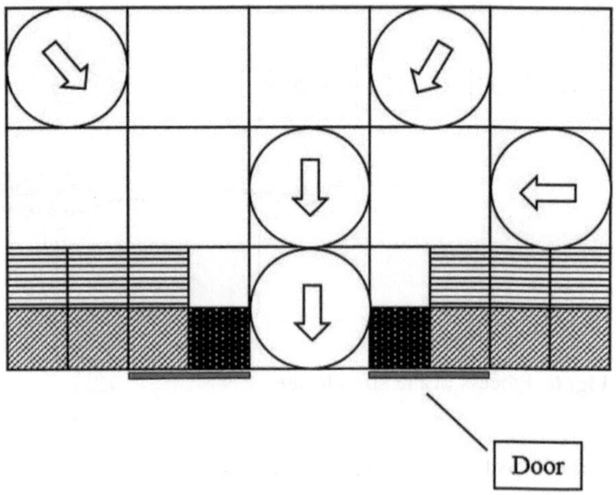

Fig. 8 Crowded situation

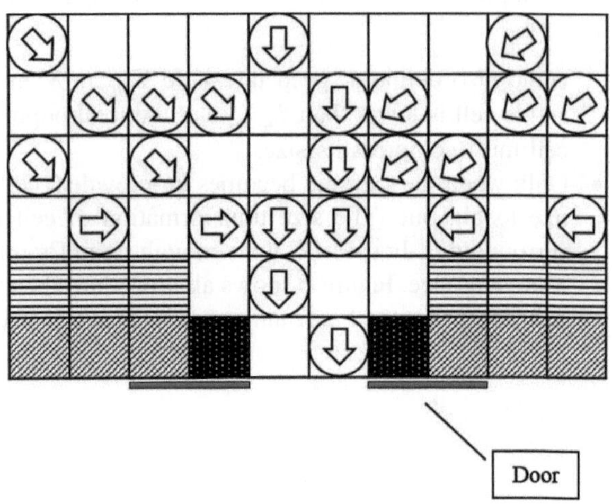

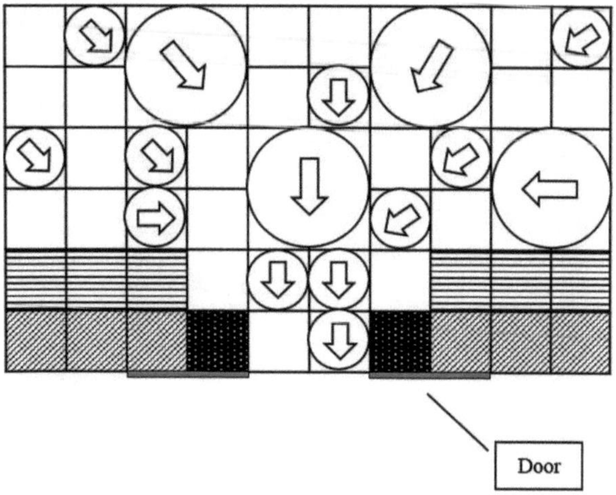

Fig. 9 Moderate crowded situation

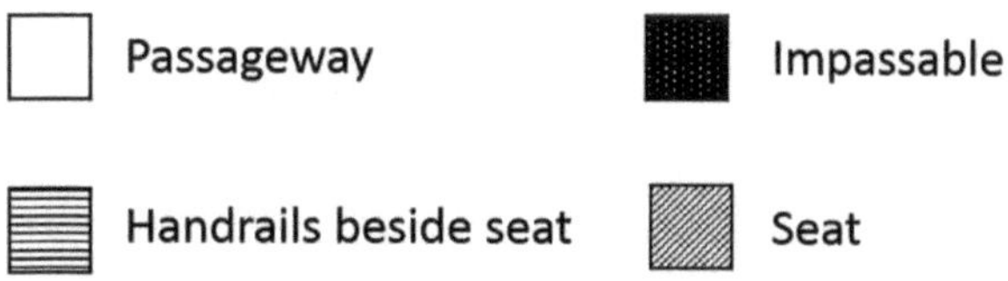

Fig. 10 Notation of typical status figures

In most situations, both large cells and small cells existing in the same carriage means some passengers will occupy more spaces than others. After some passengers get off, the population density might fall down below the threshold, in which people are likely to occupy the larger space, that is, in a whole rectangle. This process is similar to our real life where more room for people leads to a more comfortable feeling. The moderate crowded situation described above can be shown in Fig. 9.

The notations of the signals of Figs. 7, 8 and 9 are in the following Fig. 10.

3.2 Simulation

Based on the typical cases stated above, the simulation for all cases can be undertaken and their processes are achieved as shown in Figs. 11, 12 and 13, which are adopted to describe the process for the extremely crowded situation happening in subway systems. The subway door in this simulation is on the left top side. To make our simulation more clearly, some facilities in the carriage are simplified, such as seats and handrails. All white area is passageway because the target of this simulation is the observation of basic regularity of cell movements.

In this situation, the carriage is filled with passengers and the population density is mighty high. So each person has to occupy less area in order to get on or get off

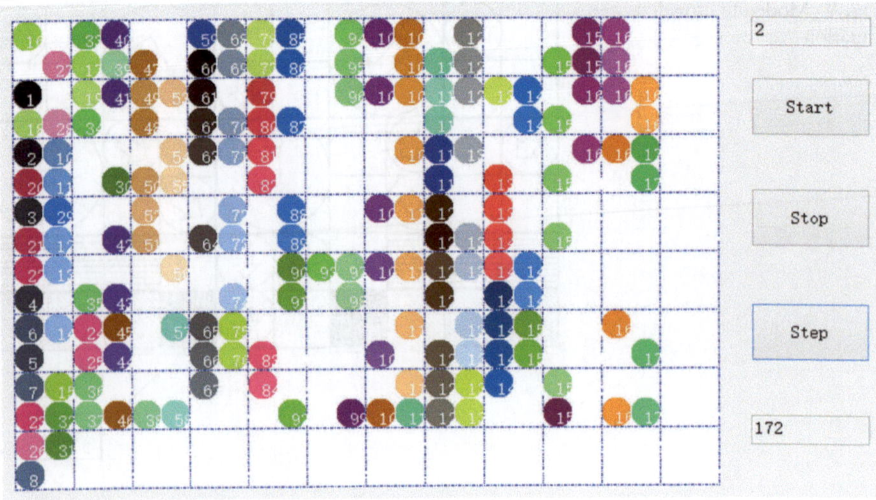

Fig. 11 Step 2 of the most crowded situation

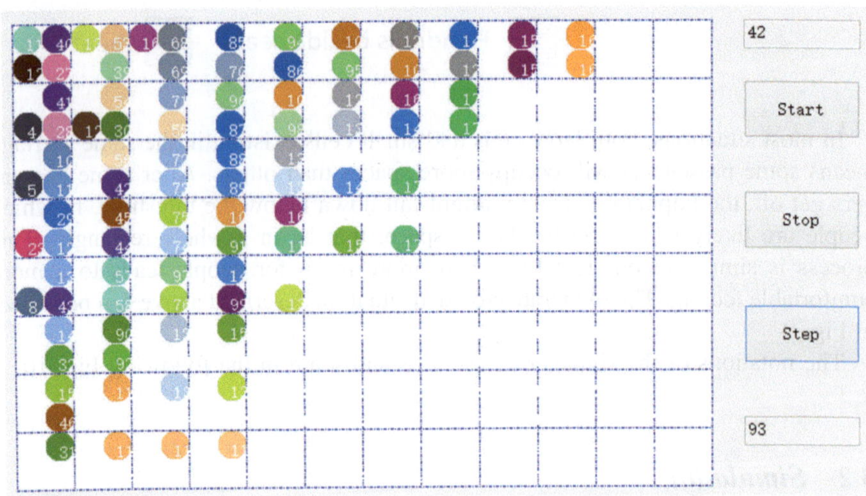

Fig. 12 Step 42 of the most crowded situation

the subway successfully. The following figures shows the process of passengers getting off and each picture represents three different stages (step 9, step 42 and step 87). The first figure shows the initial condition where all people are waiting to leave. The second figure shows that half of people have already got off. And the last figure shows almost all people have left from the carriage. As a result, the CA-VCS could well mimic the pedestrian flow when the subway is extremely crowded.

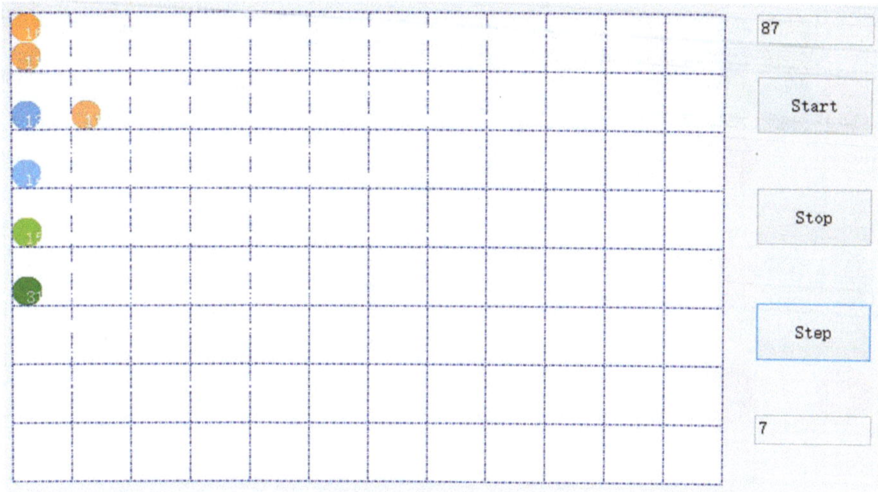

Fig. 13 Step 87 of the most crowded situation

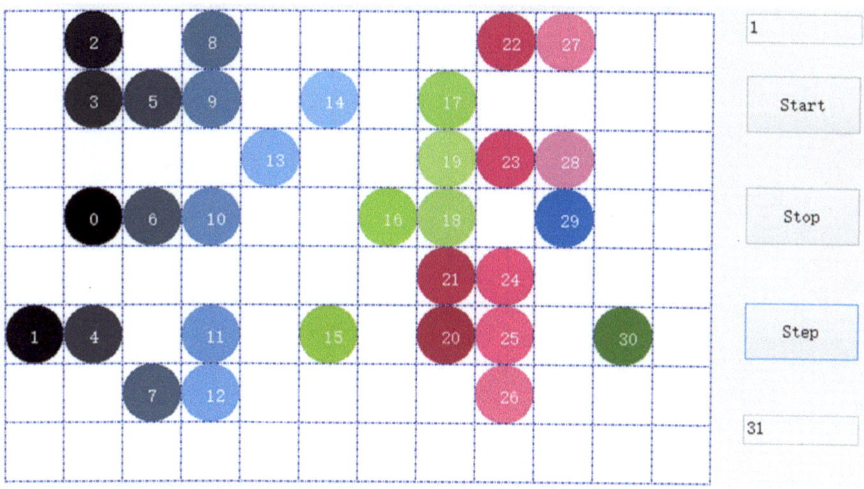

Fig. 14 Step 1 of the emptiest situation

The ideal situation, in which there are only a few passengers in the carriage, is also explored here. Figures 14, 15 and 16 shows the simulation processes to display the movement of passengers. The reason that some cells do not change their size like what they do in Fig. 18 is that the fundamental assumption of this part is the population density would keep in high level and would not change due to cell movement. The explanation for this phenomenon is that other passengers are getting on subway but not showed in the simulation because it will make a chaos if

Fig. 15 Step 11 of the emptiest situation

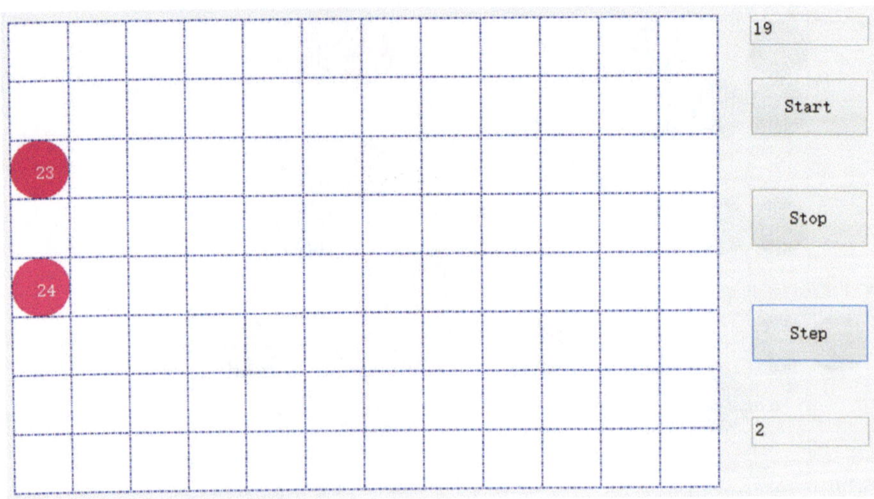

Fig. 16 Step 12 of the emptiest situation

showed and become hard for us to figure out two groups of getting on and getting off.

In this situation, there are only a few passengers in the subway, so each person could occupy more area in order to gain a more comfort feelings. Compared to the previous situation, people have more freedom to move to next place. The following three figures show that passengers will get off the subway and each picture stands for three different stages (step 1, step 11 and step 19). The meaning of each figure is

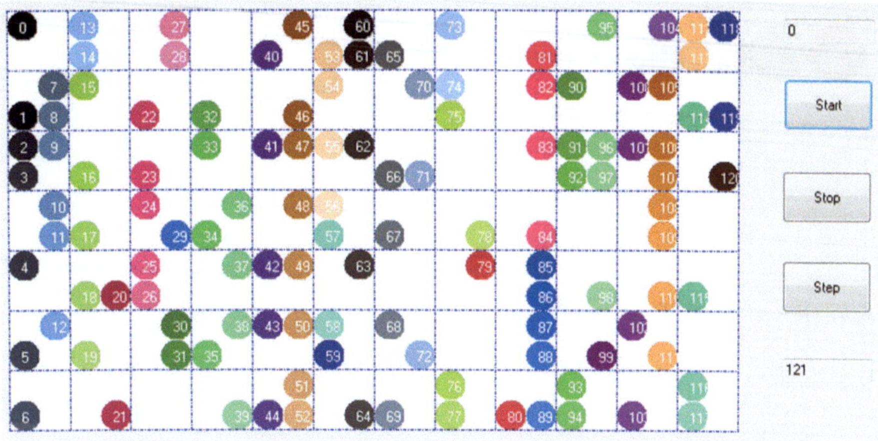

Fig. 17 Step 0 of moderate situation

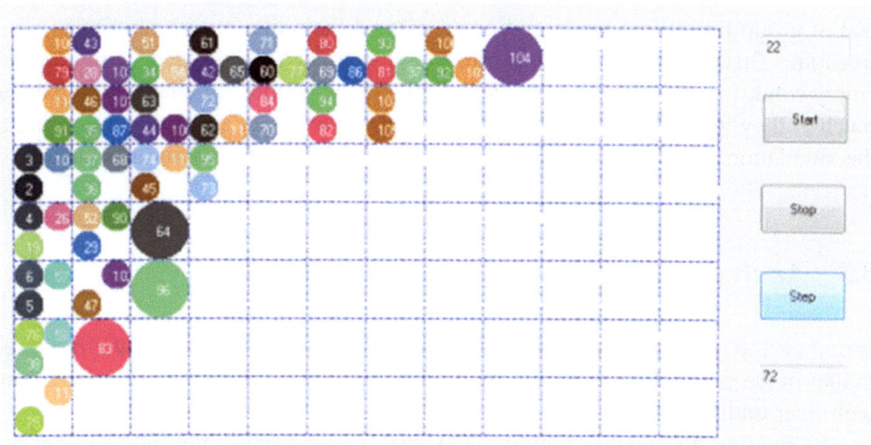

Fig. 18 Step 22 of moderate situation

similar to the crowded situation. According to the simulation result, the CA-VCS model could fit this situation well.

In most cases, the population density in subway is between empty situation and really crowded situation, which is named moderate crowded situation here. Figures 17, 18 and 19 shows the simulation process where there is someone changing their psychological size when there is enough space. And this procedure can be divided into three obvious stages (step 0, step 22 and step 69).

At the beginning, there are a lot of passengers in the carriage and each one has to occupy less area. This stage is similar to the most crowded situation. After some passenger get off the subway, the space is not crowded as before. So people at the

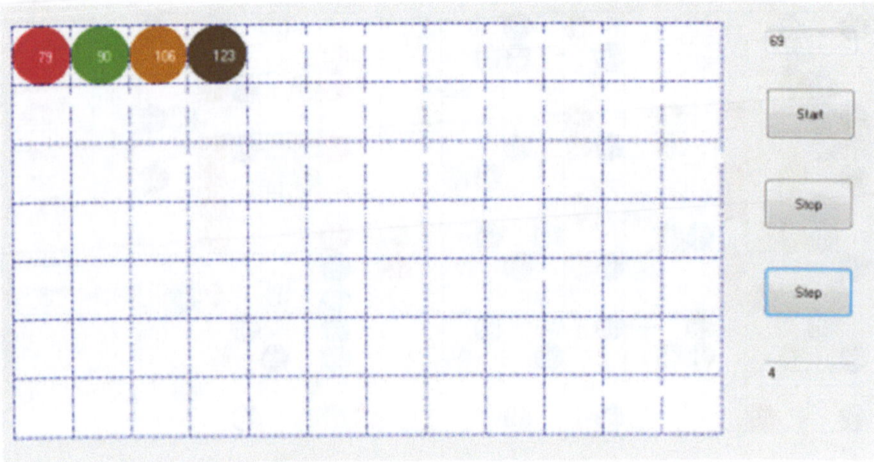

Fig. 19 Step 69 of moderate situation

end of group have more space and could expand their size in order to obtain more freedom. This stage is a transient process that links the crowded condition and empty condition. At last, most people get off and the people remained could have as much as they want. And this situation is similar to the empty situation. According to the simulation result, the CA-VCA morel could fit the most common situation.

3.3 Analysis

Based on the description and simulation above, we get to know that the proposed model in the paper, CA-VCS, can be used and expended to define more situations well than traditional ones.

The running process of traditional CA model is showed in the Fig. 20. The cell size is fixed whenever there is enough space, and it is obviously not corresponding to the real situation in subway. Because passengers would alter their psychological area due to surrounding environment. So our CA-VCS could suit more situations than traditional CA models.

According to results above, time for all passengers getting off increases due to the increase of population density, which we use step number to represent. This regularity of variance of population density is shown in the Fig. 21.

Where the step number in the Fig. 21 is the average number of about 10 tests with same parameters. The population density is the ratio of initial total number of pedestrian and the maximum number which a carriage could contain.

And this trend is corresponding to our life experience. The cell size for CA-VCS would be changeable in the simulation depending on the population density. When there are a lot of passengers in the carriage, corresponding to our peak hours, almost

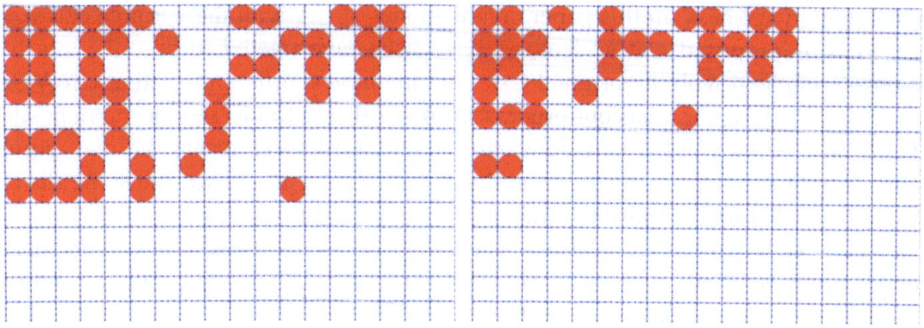

Fig. 20 Simulation process of traditional CA

Fig. 21 Step number versus population density

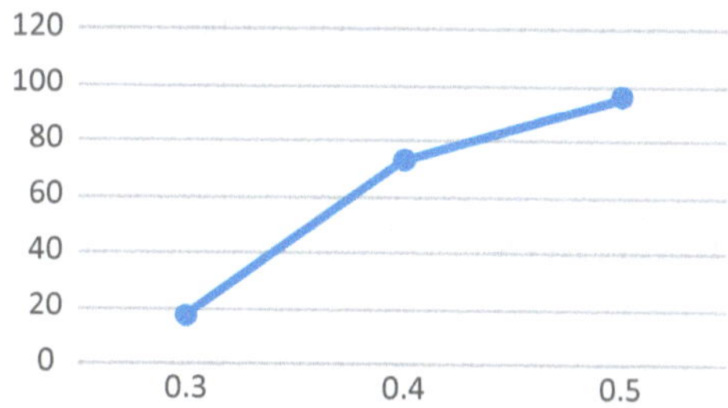

each passenger is trying to keep its minimum space in order to get on the subway. At this moment, subway clients would also push as many people as possible into the carriage. In this situation one subway could carry much more people on rush hour and even beyond the required number. And traditional model could only reach the maximum number which could not reflect the extreme situation and this is also common in most big cities. Only after some people get off the subway, the population density will decrease and some people are keen to occupy more space in order to gain higher comfortable degree. This transformation is well displayed in the simulation part 3, which includes Figs. 17, 18 and 19. In addition the traditional and modified model could produce the similar result when all latter model cells recovered to usually large size or are compressed into small size. In conclusion, the new CA-VCS could suit more conditions and expand the application ranges.

4 Conclusion

The simulation of passenger behaviors in the subway is significant, because the passenger density in most China cities subway changes a lot during different time especially in the peak hours. The traditional fixed cell size automaton model is obviously not suitable for this situation and CA-VCS model is thus proposed to make up its weakness. In this paper, a new model of CA with the variable cell size, CA-VCS, is proposed and its features have been investigated to describe the difference compared with the original CA. And the largest difference is that CA-VCA model could show the transient process from small size to big size or the opposite. Based on CA-VCS, most situations can be simulated to represent the real world, which might not happen for the traditional CA. There are, however, some disadvantages of this model existing, such as the small cell might walk farther than big one at one step. And as we all know, the CA model is not a traditional mathematic model which is strictly defined by equation. The value of parameters thus is not set by scientific calculation but expert experience. As a result, CA-VCS model is meaningful to implement the analysis of passengers in subway and more efforts will be expected to improve its performance.

Acknowledgment This work was partially supported by the National Key Basic Research and Development (973) Program of China (No. 2012CB725405), and the National Natural Science Foundation of China (No. 61273238).

References

1. Bak, Pek: Self-organized criticality in non-conservative models. Phys. A Stat. Mech. Appl. **191**(1), 41–46 (1992)
2. Benyoussef, A., Chakib, H., Ez-Zahraouy, H.: Anisotropy effect on two-dimensional cellular-automaton traffic flow with periodic and open boundaries. Phys. Rev. E 68 (2003)
3. Helbing, D., Farkas, I., Vicsek, T.: Simulating dynamical features of escape panic. Nature **407**, 487–490 (2000)
4. Zou, B., Hu, J., Wang, Q., Ke, G.: A distributed shortest-path routing algorithm for transportation systems. In: 7th International Conference on Traffic and Transportation Studies, vol. 1 (2010)
5. Blue, V.J., Adler, J.L.: Emergent Pedestrian streams and cellular automata microsimulation. In: Transportation research record 80th Annual Meeting, Washington D.C., 2001, TRB CD
6. Helbing, D., Molnar, P.: Social force model for pedestrian dynamics. Phys. Rev. E **51**, 4282–4286 (1995)
7. Yamamoto, Kazuhiro, Kokubo, Satoshi, Nishinan, Katsuhiro: Simulation for pedestrian dynamics by real-coded cellular automata. Phys. A Stat. Mech. Appl. **379**, 654–660 (2007)
8. Weiguo, S., Yanfei, Y., Binghong, W., Weicheng, F.: Evacuation behaviors at exit in CA model with force essentials: a comparison with social force model. Phys. A Stat. Mech. Appl. **371**, 658–666 (2006)
9. Weifeng, Fang, Lizhong, Yang, Weicheng, Fan: Simulation of bi-direction pedestrian movement using a cellular automata model. Phys. A Stat. Mech. Appl. **321**, 630–633 (2003)
10. Jiang, Y.-P., Philip Chen, C.L.: A novel approach for designing intelligent transportation system. In: IEEE International Conference on System, Man, and Cybernetics, October, 2012
11. Zheng, Y., Guo, W., Zhang, Y., Hu, J.: A generalized comfort function of subway systems based on a nested logit model. Tsinghua Sci. Technol. **19**(3), 300–306 (2014)
12. Wang, Y., Long, D., Shi, F.: Cellular automaton model for analyzing capacity of branch road section. J. Central S. Univ. Technol. **18**(5), 1744 (2011)
13. Dijkstra, J., Jessurun, J., Timmermans, H.: A multi-agent cellular automata model of pedestrian movement. Pedestrian Evacuation Dyn. 173–181 (2001)

Erratum to: Proceedings of the Second International Conference on Intelligent Transportation

Huapu Lu (Ed.)

Erratum to: H. Lu (ed.), *Proceedings of the Second International Conference on Intelligent Transportation*, Smart Innovation, Systems and Technologies 53, DOI 10.1007/978-981-10-2398-9

The original version of the book was inadvertently published without Acknowledgment in Chaps. 14 and 15, which has to be included. The erratum chapter and the book have been updated with the change.

The updated online versions of these chapters can be found at
http://dx.doi.org/10.1007/978-981-10-2398-9_14
http://dx.doi.org/10.1007/978-981-10-2398-9_15

© Springer Science+Business Media Singapore 2017

H. Lu (ed.), *Proceedings of the Second International Conference on Intelligent Transportation*, Smart Innovation, Systems and Technologies 53,
DOI 10.1007/978-981-10-2398-9_29

MIX
Papier aus verantwortungsvollen Quellen
Paper from responsible sources
FSC® C105338

If you have any concerns about our products,
you can contact us on
ProductSafety@springernature.com

In case Publisher is established outside the EU,
the EU authorized representative is:
Springer Nature Customer Service Center GmbH
Europaplatz 3, 69115 Heidelberg, Germany

Printed by Libri Plureos GmbH
in Hamburg, Germany